Philosophical Studies Series

Volume 160

Philosophical Studies Series aims to provide a forum for the best current research in contemporary philosophy broadly conceived, its methodologies, and applications. Since Wilfrid Sellars and Keith Lehrer founded the series in 1974, the book series has welcomed a wide variety of different approaches, and every effort is made to maintain this pluralism, not for its own sake, but in order to represent the many fruitful and illuminating ways of addressing philosophical questions and investigating related applications and disciplines.

The book series is interested in classical topics of all branches of philosophy including, but not limited to:

- Ethics
- Epistemology
- Logic
- Metaphysics
- Philosophy of language
- Philosophy of logic
- Philosophy of mind
- Philosophy of religion
- Philosophy of science

Special attention is paid to studies that focus on:

- the interplay of empirical and philosophical viewpoints
- the implications and consequences of conceptual phenomena for research as well as for society
- philosophies of specific sciences, such as philosophy of biology, philosophy of chemistry, philosophy of computer science, philosophy of information, philosophy of neuroscience, philosophy of physics, or philosophy of technology; and
- contributions to the formal (logical, set-theoretical, mathematical, information-theoretical, decision-theoretical, etc.) methodology of sciences.

Likewise, the applications of conceptual and methodological investigations to applied sciences as well as social and technological phenomena are strongly encouraged.

Philosophical Studies Series welcomes historically informed research, but privileges philosophical theories and the discussion of contemporary issues rather than purely scholarly investigations into the history of ideas or authors. Besides monographs, *Philosophical Studies Series* publishes thematically unified anthologies, selected papers from relevant conferences, and edited volumes with a well-defined topical focus inside the aim and scope of the book series. The contributions in the volumes are expected to be focused and structurally organized in accordance with the central theme(s), and are tied together by an editorial introduction. Volumes are completed by extensive bibliographies.

The series discourages the submission of manuscripts that contain reprints of previous published material and/or manuscripts that are below 160 pages/88,000 words.

For inquiries and submission of proposals authors can contact the editor-in-chief Massimo Durante via: massimo.durante@unito.it

Piero Dominici

Beyond Black Swans: Inhabiting Indeterminacy

Foreword by Edgar Morin

Springer

Piero Dominici
Philosophy, Social, Human & Educational Sciences
University of Perugia
Rome & Perugia, Roma, Italy

Foreword by
Edgar Morin
Philosophy & Social Sciences
École des Hautes Études en Sciences Sociales
Paris, France

ISSN 0921-8599 ISSN 2542-8349 (electronic)
Philosophical Studies Series
ISBN 978-3-032-09028-7 ISBN 978-3-032-09029-4 (eBook)
https://doi.org/10.1007/978-3-032-09029-4

Translation from the Italian language edition: “Oltre i cigni neri: l’urgenza di aprirsi all’indeterminato” by Piero Dominici with foreword by Edgar Morin, © FrancoAngeli 2023. Published by FrancoAngeli.

This Springer imprint is published by the registered company Springer Nature Switzerland AG
The registered company address is: Gewerbestrasse 11, 6330 Cham, Switzerland

Translated from the Italian by Kay McCarthy
Original Title: Oltre i Cigni Neri. L'urgenza di aprirsi all'indeterminato
Published by FrancoAngeli, 2023

I would like to dedicate this book, this enjoyable endeavour, to Edgar Morin, my Teacher and fraternal Friend, over time, beyond time! "Your old(er) brother", as he often repeats, is also extraordinary for his humanity and generosity. An essential reference point, not only during the lengthy and complex scientific research carried out over all these years. I also thank him for the significant foreword which, without the least hesitation, he wished to write for this seventh monograph of mine.

I need to point out that although this work was begun prior to 2020 (including the foreword by Edgar Morin), due to the global, systemic emergency which overwhelmed us all, it was necessary to reopen and revise several of its parts. As I hope you will grasp when reading it, this volume is the result of years of study and research, as well as of continuous experiences, projects, travels, and encounters that have never failed to enrich me.

It is also dedicated to the strongest of "motivations" that has kept me at university, a reason that reaches far beyond my love for thinking, studying, and research: the possibility of meeting and engaging in dialogue, discussions, and debates with young people and in a relationship based on sharing. Rhetoric aside, I continue to receive much more than I give (or try to give) and share with commitment and passion.

This work is dedicated to them as well, also and above all because societies and democracies, at present going through a markedly difficult phase of redefinition and transition, are in desperate need of People, not "individuals", of critical, flexible minds, not "mere executors of functions and rules". To be people and become citizens, as well as to participate in the construction of a truly open society, of a mature, complete democracy, these are processes that have grown increasingly complex and make no concession to shortcuts, simplifications, reductionism, or merely formal, cosmetic solutions.

P. D.

Foreword by Edgar Morin

I have found extremely convincing the presuppositions, approach, and the epistemological and cognizant endeavour that Piero Dominici has undertaken to develop in this latest work. Piero Dominici has the merit of showing us the importance and potentiality of a systemic treatment of the ongoing transformation, and of taking up fundamental questions in terms of approach and method that have long been the objects of my own reflection and work (see *Lost Paradigm* and *Method*), from an original and critical perspective. The comprehension of what he defines as the "complexity of complexity" fends off reductionist and deterministic temptations and capitulation to the flattery of hegemonic narratives and explanations that go too far in simplifying the issues of our times, not least those of democracy and globalization. He carefully keeps together a system of variables and contributing causes, highlighting the correlations and connections, even when he deals with thorny questions linked to digital transformation and the civilization of automation and artificial intelligence. These are complex dimensions and evolutions determining yet another change in paradigms, languages, cultures, lifeforms, and relational systems. Dominici's research leads toward a radical rethinking of education, training, and research, reunifying several historical, primordial *fractures*. Among these diverse fractures, Dominici hones in on the current rift between culture and technology, triggered, like others, by the equally important and devastating split between studies in the humanities and the sciences. Fractures and "logics of separation" that show, in an even more evident and embarrassing manner, the total inadequacy of our schools and universities, precisely during this new phase of anthropological transformation in which we must search for new pathways in education and, above all, for a *new thought system.*[1]

[1] See several of my own works, which are also cited in this book: Morin, E. (1973), Le paradigme perdu (The lost paradigm). Paris: Seuil; Morin, E. (1977–2004), La Méthode, (Method) Vol I-VII. Paris: Seuil; Morin, E. (1990), Introduction à la pensée complexe (Introduction to complex thought), Paris: Seuil; Morin, E. (1991), Les sept savoirs nécessaires à l'éducation du future, English translation: Seven complex lessons in education for the future. UNESCO (2001); Morin, E. (1991) La tête bien faite (Well-made heads) Paris: Seuil; Morin, E. (2015), Penser global.

Dominici's thought and research extend, with merit, beyond the rhetoric of the so-called digital revolution, underlining—one of the few, if not the only one, to do so in these years—how "the digital" has led people, organizations, social systems, and life itself toward a condition of "augmented complexity", not toward simplification.

In this manner, research becomes an essential "instrument" for going beyond the common (not to say hegemonic) vision that links the ongoing paradigm shift to a mainly technical and technological transformation: in this case, the author argues that educational and training processes are in need of a radical breakthrough, far beyond the idea of an extension or an adaptation to the nature of technological change and to the so-called digital revolution. Cognitive, linguistic, communicative, social, cultural, existential, and anthropological *fractures* that find themselves anchored to educational, pedagogical, and cultural models, which, as Dominici also affirms, carry out a series of functions vital for the very survival of communities and social systems.

If we do not carry through with this radical transformation of education and training, we will once again find ourselves in the condition of merely undergoing the technological transformation that is occurring: a transformation—insists Dominici—that is above all *anthropological* and which calls into question the identities and subjectivities at stake, while further determining new asymmetries and inequalities on a global scale.

The ongoing global mutation, which is profound, radical, and invasive, finds us totally unprepared to manage such a dangerous transitional phase; this is a crisis, as I have often affirmed in the past, which is first and foremost a "thinking crisis", other than a cultural one.

For some time, therefore, Dominici, in perfect harmony with the complexity paradigm and approach, has been carrying forward the challenge of trying to put these fundamental "fractures" back together from within the educational and training institutions, whose inadequacies he has made more than evident. Fractures that in the meantime have become existential as well, impacting the experiences of people and groups, within and without the organizations, within and without the human ecosystems. These *fractures*, which are also epistemological, were defined by Dominici years ago as "false dichotomies": they continue to feed into the social production of knowledge and culture and into social life itself, in all its forms and representations. Maximum appreciation, consequently, goes to this continual and incessant effort to define the complex nature, ambiguous and unpredictable through and through, of complexity and of life (biological, social, and human), even while setting aside an important amount of leeway for the invisible and the undefinable, since not all is visible and not all is measurable. I fully agree with him when he denounces, with great clarity, as in his preceding publications, the error that we are making in wanting at all costs to transform what is *qualitative* (the Human) into what is *quantitative* and measurable, thus "objective" and "scientific". He decries

---

L'Humain e son universe (Thinking global. The human and its universe) Maison des Sciences de l'Homme; Morin, E. (2020) Changeons de voie. Les Leçons du coronavirus (Let's change our ways. The lessons of the Coronavirus) Paris: Denoël: See also: Morin, E., Ciurana, É.-R., Motta, D.R. (2003), Eduquer pour l'ère planetaire (Educating for the planetary era) Paris: Balland.

the negative consequences of a "standardization culture" that has, at this point, taken over our organizations and public life. Thence the vital importance, for Dominici, not to keep perpetuating the "ultimate error", confusing the "complex" with the "complicated", complex systems with complicated ones, continuing to trace life through reductionist and deterministic analyses, through an infinite sequence of data and formulas, deluding ourselves that we are capable of reducing variety and complexity, and at the same time, continuing to keep culture and technology separate.

One concept that is emphasized repeatedly in this book is the importance of "inhabiting hypercomplexity", given that this (hyper)complexity—according to the author—cannot be managed in any way, not even with the support of technology or the new hyperconnected environments. This leads to the urgency of recovering what Dominici has defined as the "complex dimensions of educational complexity", *educating and training toward unpredictability,* building a veritable "culture of error", from the first years of life and study on. Indubitably, this is another of the questions on which I find myself in complete agreement with the theses expressed by this Italian sociologist and philosopher.

In this sense, I would like to propose an extract:

> [...M]ore and more frequently, change lurks in the zones of *tension* and *conflict*, in our weaknesses and inadequacies, in the anomalies,[...] the fluctuations and the dilemmas that characterize knowledge, social action, [and] complex (adaptive) systems [...]; change even lurks in our incompleteness, which [...] allows us to be creative and to dip into our imagination, seeking alternative pathways, abandoning, if necessary, the paths already trodden; more and more frequently, change lurks in the moments of uncertainty, in those errors [...] and in those vulnerabilities that we often ignore and/or try not to see [...]
>
> Change (and innovation), which risk, however, being opportunities "for a select few". As I often repeat, it is necessary to question the fields of knowledge, the borders between them, and the consolidated practices, reconsidering the strategic force of emotions and of individual and collective imagination(s); in other terms, [...] it is also necessary to have the courage to shatter equilibriums, *breaking the chains of tradition*, abandoning the certain for the uncertain; choosing, at least temporarily, to run *the risk of being vulnerable*. Inhabiting the borderlines, the unexplored territories, overstepping those limits and those logics of separation (typical of educational and training institutions) that prevent us from grasping the deepest meaning of the vital, the social, and the relational, and of understanding their complexity and ambivalence. Dimensions, in effect, that are complex, never reducible/attributable to mathematical formulas and/or sequences of data.

The civilization of intelligent machines, in which artificial and natural coincide, is marked by a progressive growth of the dimensions of *technological control*, which marginalizes the Human, debases thought, and diminishes the *sphere of responsibility*, as clarified by Dominici, who redefines the challenges of complexity precisely by moving in the direction of *rethinking/redefining the centrality of the Person and the Human.* This further signifies—in his words—rethinking the relational spaces and areas of communication within the educational and training institutions, relaunching education from a perspective that cannot but be systemic and transdisciplinary. We continue, in any case, to ignore an important aspect: inescapably, the human factor is/will be determining, given that *it is* behind every process, behind every mechanism, behind every form of artificial intelligence. The possibilities and prospects for the *Planetary Humanism* I have long spoken about are still with us. Along these lines of discourse, I find his idea of educating and training "hybrid figures", first proposed back in the mid-1990s, to be very interesting. These figures have in common a logical, methodological, and epistemological basis for their preparation, which falls outside the boundaries of disciplinary pathways; figures that not only know how to manage/control technology and the new hyperconnected environments, taking advantage of their practical-applicative potential to the highest degree, but are capable of *inhabiting hypercomplexity*.

Dominici understands and ably explains that it is time to "rethink how we think", the true foundational dimension of the Human: a crucial dimension that brings along that of "rethinking the *human* factor and its interaction with *technique* and *technology",* sealing the fracture between culture and technology and therefore between studies in the humanities and the sciences. As we swing back and forth between new utopias and dystopias, we must be aware that the future—as he affirms—belongs to those who will succeed in healing the *fracture between the human and the technological*, to those who will be able to redefine and rethink the complex relationship between *natural and artificial,* of those who will understand how to unite (not separate) knowledge and competences, of those who will know how to merge the humanities with the sciences, both in the sphere of education and training and in defining professional profiles and competences.

I further agree with Dominici when he affirms: "…technology (= culture) has come in forcefully to take part in the *synthesis of new values and evaluation criteria.* We are going through a process of *anthropological transformation* (1995), midway through a radical mutation of paradigms and languages. Hurled into hypercomplexity, a hypercomplexity whose profound epistemological and ethical implications we have not yet understood, little aware that all we can do is to attempt to *inhabit* it. A civilization, increasingly marked by paradoxes and contradictions, by distances and asymmetries, that must come to terms with the absence of a thought system and of a theoretical-interpretative model capable of comprehending the hypercomplexity of the ongoing upheavals. The traditional borders between nature and culture, between the natural and the artificial, between humanistic and scientific studies, between fields of knowledge, between social experiences, between vital worlds, have been completely done away with. The numerous epistemological implications are as yet underestimated, under the misleading conviction that rationality, technique and technology and the infinite availability of data/information will be able once again to rebalance the loss of equilibrium and control. Order and chaos coexist in this *era of interconnected ecosystems* as never before; they cohabit; they are both present, however and forever, feeding back into the systemic framework of the complexity of the living, and even more so, of the social, which continues to reveal its quality of never being thoroughly comprehensible or intelligible."

Opportunely, the author deals with the dimensions of a subjective, vital, relational, organizational, and social complexity—but also linguistic and communicative—that, in his words, *is not objectifiable* in any kind of mathematical formula nor in any data or sequence of data, even if infinite. We are in the throes of this *anthropological transformation*, which expresses itself, as Dominici defines it, in an overturn of the complex interaction between biological and cultural evolution; this is a profound question and, above all, in terms of a "communication culture", it is rendered even more complex and problematic by the absence of a thought system and of a theorical-interpretative model capable of observing, recognizing, and (attempting to) understand hypercomplexity.

The other crucial issue that is addressed in this work is related to the *grand illusions of the hypertechnological civilization,* illusions that intersect, evidently, themes of freedom and democracy. He affirms acutely that "the hypertechnological, hyperconnected civilization, with its intelligent (?) machines, continues to fuel an age-old and counterproductive illusion: that of the *possibility of expelling/eliminating error* (prerequisite of any kind of knowledge, of life, and of freedom itself) and unpredictability, not only from education and training, but even from complex systems themselves, other than from the social and organizational processes that characterize them." In perfect continuity with that still dominant conviction (worldview), which puts techno-economic progress on a par with human and social progress, fueling what I have described as the "euphoric delirium of transhumanism", through which human beings convince themselves of being masters of their own destiny, of managing to obtain immortality, no less, and indeed, as is effectively argued in the book, to be able to control and pre-determine everything through automation and artificial intelligence.

Themes and issues of crucial importance that render the work of Dominici important not only for studies and research on complexity and democracy, but also for the quest for a New Humanism, one capable of going beyond the premises of the previous cultural and historical experience. Making us capable of grasping, better and more in-depth, lessons linked to the coronavirus emergency and the challenges that have derived from it: "The unpredictable future is in gestation today. We hope that it will bring forth a regeneration of politics, protection for the planet and the humanization of society: it is time to change directions".[2]

[2] See E.Morin (2020), *Changeons de voie. Les leçons du coronavirus*, Italian translation., *Cambiamo strada. Le 15 lezioni del coronavirus*, Raffaello Cortina Editore, Milan, 2020, p.25.

Because, as is so well explained in this book, in the society of artificial intelligence, automation, and simulation, what is being put on the spot is our very humanity, our "being human", our freedom, and our responsibility to learn and to improve our lives, even — and above all—committing errors. And what is hoped and wished for is that, sooner or later, human beings will realize that the growth of their power goes hand in hand with the growth of their weakness, learning to cohabit with the uncertainty that has always accompanied the grand adventure of humanity.

Paris, France Edgar Morin

Competing Interests The author has no competing interests to declare that are relevant to the content of this manuscript.

About This Book

This book describes the urgent need of modern humanity to renew and reinforce an open attitude to the complexity of life, above all by embracing its intrinsic indeterminacy, rather than attempting futilely to control its evolution. Oblivious to this ever-more urgent necessity, seduced by the speed and virality of digital pattern recognition, computing, and artificial simulation of human thought, society has reverted to a linear, deterministic concept of reality, under the belief that everything can be measured and managed, and that error and unpredictability will soon be eliminated from our lives and organizations. Consequently, choices and responsibilities have been delegated to technology, artificial intelligence, and algorithms, even in educational institutions, which are now preoccupied with teaching mere skills and know-how, thus committing the fatal error of confusing artificial, mechanical, complicated systems with living, complex, adaptive systems.

This volume is intended not only for complexity/social scientists, philosophers, and students, but for the curious from all walks of life. It calls for learning to inhabit complexity, while recognizing and participating in its interdependent, interconnected, interactive systems of relationships. Dominici reveals the futility of endeavoring to control the uncontrollable or observe the unobservable, showing how self-organization and emergence, triggered from the smallest and most modest elements, impact the entire system.

Contents

List of Figures

Chapter 1
Black Swans and Obsolete Paradigms. The Impossibility of a Working Synthesis and *the Urgency of Inhabiting Indeterminacy*

One cannot simply see something new and different immediately. First, the whole style of thinking must change, the entire intellectual atmosphere must vacillate and the force of a thought endowed with directed promptness must weaken. A peculiar kind of intellectual restlessness and a change need to arise along with a shift in the feelings of collective thinking, which create at the same time only the possibility and the need to see something new, different.
Ludwik Fleck
We naturally associate democracy, to be sure, with freedom of action, but freedom of action without freed capacity of thought behind it is only chaos.
John Dewey
The reform of thinking is a key anthropological and historical problem. This implies a mental revolution even more important than the Copernican one. Never in the history of humanity have the responsibilities of thought been so immense. The heart of the tragedy lies also in thought.
Edgar Morin
The person is not being, but the movement of being towards being

Emmanuel Mounier, 1932

Abstract Complex (living) systems, despite the obsolete and inadequate paradigms which propagate the fatal error of equating them with systems that are merely complicated (that is, artificial, mechanical systems), require us to adopt a socio-systemic, relational paradigm in order to learn to inhabit the intrinsic indeterminacy of human life and all other living systems, in which "black swans" and other unexpected events are understood to be connotative features of life and complexity. In fact, *life itself is emergence; it is permanently relational,* an infi-

P. Dominici, *Beyond Black Swans: Inhabiting Indeterminacy*, Philosophical Studies Series 160, https://doi.org/10.1007/978-3-032-09029-4_1

nite, *non-linear sequence of dynamic processes,* where the emergent manifests itself in all possible and unimaginable, unpredictable ways. It is necessary to dissolve our obsession with total control and rationality, and to break free of the cognitive cages from which we attempt to simplify what cannot be simplified, observe the unobservable, and even aspire to eliminate error from our lives and social systems.

Keywords Complex systems · Living systems · Emergence · Black swans · Cognitive cages · Persistent paradox · Organism-society · Mechanism-society · Obsolete paradigms · Intrinsic indeterminacy

For almost thirty years, I have been studying and dealing with (researching) "complex systems", with (*hyper*)complexity and, more specifically, with a very particular type of complexity, capable of making ambiguities, ambivalences, dichotomies, contradictions, conflicts, asymmetries, imaginaries, even, paradigms, *coexist.* This is a kind of hypercomplexity (2000) which, with the advent of digital technology and artificial intelligence systems, has been extended and amplified further (far from simplification!) to all areas of *living* and *non-living*, also causing all the boundaries to dissolve alongside possible distinctions between "nature" and "culture", between *natural* and *artificial.* It is a hypercomplexity whose variables and dimensions are not immediately "observable" (an unavoidable epistemological and scientific question), and which is marked by "emergent properties", capable of self-organization, and of being *generative* without knowing/going through moments of balance and stability (1996). A hypercomplexity, a constitutive element of organic aggregates and, specifically, of social and human systems, characterized by the systemic and radically interdependent nature of the relations of the parts/entities that constitute/structure it. A (hyper)complexity which, to be described and investigated only, needs a systemic as well as a multi/inter/transdisciplinary approach, above all, a (hyper) complexity requiring and implying the coexistence and use of thoughts, methods, methodologies and epistemologies. An "other" type of gaze and thinking, a new epistemology (1995–1996) above all. I am referring, obviously, to *social complexity*[1] and, by way of extension, to that of the Human.

[1] With regard, in particular, to social and cultural complexity, within the context of an immense and multidisciplinary scientific literature, see: Crozier M., Friedberg E. (1977), *L'Acteur et le Système. Les contraintes de l'action collective*, Italian translation, *Attore sociale e sistema. Sociologia dell'azione organizzata*, ETASLibri, Milan 1978.; Luhmann N. (1975), *Macht*, Italian translation, *Potere e complessità sociale*, Il Saggiatore, Milan 1979; Elias N., *Die Gesellschaft der Individuen*, Frankfurt, Suhrkamp, 1987, Italian translation *La società degli individui*, Il Mulino, Bologna 1990; Coleman J.S. (1990), *Foundations of Social Theory*, Italian translation, *Fondamenti di teoria soci-*

We are trying to *inhabit* an interconnected/hyperconnected society, a new global ecosystem which, in addition to living on emergence (hypercomplexity) and emergencies (also in the sense of *perilous public situations*), self-feeds/self-produces its change and (non-linear) evolution by means of emerging variables and situations which we humans, punctually, continue to fool ourselves we can control, manage and predict fully. *Emerging* variables and *properties*, as well as unexpected

ale, Il Mulino, Bologna 2005; Hannerz H. (1992), *Cultural Complexity. Studies in the Social Organization of Meaning*, trad. It., *La complessità culturale. L'organizzazione sociale del significato*, Il Mulino, Bologna 1998; Ardigò A. (1989), *La sociologia dell'attore sociale di fronte alla sfida dell'ipercomplessità societaria*, in, Studi di Sociologia—Anno 27, Fasc. 3 (luglio-settembre 1989), pp. 267–280; Ardigò A. (1990). *L'ipercomplessità come problema epistemologico e di organizzazione sociale. L'approccio del sociologo dopo le cibernetiche e la socio-sistemica*, in, Ardigò A., Mazzoli G. (ed.), *L'ipercomplessità tra socio-sistemica e cibernetiche*. FrancoAngeli, Milan, pp. 13–40; Crespi F., *Manuale di Sociologia della cultura*, Laterza, Rome-Bari 1996; Belardinelli S., *Il progetto incompiuto. Agire comunicativo e complessità sociale*, FrancoAngeli, Milan 1996; Ferrarotti F., *La sociologia alla riscoperta della qualità*, Laterza, Bari 1989; Ferrarotti F. (1991), *Trattato di sociologia*, UTET, Turin; Berger P.L., Luckmann T. (1966), *The Social Costruction of Reality*, Italian translation, *La realtà come costruzione sociale*, Il Mulino, Bologna 1969; Izzo A., *L'invincibile perplessità. Fondamenti, storia e problemi della sociologia della conoscenza*, Armando, Rome 1999; Cipolla C. (2014), *Oltre il "peccato originale" della selezione naturale. Alcune proposte teoriche nella società digitale*, FrancoAngeli, Milan; Ferrara A., *Pensare la società. L'idea di una filosofia sociale*, Carocci, Rome -Bari 2001; Bourdieu P. (2001), *Science de la science et réflexivité*, Italian translation, *Il mestiere di scienziato*, Feltrinelli, Milan 2003; Hughes J.A., Sharrock W.W. (1997 [3]), *The Philosophy of Social Research*, Italian translation, *Filosofia della ricerca sociale*, Il Mulino, Bologna 2005; Hedström P. (2005), *Dissecting the Social. On the Principles of Analytical Sociology*, Italian translation, *Anatomia del sociale. Sui principi della sociologia analitica*, Bruno Mondadori, Milan 2006. Hughes J.A., Sharrock W.W. (1997 [3]), *The Philosophy of Social Research*, Italian translation, *Filosofia della ricerca sociale*, Il Mulino, Bologna 2005.

From among our publications see Dominici P. (1995–1996). *Per un'etica dei new-media. Elementi per una discussione critica*, Florence: Firenze Libri Ed. 1998; (2003). *La comunicazione nella società ipercomplessa. Istanze per l'agire comunicativo e la condivisione della conoscenza*, Rome: Aracne Ed. 2005; (2005), *La comunicazione nella società ipercomplessa. Condividere la conoscenza per governare il mutamento*, Milan: FrancoAngeli, 2011 (nuova ed.); Dominici P. (2008), "Sfera pubblica e società della conoscenza" in Morcellini M., Mazza B. (a cura di), *Oltre l'individualismo. Comunicazione, nuovi diritti e capitale sociale*, Franco Angeli, Milano; (2009), *La società dell'irresponsabilità*, FrancoAngeli, Milan 2010; (2014). *Dentro la società interconnessa. Prospettive etiche per un nuovo ecosistema della comunicazione*, Milan: FrancoAngeli; (2014). *La modernità complessa tra istanze di emancipazione e derive dell'individualismo*, in "Studi di Sociologia", n°3/2014, Milan: Vita & Pensiero; (2015). *Communication and Social Production of Knowledge. A new contract for the Society of Individuals*, in "Comunicazioni Sociali", n°1/2015, Milan: Vita & Pensiero; -*The hypertechnological civilization and the urgency of a systemic approach to complexity. A New Humanism for the Hypercomplex Society*" in, Neto A.B.F., Nieto E.C., Chiodi G., Petroccia S., (Eds.)., *Governing Turbolence. Risk and Opportunities in the Complexity Age*, Cambridge Scholars Publishing, Cambridge 2017. Again with regard to these issues, see also the entry/essay—*La complessità della complessità e l'errore degli errori*, in Enciclopedia Italiana TRECCANI, Treccani, Year 2018 http://www.treccani.it/magazine/lingua_italiana/speciali/digitale/5_Dominici.html

perturbations, are a matter, as we know, of typical dimensions and elements characteristic of *complex systems*, veritable "organisms" that are structurally dynamic, *dissipative*,[2] unpredictable and, above all, *irreversible* in their dynamics.

[2]It is useful to clarify further some fundamental assumptions: complex systems are "organisms" and, as such, they are dissipative structures that are constituted by means of self-organizational processes, originating from a condition of initial disorder—which already implies dissipation. Furthermore, the system-organism (continuously) acquires "energy" from the "external" environment/ecosystem of reference and, at the same time, triggers/causes/determines "entropy" in the same environment/ecosystem of reference, with which, evidently, there is not only a simple relationship of dependence/adaptation or prevalence of the environment (Darwin) on the system-organism. The complex relationship between the system-organism—with all the dimensions/issues relating to *homeostasis*—and the reference environment/ecosystem is one of radical interdependence and reciprocal conditioning. We shall return to this later. Among the "classics", see: Schrödinger E. (1944), *What is Life? The Physical Aspect of the Living Cell*, Italian translation, *Che cos'è la vita?*, Adelphi, Milan 1995 (There is also his concept of "negentropy" or negative entropy, or syntropy, capable of bringing systems back to order. For Schrödinger, "matter feeds on negative entropy" and again "living matter escapes decay moving towards states of equilibrium"...); Morin E. (1973), *Le paradigme perdu: la nature humaine*, Italian translation, *Il paradigma perduto. Che cos'è la natura umana?*, Feltrinelli, Milan 1974; Bertalanffy von L. (1968), *General System Theory: Foundations, Development, Applications*, Italian translation, *Teoria generale dei sistemi*, Isedi, Milan 1975; Monod J. (1970), Le hazard et la nécessité, Italian translation *Il caso e la necessità. Saggio sulla filosofia naturale della biologia contemporanea*, Milan, Aronoldo Mondadori Ed. 1997; Morin E. (1977), *La Méthode I. La Nature de la Nature*, trad. it. *Il metodo 1. La natura della natura,* Raffaello Cortina Editore, Milan 2001; Morin E. (1980), *La Méthode II. La Vie de la Vie*, trad. it., *Il metodo 2. La vita della vita*, Raffaello Cortina Editore, Milan 2004; Prigogine I. (1980), *From Being to Becoming: Time and Complexity in the Physical Sciences*, Italian translation di Bocchi G. e Ceruti M., *Dall'essere al divenire,* Einaudi, Turin 1986; Miller J.G. (1971), *Living system*, tr. it., *Teoria generale dei sistemi viventi*, FrancoAngeli, Milan 1986[3]; Cramer F. (1988), *Chaos und Ordnung. Die Complexe Struktur des Lebendigen*, Italian translation, *Caos e Ordine. La complessa struttura del vivente*, Turin: Bollati Boringhieri 1994; Prigogine I. (1996), *La fin des certitudes. Temps, chaos et les lois de la nature*, Italian translation, *La fine delle certezze. Il tempo, il caos e le leggi della natura*, Turin: Bollati Boringhieri, 1997; Maturana H., Varela F. (1972), *Autopoiesis and Cognition. The Realization of the Living*, Reidel Italian translation, *Autopoiesi e cognizione. La realizzazione del vivente*, Venezia, Marsilio, 1985. On these issues and, specifically, on the concept of "autopoietic organization", see also Maturana H.R., Varela F.J. (1985), *The Tree of Knowledge*, Italian translation, *L'albero della conoscenza*, Milan: Garzanti 1987; Prigogine I.—Stengers I. (1984), *Order out of Caos*, New York: Bentham Books; Heisenberg W., *Physics and Philosophy: the Revolution in Modern Science*, Prometheus Books, Buffalo N.Y., 1958; Margulis L., *Symbiosis in Cell Evolution in the Archean and Proterozoic Eons*, Freeman, New York 1993; Nicolis G.—Nicolis C. (2007), *Foundations of Complex Systems*, Singapore: World Scientific; Goodwin B., *How the Leopard Changed Its Spots. The evolution of Complexity,* Princeton Science Library, Princeton 1994; Israel G., *La macchina vivente. Contro le visioni meccanicistiche dell'uomo*, Bollati Boringhieri, Turin, 2004; Bertuglia C.S., Vaio F., *Non linearità, caos, complessità*, Bollati Boringhieri, Turin, 2003; Israel G., *The Science of Complexity. Epistemological Problems and Perspectives*, in "Science in Context", 18, Anno 2005, pp. 1–31. See also, paying particular attention to them, studies and research carried out by the Italian physicist Giorgio Parisi, (an extraordinary person and not only as a scientist, recently awarded the Nobel Prize for Physics) both with reference to complexity and the theory of "disorderly systems": Parisi

It is a matter of an interconnected society/a global ecosystem comprising infinite, closely interdependent and hyperconnected ecosystems and subsystems, which would require, contrary to what actually happens, a systemic vision and an oft-mentioned *paradigm shift*. Moreover, no action and/or strategy is carried out considering the long term and/or, in any case, with a view to building a "culture of responsibility and prevention". In other words, no one works on the creation of a "culture of complexity" (1995–1996) which can only be configured as a "culture of responsibility". To all this, we need to add the continuous maintenance, within educational and training institutions, of an *overall architecture of fields of knowledge and skills* which, marked—as always—by a logic of the separation between, and the impermeability of the disciplinary sectors and a logic of control/power, makes any attempt at changing the methodologies, epistemology and approaches to social and human complexity, unfeasible. Humanistic, critical, systemic training continues to be deemed useless. We continue not to consider the social sciences and their dialogue with the so-called fundamental, exact sciences (specialization vs the multi/inter/transdisciplinarity, which I had referred to in the mid-nineties with the term "false dichotomies"). All the most creative disciplines continue to be seen as "useless", while recognition and "value" are assigned only to evident (?), empirical, immediate and measurable, quantitative and statistical results, relying, moreover, entirely on the (simple) solutions presented by technology. The months of 2020 spent in *lockdown*, in quarantine, in isolation, did not teach us and probably in the future will not really teach us anything, because we live in *a permanent state of inadequacy*, incapable of *systemic thinking*, of creating an adequate *system of thought* to deal with the *hyper complexity* of the change taking place. We continue to believe and entertain the delusion that we know how/are able to predict and control, even *pre-determine everything*, instead of learning *to inhabit error*,[3] the doubts

G., *Field Theory, Disorder and Simulations*, World Scientific Pub. Co Inc., Singapore 1992; Parisi G., *La complessità da un punto di vista di un fisico*, in Rendiconti Accademia Nazionale delle Scienze—*Memorie di Scienze Fisiche e Naturali* 119° (2001), Vol. XXV, pp. 7–14; Parisi G., Urbani P., Zamponi F., *Theory of Simple Glasses*, Cambridge University Press 2020. See also his latest book, more general in intention Parisi G., *In un volo di storni. Le meraviglie dei sistemi complessi*, Rizzoli, Milan 2021; see also the recent work of: Benasayag M., Cany B. (2021), *Les Nouvelles Figures de l'Agir*, Italian translation, *Corpi viventi. Pensare e agire contro la catastrofe*, Feltrinelli, Milan 2022.

And more recent works such as: Al-Khalili J., McFadden J. (2014), *Life on the Edge. The Coming of Age of Quantum Biology*, Italian translation, *La fisica della vita. La nuova scienza della biologia quantistica*, Bollati Boringhieri, Turin 2015; Catania K. (2020), *Great Adaptations*, Italian translation, *Adattamenti meravigliosi. Sette irresistibili misteri dell'evoluzione*, Bollati Boringhieri, Turin 2021; on "variability" as a primary property of the living, see: Kupiec J-J., *La concezione anarchica del vivente*, Milan: Eleuthera 2021; Zimmer C., *Life's Edge: The Search for What it Means to Be Alive*, Palgrave Macmillan, 2021; Josup M., Holme P., Kanazawa K. Et al., *Social Physics*, in Physics Reports, 948 (2022), 1–148, Elsevier.

[3] N.B. The use of an asterisk indicates a definition proposed by the author.

and the unpredictability that distinguish and shape complex systems during their non-linear evolution.[4]

The extraordinary scientific discoveries and technological innovations of the past few decades, besides having endowed us with increasingly efficacious opportunities by which to determine the mechanisms of biological evolution and providing opportunities (at least for now, *for a few*) and unimaginable scenarios, which, with incalculable potential, have ushered us, definitively, into a time of *maximum unpredictability, uncertainty and obsolescence* (of knowledge and skills), into systemic and, by now, existential dimensions as well. Furthermore, the systemic relationship, no longer simply dialectical, between knowledge, risk and trust has become stronger within the framework of total interdependence.

This time, the paradigm shift is so profound and irreversible that it obliges us to rethink/redefine everything and question the very idea and concept of "science" in all its fundamental, epistemological and methodological pillars. Above all, we need to question the principle of "observability",[5] variables and phenomena, and the conviction/belief/axiom that all *reality*, including the Social and the Human, are reduc-

---

[4] Over the years, I have returned several times, with studies and research, to these questions. See, in particular: Dominici P. (1996). *Per un'etica dei new-media. Elementi per una discussione critica*, Florence: Firenze Libri Ed. 1998; (2005), *La comunicazione nella società ipercomplessa. Condividere la conoscenza per governare il mutamento*, Milan: FrancoAngeli, 2011 (nuova ed.); *La società dell'irresponsabilità*, FrancoAngeli, Milan 2010; (2014). *Dentro la società interconnessa. Prospettive etiche per un nuovo ecosistema della comunicazione*, Milan: FrancoAngeli; Dominici P. (2016a), "La filosofia come "dispositivo" di risposta alla società asimmetrica e ipercomplessa", in Candiotto L., Gambetti F., *Il diritto alla filosofia. Quale filosofia nel terzo millennio?*, Diogene Multimedia, Bologna; (2016b). *L'utopia Post-Umanista e la ricerca di un Nuovo Umanesimo per la Società Ipercomplessa*, in "Comunicazioni Sociali", n°3/2016, Milan: Vita & Pensiero; *Il grande equivoco. Ripensare l'educazione (#digitale) per la Società Ipercomplessa* [*The Great Mistake. Rethinking Education for the Hypercomplex Society*], in "Fuori dal Prisma", Milan: Il Sole 24 Ore, 2016; *L'utopia Post-Umanista e la ricerca di un Nuovo Umanesimo per la Società Ipercomplessa*, in "Comunicazioni Sociali", n°3/2016, Vita & Pensiero, Milan 2016; *The Hypercomplex Society and the Development of a New Global Public Sphere: Elements for a Critical Analysis*, in, RAZÓN Y PALABRA, Vol. 21, No. 2_97, Abril-junio 2017—ISSN: 1605–4806, pp. 380–405; *L'ipercomplessità, l'educazione e la condizione dei saperi nella Società Interconnessa/iperconnessa*, in "Il Nodo. Per una pedagogia della Persona", Anno XXI, n°47, Falco Editore, Cosenza 2017, pp. 81–104; *Dentro la Società Interconnessa. La cultura della complessità per abitare i confini e le tensioni della civiltà ipertecnologica,* FrancoAngeli, Milan 2019; *Controversies on hypercomplexity and on education in the hypertechnological era*, in, A. Fabris & G. Scarafile, Eds, *Controversies in the Contemporary World*, Amsterdam-Philadelphia: John Benjamins Publishing Company, 2019; Dominici P., *La complessità della complessità e l'errore degli errori*, in Enciclopedia Italiana "Treccani", Istituto della Enciclopedia Italiana Treccani, Anno 2019 http://www.treccani.it/magazine/lingua_italiana/speciali/digitale/5_Dominici.html; Dominici P., "*Educating for the Future in the Age of Obsolescence*", in CADMUS, Volume 4—Issue 3, November 2020, pp. 93–109. To broaden your view further see Godfrey-Smith P. (2021), *Theory and Reality. An Introduction to the Philosophy of Science*, Italian translation, *Teoria e realtà. Introduzione alla filosofia della scienza*, Raffaello Cortina Ed., Milan 2022.

[5] On these questions see, in particular: Dominici P., op.cit., 1996–2020; Blastland M. (2019), *The Hidden Half. How the World Conceals its Secrets,* Italian trans. *La metà nascosta. Le forze invisibili che influenzano ogni cosa,* Bollati Boringhieri, Torino 2021.

ible and attributable to objective variables, dimensions and measures, as well as translatable into quantitative data and statistical and mathematical models. *From qualitative to quantitative*: this is the riskiest and most misleading of reductions and simplifications and not only within the field of scientific research. Because not everything can be simplified *and/or reduced*, not everything is manageable (quite the opposite!), but technologies of connections and the infinite and inexhaustible availability of data are flooding us with extremely *risky* certainties and securities. This was certainly brought home to us not too long ago, during the global pandemic, when each and every *hyperconnected, hyper-accelerated and technologically controlled/supervised—but scientifically (dis)organized*—nation-state was simultaneously thrown into an un precedented state of crisis by a mere virus.

1.1 For a Socio-Systemic and Relational Paradigm

In this sense, the social sciences, based also on scientific research methods and on a *socio-systemic approach*, are interested not only in the fundamental dimensions required to describe, measure and observe but also, and, above all, in those regarding *understanding and living*. These proceed by conjectures and refutations, resorting to both qualitative and quantitative methodologies and tools of detection, in an effort to recover/recuperate the complex, unpredictable and heterogeneous dimensions of the Social, but also to acquire sense and meaning by recovering heuristics and epistemologies, too often overlooked or underestimated. In particular, sociology, which I intend and define as the *social science of complexity*, does its utmost to investigate, understand and *inhabit the hybrid zones of change,* by studying the static and dynamic aspects of social systems, focusing on social praxis and, above all, trying to evaluate the impact of new paradigms, sometimes even at loggerheads with each other, in particular those concerning human relationships within different socio-cultural contexts. The relevance and necessity of this relational socio-systemic paradigm find their strategic centrality precisely in the complex dimensions of interaction and (systemic) "relationships". These dimensions are of vital importance and are capable of bringing not only knowledge and skills, but also, and, above all, this archetype closer to other scientific paradigms. It is a matter of an approach and a research method that might turn out to be decisive in the task, far from simple, of mending the tears, bridging the gaps, of healing the lacerations and fractures between the different disciplinary fields and experiences.

Sociology, in particular, is a social science open to dilemma, to open dialectics that fail to find any superior synthesis, to mutual contamination between varieties of knowledge and disciplines. It is, above all *a province of meaning* "open to the coexistence of different paradigms" (1996). In other words, it is *a social science open*, inevitably, to dialogue with other disciplines (there can be no sociology/social science without the fundamental contribution of philosophy, history, economics, statistics, social statistics, ethnography, anthropology, social psychology, even biology and physics etc. and, today, the sciences of computers, neurology and cognition).

This is due to the fact that sociology is vitally in need of cross-contamination for the precise purpose of clarifying its approach and, more so, of endowing it with practical operationality. It is social science which, with the essential contribution of Philosophy—also in terms of continuous orientation and theoretical re-orientation/re-positioning –of Epistemology and research methodology in particular, seeks to know, understand and inhabit the hybrid areas of change.

Ambivalence, ambiguity, conventionality and heterogeneity are just some of the structural characteristics of Sociality as well as of social and human systems. Sociology, as a social science in a state of constant (precarious) equilibrium hovering between the qualitative and the quantitative, without renouncing methodological rigour, which is the only antidote to the Weberian *impossibility of a truly non-evaluative knowledge of reality*, is open to the coexistence of paradigms, as mentioned above, and should be based only on an approach that is necessarily systemic and socio-relational, capable of identifying and recognizing levels of connection, correlation and *systemic and non-linear causality*. The single social actor is already a "complex system" as such, a matter of "relationship" stemming from the systemic dimension of processes and phenomena which generate norms, values, beliefs, social structures, networks, systems of value and cognitive orientation.

On the other hand, we are going through and trying to inhabit an era marked by profound changes and processes of *complex synthesis*, whose epistemological and ethical implications, in addition to throwing open completely new perspectives and trajectories, we are unable to evaluate fully, despite having at our disposal increasingly sophisticated tools for measuring, detecting and processing data. We are experiencing an accelerated phase of radical change of paradigms and an *anthropological transformation* which is assuming the form of a *progressive overturn of the complex interaction between biological and cultural evolution*[6] (Fig. 1.1).

Divisions, breakdowns, the logic of separation and confinement, fractures, "false dichotomies",[7] are traits that reveal and lay bare all their inadequacy and ineffectiveness. Boundaries and limits, between nature and culture, between the *natural* and artificial, which have broken down gradually and completely. Science, and technology which—flanked by the increasingly strategic and vital role of communications, especially within the public sphere and public opinion—have triggered an unprecedented paradigmatic shift with reference, above all, to perspectives concerning the advent of artificial intelligence, have *opened* a new "epistemological fracture", to which we shall return later. This fracture is reconfiguring the entire phenomenal reality and conditioning the actors/organisms that are its more or less, active and/or

---

[6] See Dominici P. (1995–1996), *Per un'etica dei new media. Elementi per una discussione critica*, Firenze Libri Ed. Florence 1998. See also Dominici P. (2003). *La comunicazione nella società ipercomplessa. Istanze per l'agire comunicativo e la condivisione della conoscenza*, Rome: Aracne Ed. 2005; (2005), *La comunicazione nella società ipercomplessa. Condividere la conoscenza per governare il mutamento*, Milan: FrancoAngeli, 2011 (nuova ed.); Dominici P. (2008), "Sfera pubblica e società della conoscenza" in Morcellini M., Mazza B. (a cura di), *Oltre l'individualismo. Comunicazione, nuovi diritti e capitale sociale*, Franco Angeli, Milan.

[7] *Ibidem.*

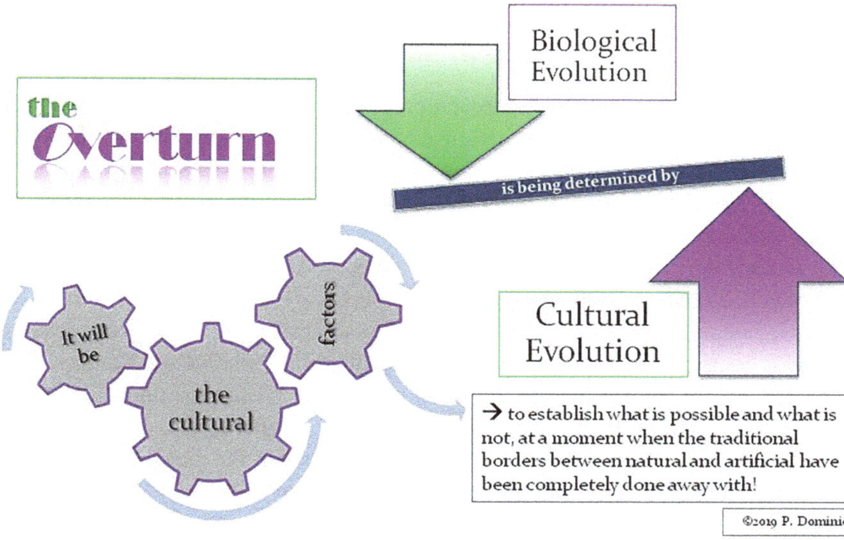

Fig. 1.1 The overturn of the complex interaction between biological and cultural evolution: an anthropological transformation

participating protagonists. The *boundaries and limits* in question are destined to transform themselves more and more into *hybrid zones* of contamination which, at least for the moment, find our educational and training institutions, as well as public and private organizational cultures, completely unprepared to address variability, ambivalence, deviance, diversity but also, and, above all, to deal with the unexpected, the unpredictable, the indeterminate and the unexpected. These are some (some of the many) of the structural ontological, connotative characteristics of complexity and complex systems.

Complexity, on the other hand, is an essential characteristic of *organic aggregates*, in other words of biological, social, relational, human systems, always characterized by systemic relationships and numerous levels of interconnection/ interdependence: open and *dissipative* systems,[8] comprising"/units/components in

[8]complexity and "complex systems", fundamental contributions have arrived, over time, from all disciplinary and research fields: from biology to physics, cybernetics to philosophy, economics to sociology, chemistry to mathematics, from the sciences of education to psychology, from medicine to organisational studies, to all social, human and life sciences. I shall leave aside the arts and the more creative disciplines which, although not based on "observability" and the "scientific method", have proved to be extremely important as far as this complex evolution is concerned. Within the literature of scholarship, which, to say the least, is immense besides being multidisciplinary, see, in particular, among the "classics": Anderson, P., *More is Different*, Science, 04 Aug 1972, Vol 177, Issue 4047, pp. 393–396; Prigogine I.—Stengers I. (1979), *La Nouvelle Alliance. Métamorphose de la science*, Italian translation, *La nuova alleanza. Metamorfosi della scienza*, Turin: Einaudi, 1981; Prigogine I. (1980), *From Being to Becoming: Time and Complexity in the Physical Sciences*, Italian translation di Bocchi G. e Ceruti M., *Dall'essere al divenire,* Einaudi, Turin 1986; Cramer F. (1988), *Chaos und Ordnung. Die Complexe Struktur des Lebendigen*, Italian translation, *Caos e*

themselves complex; systems extremely sensitive to external *perturbations* and marked by irreversible dynamics at what has been defined the *"edge of chaos"*,[9] *capable of self-organization*—both as single constituent "units" and as a "system"—well-structured and, as mentioned previously, made up of "parts" which, in their multiple and systemic interactions, condition the behaviour and the (non-linear) evolution of the systems themselves. (Complex) systems which, among many peculiar and founding dimensions, are not analysable/observable/interpretable/understandable, as if we were "external observers" and/or neutral ones. Instead, as far as the *world of objects and things* is concerned, we are speaking of *complicated* (mechanical and artificial) and non-complex *systems,* which can be described in terms of mathematical formulae. These are systems (immediately) "observable" in all their constituent dimensions, measurable, therefore predictable (manageable) and, above all, broken up into their constituent parts in order to understand their behaviour and workings. Although, for the sake of completeness, there would be much to write about *deterministic chaos*, but also about the profound impact of artificial intelligence on the architecture of the fields of knowledge, practices and experiences. I have spoken of this in the past—as already mentioned—in terms of a "new epistemological fracture",[10] on a par with the *discovery of chaos* and its related theory.

Ordine. La complessa struttura del vivente, Bollati Boringhieri, Turin 1994; Cini M., *Un paradiso perduto. Dall'universo dele leggi naturali al mondo dei processi evolutivi*, Feltrinelli, Milan 1994; Prigogine I. (1996), *La fin des certitudes. Temps, chaos et les lois de la nature*, Italian translation, *La fine delle certezze. Il tempo, il caos e le leggi della natura*, Turin: Bollati Boringhieri, 1997; Prigogine I.—Stengers I. (1984), *Order out of Caos*, New York: Bentham Books; Heisenberg W., *Physics and Philosophy: the Revolution in Modern Science*, Prometheus Books, Buffalo N.Y., 1958; Heisenberg W. (1959, 9th ed.), *Wandlungen in den Grundlagen der Naturwissenschaft*, Italian translation, *Mutamenti nelle basi della scienza*, Bollati Boringhieri, Turin 1978; Heisenberg W., *I principi fisici della teoria dei quanti*, Bollati Boringhieri, Turin 1976; Heisenberg W. et al., *Discussione sulla fisica moderna*, Bollati Boringhieri, Turin 1980; Bar-Yam Y., *Dynamics of Complex Systems*, Addison-Wesley, Reading, Massachusetts 1997; Kauffman S.A. (1993), *The Origin of Order. Self-organization and Selection in Evolution*, Oxford University Press, Oxford; Capra F. (1975), *The Tao of Physics*, Italian translation, *Il Tao della fisica*, Adelphi, Milan 1982; Capra F. (1996), *The Web of Life*, Italian translation, *La rete della vita. Una nuova visione della natura e della scienza*, Rizzoli, Milan 2001; Heisenberg W., *Lo sfondo filosofico della fisica moderna*, a cura di G. Gembillo e E.A. Giannetto, Sellerio, Palermo 1999; Heisenberg W., *Indeterminazione e realtà*, ed. G. Gembillo e G. Gregorio, Guida, Napoli 2002 Ageno M., *Le origini dell'irreversibilità*, Bollati Boringhieri, Turin 1999 e Grassi W., *Farfalle e uragani. Complessità: la teoria che governa il mondo*, Hoepli, Milan 2018; Complexity: the theory that governs the world, Hoepli, Milan 2018; interesting, again from this perspective, the concept of "synergetics", proposed by the physicist Hermann Haken, to define the "science of combined effects" which studies systems comprising numerous subsystems in a relationship of continuous and reciprocal interaction. See Haken H. (1977), *Synergetics: An Introduction. Nonequilibrium Phase-transitions and Self-organization in Physics, Chemistry, and Biology*, Heidelberg: Springer (new ed. 1983).

[9] Packard, N. (1988) "Adaptation Toward the Edge of Chaos". University of Illinois at Urbana-Champaign, Center for Complex Systems Research.

[10] See Dominici P., (1996, 2003–2005, 2014–2020), *op.cit.*

In the meantime, we humans, unaware of our limits and fragility, continue to try to control, to cage, to mould/model all the complexity of the human and the social, the *vitality of the spirit* and of the "unobservable", confining them to mathematical models and formulae, to infinite sequences of data and numbers, to molecules, synapses, hormones and chemical reactions. And, in doing so, we also try to *visualize* what cannot be visualized: the mere idea that something can escape our "gaze" and our control has always frightened us. And it is precisely this attitude—this *organized/disorganized obsession* with control—which, paradoxically, condemns us to unpreparedness and to waiting eternally for a black swan (ancient metaphor and classical *post-rationalization)* unaware of how emergencies, dynamism and infinite variability are inherent elements of complex systems. Moreover, historically speaking, the more we humans try—and delude ourselves—in every situation and under every circumstance, to maintain and/or exercise control over phenomenal reality (that is, the reality of phenomena) and our existence, not to mention control/power over Nature, the more this control eludes us to appear and be confirmed as a *chimaera,* not only of industrial modernity. In actual fact, we are dealing with complex trends and dynamics that emerge, in an even more evident and radical way, in the hypertechnological and hyper-connected civilization forced to deal with a sort of "New Nature" (1996), *within* which "natural" and "artificial" are confounded completely.

An organized/disorganized obsession (as I call it) besides all else, requires and produces further breakups and fragmentations, divided into systems (apparently?) more and more detached from the overall picture and the general situation.

The Human, the Social, the Vital and, more in general, "complex systems" *cannot be reduced or simplified, much less made wholly measurable,*[11] *predictable and*

[11] To further explore this hypothesis of study and research on which I as well as the numerous intimately related issues, have been working for over twenty-five years, I shall limit myself to dealing with issues from a perspective that can only be multi/inter/transdisciplinary: Schrödinger E. (1944), *What is Life? The Physical Aspect of the Living Cell*, Italian translation, *Che cos'è la vita?*, Adelphi, Milan 1995; Schrödinger E. (1958), *The Physical Basis of Consciousness, Science and Religion, The Mystery of the Sensual Qualities*, Italian translation, *L'immagine del mondo*, Bollati Bopringhieri, Turin 1963 e 2017; Damasio A.R. (1994), *L'errore di Cartesio. Emozione, ragione e cervello umano*, Adelphi, Milan 2009; Morin E. (1980), *La Méthode II. La Vie de la Vie*, trad. it., *Il metodo 2. La vita della vita*, Raffaello Cortina Editore, Milan 2004; Dominici P., *For an Inclusive Innovation. Healing the fracture between the human and the technological*, in, European Journal of Future Research, Springer, 2017; *Controversies on hypercomplexity and on education in the hypertechnological era*, in, A. Fabris & G. Scarafile, *Controversies in the Contemporary World*, John Benjamins Publishing Company, 2019; "*La complessità della complessità e l'errore degli errori*", *in* TRECCANI—sezione "Lingua Italiana"—Istituto Enciclopedia Italiana "Treccani"—December 2018. See also: C.J. Preston (2018), *The Synthetic Age*, Italian translation, *L'era sintetica. Evoluzione artificiale, resurrezione di specie estinte, riprogettazione del mondo*, Einaudi, Turin 2019; M. Tegmark (2017), *Life 3.0. Being Human in the Age of Artificial Intelligence*, Italian translation, *Vita 3.0. Essere umani nell'era dell'intelligenza artificiale*, Raffaello Cortina Ed, Milan 2018; See Éric Sadin (2018), *L'Intelligence artificielle on l'enjeu du siècle*, Italian translation, *Critica della ragione artificiale*, LUISS University Press, Rome 2019; H. Fry (2018), *Hello World. How to Be Human in the Age of the Machine*, Italian translation, *Hello World. Essere umani nell'era delle machine*, Bollati Boringhieri, Turin 2019. Of the classics see: H. Jonas (2001), *The Phenomenon of Life*, Italian translation, *La cibernetica e lo*

manageable, also because, besides being extremely sensitive to external perturbations that modify, irreversibly, the initial conditions, not all the structural/constitutive/interdependent dimensions that characterize them are "observable". This is one of the key points of the "management/control of complexity" issue. On the contrary, instead of separating and breaking things down into parts, we should try to observe and recognize their complexity, re-assemble what appears to be separate (education and training), recognize and highlight their systemic connections and relationships (knowledge), *their vitality of spirit*, the essential in them, which, as *The Little Prince* held, is 'invisible to the eye'.[12] We need to acquire awareness, something far from being attained and metabolized as yet, that—as mentioned above- as not everything is "observable" it cannot be "measured" and "foreseen" We continue to define "boundaries" and to feed fractures that do not lead to knowledge, but to that sense of reassurance (necessary and functional) with respect to the uncertainty of life and the indeterminacy of reality. Beyond the significant delay with which we become aware of it, it is time to inhabit these border areas and to try to heal these fractures, aware of our incompleteness and the urgent need to learn how *to inhabit hypercomplexity*, chaos and emerging emergency.

And indeed... *Order and Chaos*[13]: it is no longer sufficient to try to distinguish between them to restore the balance and control that have been lost. Seeing that

---

scopo: una critica, ETS, Pisa 1999. Finally, I recall another classic—in my opinion—extraordinary: Koyré A. (1961), *Les philosophes et la machine. Du monde de l'"à-peu-près" à l'universede la précision*, Italian translation, *Dal mondo del pressappoco all'universo della precisione*, Einaudi, Turin 1967; Dominici P., *Dentro la "civiltà senza errore". Sintesi complesse: l'Umano, il legame sociale e la civiltà ipertecnologica,* in A. De Cesaris (ed.)*Vite digitali. Essere umani nella società del XXI Secolo*, FrancoAngeli, Milan 2020; Al-Khalili J., McFadden J. (2014), *Life on the Edge. The Coming of Age of Quantum Biology*, Italian translation, *La fisica della vita. La nuova scienza della biologia quantistica*, Bollati Boringhieri, Turin 2015; Lanza R., Pavšič M., Berman B. (2020), *The Grand Biocentric Design*, Italian translation, *Il grande disegno biocentrico. Come la vita crea la realtà*, Il Saggiatore, Milan 2022.

[12] See Saint-Exupery Antoine de, *The Little Prince*, Reynal and Hitchcock/Gallimard, Paris 1943. On the topic of "invisibility", on the history of the relationship between humanity and the invisible, see the fascinating account by Ball P. (2014), *Invisible. The dangerous allure of the Unseen*, Italian translation, *L'invisibile. Il fascino pericoloso di quel che non si vede*, Einaudi, Turin 2016. I wish to mention a great "classic" on these questions, in particular on the problematization of the opposition (?) between "visibility" and "invisibility", their intimate relationship and interdependence: Merleau-Ponty M. (1964), *Le visible et l'invisible*, Italian translation, *Il visibile e l'invisibile*, Bompiani, Milan 1969.

[13] I shall return to the theme of "chaos" several times, and with the necessary bibliographic references and in-depth studies. Here I limit myself to recalling the interesting cycle of lectures held by Ilya Prigogine at the State University of Milan and the chair of Philosophy of Science, then held by Giulio Gioriello (12–14 February 1992), and published in: Prigogine I. (1992), *Le leggi del caos*, Laterza, Bari-Rome 1993 (see also ed. 2003). See the excellent: Cramer F. (1988), *Chaos und Ordnung. Die Complexe Struktur des Lebendigen*, Italian translation, *Caos e Ordine. La complessa struttura del vivente*, Turin: Bollati Boringhieri 1994. See also Ruelle D. (1991), *Hasard et Chaos*, Italian translation, *Caso e Caos. La scoperta della complessità ai confine tra matematica e fisica*, Bollati Boringhieri, Turin 1992; Mitchell Waldrop M., *Complexity. The Emerging Science at the Edge of Order and Chaos*, Simon & Schuster Paperbacks, New York 1992; Strogatz H.S., *Sync: How Order Emerges from Chaos in the Universe, Nature, and Daily Life*, Hachette Book Group, 2004.

even in complex systems, order and chaos coexist, they respond retroactively within the systemic framework of the *complexity of the living* and, even more so, of the social, which continues to reveal itself, in all its dynamic and unstable being, which is never observable, measurable, predictable, never fully comprehensible and intelligible, except by resorting to reductionist and deterministic explanations. Furthermore, the "complex systems" which can be described, represented and visualized as *networks comprising nodes and connections*, have various peculiar characteristics and dimensions. These are (1) *systems featuring non-equilibrium thermodynamics*, within which forces/variables/interactions/actors that make them constantly dynamic and unstable act; they are (2) characterized by "emergent properties", unobservable from the outset, linked to the entire network and not to single variables/factors, which manifest themselves exclusively during their non-linear evolution; (3) consequently, they turn out to be unpredictable, they are traversed by chaotic dynamics that highlight the predictive limits of Science itself; (4) *complex systems* reveal themselves and develop non-linearly and in variable forms; (5) they have a capacity for self-organization and self-generation; (6) they are "open" systems, i.e. sensitive to external perturbations of the environment/ecosystem and to any kind of feedback; (7) they are divided into closely interconnected, interdependent *subsystems* (Simon's hierarchy principle); (8) finally, complex systems, especially when it comes to the social and the human, do not necessarily share the same space-time (indeed): this implies that the relative interactions manifest themselves and are recognizable even at extremely significant distances.

The hypertechnological and hyperconnected civilization—increasingly conceived, designed and built as a "mechanism" and not as an "organism"—continues, erroneously, to be built on the assumption of the progressive *marginalization of the Human* and its relative systemic space of relations and responsibility. It is marked by a progressive, exponential, unstoppable growth of the dimension of the *technologically controlled*, capable—at least, apparently—of circumscribing/reducing the probabilities of errors and unexpected events. This kind of civilization is based on programming, automation and (hyper)simulation of processes and dynamics, which bring with them a series of *risky illusions* (the result of forms of *reductionism* and *determinism*), in the face of a growing, increasingly more evident and recognizable hypercomplexity, which characterizes the ongoing change that educational and training institutions are largely unable to handle. Given the quantitative and qualitative exponential growth of the myriad inter-relations that feed the local and global phenomena and processes, we have been witnessing, for some time now, the hegemony of reductionist and deterministic notions and a paradoxical return to neopositivist concepts of reality.

Dynamics and processes that assume concrete form, on the one hand, in the sometimes obsessive quest for *simplification at all costs* (while it should be remembered that the opposite of complexity is not simplification, but reductionism), even when it is dangerous to simplify (education, communication, democracy), and, on the other, in what I defined, before the issue came to the fore, the *grand illusions of the hypertechnological civilization*: the illusions of rationality, control, predictability and measurability; and, finally, the most dangerous and deceptive illusion of all, that of being able to *eliminate error*, along with unpredictability, from organizations and from our lives: that is, eliminate what makes us "human beings", more than anything else, "free human beings".

Numerous illusions which, supported and strengthened further by the systematic "carte blanche" granted to technique/technology, find various operative interpretations by recurring to reductionist and deterministic approaches, based—concretely—on the exclusive involvement of specific disciplinary fields, knowledge and technical skills: that is, on those which seem best able to support and guarantee, even at the level of hegemonic narratives, those very same grand illusions.

In this sense, it is essential to start again from the *care* of thought—perhaps, in some ways, the most important *emergent property* found among complex systems[14]—from the urgency of "rethinking how we think" and of defining a thought system suited to the (hyper)complexity of the change taking place.[15] It is time to rethink the very idea and definition of "Science", accepting the epistemological and methodological challenge of *observing* and *investigating* (*besides attempting to know*) even the *non-observable*, and of deeming significant, within the pathways/prospects of scientific research, even what is *not measurable in quantitative terms.*

In other words, it is also necessary to start again from renewed awareness that the idea/view/perspective of transforming/translating/simplifying and reducing the "qualitative" (the complexity of life, of the social, of the Human) turning it into

---

[14] On matters relating to the brain, the mind and thought and, consequently, artificial intelligence, see, in particular, a great classic, highly significant with reference to studies and research regarding complexity: Minsky M. (1985), *The Society of Mind*, Italian translation, *La Società della Mente*, Adelphi, Milan 1989. Minsky argues that the human brain is a vast organized society, made up of many different parts. Crammed inside the human skull are hundreds of different kinds of motors and organizations, wonderful systems that have evolved and accumulated over hundreds of millions of years. Some of these systems, such as the parts of the brain that make us breathe, function almost independently. But in the majority of cases these parts of the mind have to coexist with the others, in a relationship that is sometimes one of collaboration, but more often one of conflict. It follows that our decisions and actions hardly ever have simple and unambiguous explanations but are generally the result of the activities of large societies of process in a continuous relationship of challenge, contrast or mutual exploitation. The great possibilities of intelligence derive from this enormous diversity, and not from a few simple principles (p. 21). See also the following interesting works: Searle J.R., *Minds, Brains and Programs*, Harvard University Press, 1984; Searle J.R. (1998), *Mind, Language and Society*, Italian translation, *Mente, Linguaggio e Società*, Raffaello Cortina Ed., Milan 2000; Noë A., *Perché non siamo il nostro cervello*, Raffaello Cortina Ed., Milan 2010. This is a text which, in addition to analyzing issues relating to thought and the brain, also reflects on the question of "consciousness", in an attempt to go beyond Descartes and his related idea/vision/conception which makes it coincide with something happening inside the Subject, excluding any connection with the environment of reference and the phenomenal world. See also: Crawford K. (2021), *Atlas of AI. Power, Politics and the Planetary Costs of Artificial Intelligence*, Italian translation, *Né intelligente né artificiale. Il lato oscuro dell'IA*, Il Mulino, Bologna 2021.

[15] On these fundamental questions, see: Dewey J. (1933), *How We Think*, Italian translation, *Come pensiamo*, Florence: La Nuova Italia, 2006; Emery F.E. (ed.) (1969), *Systems Thinking*, Italian translation, *La teoria dei sistemi. Presupposti, caratteristiche e sviluppi del pensiero sistemico*, FrancoAngeli, Milan 2001; Ceccato S. (1972), *La mente vista da un cibernetico*, Mimesis, Milan - Udine 2017.

"quantitative", quantitative data, is deceptive, misleading, even presumptuous.[16] To trace and recognise the *links*, connections and interdependencies existing between the parts, between processes, between phenomena and between People is the ultimate challenge, an educational challenge, above all! It involves seeking Error (source of knowledge and vitality), doubt, and, concretely, accepting unpredictability, the Other, the Human and Life.

1.2 *Black Swans* and Inadequate Paradigms

Life is emergence; it is permanently relational, an infinite, *non-linear sequence of dynamic processes* where the emergent manifests itself in all possible and unimaginable, unpredictable ways. Our lives are an "infinite sequence of numerous black swans",[17] according to the age-old metaphor used already by the ancients: "black swans" with numerous nuances, marked by an innate ambivalence and apparently inveterate contradictions. In this regard, one has the clear sensation that, often, at all levels of organizational action and social practice, those who, in the presence of situations/dynamics that have escaped their control, insist on the issue/metaphor of a "black swan",[18] of a unique and unforeseeable or, in any case, highly improbable event, are merely operating/constructing a ("classical") attempt at *post-rationalization,* capable of reassuring others and themselves with respect to the fact that, despite sporadic episodes, everything remains under control, manageable and

[16] These are fundamental issues and topics we have often reiterated in various scholarly and generalist publications: Dominici P., (1995–1996), (2003), (2009), (2015), (2016), (2017–2022). And it was precisely on these topics that, after years, as I "closed" the present monography, that, felicitously, I came across —and now mention with pleasure—an important work by the physicist Federico Faggin, a scholar of neuronal networks and inventor of the microprocessor, who, while investigating conscience and the mechanisms that characterise it, arrived at analogous, generative conclusions, confuting the idea/view—widespread and accepted within the scientific community-that mechanistic and reductionist paradigms are capable of explaining the Human and Life: Faggin, F, *Irriducibile. La coscienza, la vita, I computer e la nostra natura*, Mondadori, Milan, 2022.

[17] See Dominici P., *op.cit.*, 1996–2020.

[18] I am not referring, obviously, to Nassim Nicholas Taleb and his famous work *The Black Swan*. The well-known essayist, academic and stock broker of Lebanese origin, also the author of *Giocati dal chance* (and other essays), warning us against *irrefutable convictions*, only because they are confirmed by "empirical evidence" (on these topics, the scientific literature is immense, as well as multidisciplinary), which he defines the "Black Swan" (written, on purpose with a capital S): it is an (a) "isolated event", outside "normal expectations" (I add: once again, the weight of social processes of social representation, of social perception and of the "social acceptability" of risk, of the unexpected, of the improbable is well understood See also Dominici P., op.cit., 2010); an event with an (b) "enormous impact" which, inevitably, drives us humans to (c) elaborate/provide justifications for its occurrence, "to make it explainable and predictable" (see p. 11). I add: in the illusion of being able to make events explainable and, above all, predictable. Taleb N.N. (2007), *The Black Swan*, Italian translation, *Il Cigno Nero, come l'improbabile governa la nostra vita,* Il Saggiatore, Milan 2007. See also: Boorstin D. J. (1994), Cleopatra's Nose. Essays on the Unexpected, Vintage Books, New York..

predictable. And so, the age-old "illusion of control"—of *assuming mastery* over *hazard* and one's own destiny—continues to occupy a position of absolute hegemony, not only in terms of organizational and social imaginaries, despite the fact that evidence has laid bare its profound inconsistency and inadequacy. On the contrary, social and human life is characterized by infinite, non-linear sequences of black swans. Error, unpredictability, complexity and *systemic dynamism* are its constituent elements.

The systemic global emergency of 2020 (which has been in continuous evolution since then), the umpteenth "emergency" with global characteristics, certainly more important, systemic and invasive than others, has highlighted yet again a series of issues and systemic *vulnerabilities* in terms, above all, of the approach and methodology, the choices and strategies to adopt. And so, in the face of a more and more evident and recognizable form of *hypercomplexity* which continues to characterize the transformation we are currently experiencing, finding us (beginning with our educational and training institutions) drastically ill-prepared; in the face of a massive extension of the interdependencies/interconnections/interactions/conditioning/feedbacks that innervate dynamics and processes, the above-mentioned comeback of dominating analyses and explanations based on neo-positivist, linear, reductionist and deterministic visions of reality has been—almost paradoxically—gaining traction for some time now. As often happens in the presence of global, systemic emergencies, the (hyper)complexity of the ongoing processes and dynamics and the urgency of a *systemic approach* to this same complexity, often confused and abused, have been evoked from many quarters, not only at semantic and terminological levels. Unfortunately, once again, it has been a matter of evocations used more as slogans than as pathways towards renewed awareness and finally, systemic and transnational strategies! As is the case regularly during emergencies and/or in the face of crises/disasters of any kind, everyone rediscovers Science and scientific knowledge, education and in-field experience and recognises their value while acknowledging the crucial, strategic centrality of education, freedom and responsibility (individual and collective, which I repeat are relational concepts) as well as of the well-known "cultural question". In other words, everyone is rediscovers and acknowledges—as always happens in these situations—the fundamental role played by social and cultural factors/variables. With these, they also recognize the weight of/the role that *fear*, fears, emotions, feelings, *the culture of indifference*, the cultures of risk and emergency have always played, today more than ever, fueled by an ecosystem of information and communication designed, above all, to be fast and viral, according to the well-known logic of control and surveillance.

To these considerations, we need to add, with Bauman:

> Uncertainty is the natural habitat of human life—although it is the hope of escaping uncertainty that is the engine of human pursuits. Escaping uncertainty is a paramount ingredient, even if only tacitly presumed, of all and any composite images of happiness. This is why 'genuine, proper and complete' happiness always seems to reside some distance ahead: like a horizon, known to retreat whenever you try to come nearer.[19]

---

[19] Bauman Z., *The Art of Life*, Polity Press, Cambridge 2008.

Hence, amid paradigm shifts and anthropological transformations (1996), we are witnessing/experiencing—although with little awareness—the *overturn* of the complex interaction between biological and cultural evolution,[20] a profound issue also and above all in terms of the "culture of communication" (1996), rendered even more complex and problematic by the absence of a thought system of and a theoretical-interpretative model capable of observing, acknowledging and understanding the hypercomplexity and the irruption, at times overbearing, of chaos.[21]

So it happens that this *new social complexity* defines the structural conditions that permit the affirmation of the kind of *reflective knowledge* necessary to deal with the thinking crisis, and cognitive paradigms as well as the inability to sustain acceptable solutions. The systems that orient cognition and values turn out to be inadequate with regard to a constantly evolving social reality, made of complex systems—"organisms" and not "mechanisms"—which, in turn, are marked by extreme sensitivity to *perturbations* (*environmental and ecosystemic*), by initially *non-observable* properties and dimensions, capable of self-organization, of developing in an irreversible and anything but linear, predictable manner.[22]

If until some time ago, the problem was precisely that of recognizing and acknowledging the complex and irreducible "nature" of complexity, at present the question seems to be (is) that of continuing to underestimate the grounding and founding dimension of thought.

From this point of view and aware of this continuous and systemic underestimation of determining factors and variables, I can simply continue to repeat "hard times, very hard times for thinking", and not only.

Hard times, very hard times for thinking, for those who propose a mode and/or a system of cogitation, for those who "think" in an attempt to emerge from certain narratives that are overly reassuring or, on the contrary, catastrophic, for those who think actively (this is not an oxymoron, as thinking *is* (an) action) for those who, in order to start anew (the mechanistic metaphor used is "restart") and really think of matters in the "long term", should/should have become radically involved in education and research regarding education and didactics. These are *hard times*, very hard times—and, evidently, I am not referring only to the past extremely difficult and critical years of the recent global emergency, – which had given new life and strength to old simplistic, reductionist, deterministic narratives, including those concerning the digital and hyper-connected civilization.

Hard times, very hard times, for those who continue to insist on the long term, on the urgency of creating a culture of responsibility and prevention, starting again, and I insist on this once more, from education, training, research, freeing them from mono-disciplinary logic, that of the internment of knowledge and the creation of false dichotomies. Hard times, very hard times, for those who reflect on and study the *crisis of thinking* (not of words and languages), which has been continuing for

---

[20] Dominici P. (1995–1996) and following., *op.cit.*

[21] *Ibidem.*

[22] See, in particular, Dominici P., 2005, 2014, 2019, *op.cit.*

quite some time now, in all its nuances and declinations. This is a dimension that has been underrated, totally, for some decades now, also because the philosophical, the theoretical (theory vs research/practice, another false dichotomy... the elements of which feed each other), the abstract, even speculation, as well as visibility and perceivability, are seem and not only by public opinion, as totally "useless", destined to "lose", in any case and always, when making comparisons, however misleading and deceptive, with everything that is image, with everything that can be visualized, represented (somewhat akin to what has happened to school textbooks over the years) in terms of data and statistics which—for the sake of clarity and making it explicit every time—I consider fundamental, and not just when tackling a pandemic.

However, the last *pandemic*, the last series of pandemics, underscored an entire set of individual, subjective, personal, systemic, structural inadequacies and vulnerabilities which due to a series of social, historical and cultural reasons (to which we have returned often), we continue to address by applying reductionist and deterministic approaches. Perhaps — and for the writer of this book, evidently, the word 'perhaps' could be eliminated—the present is a period of a (crossing/passing/transitory) thinking crisis.—the basic dimension that Edgar Morin himself has often referred to—and of the absence of a "thought system",[23] which has been a long time coming.

It is at this level of analysis that we are able to understand all this talk of the paradigm shift, anthropological transformation, the importance of contamination between knowledge and skills, on the importance of philosophy and humanistic-political-social education, the urgency of *healing the fracture between the Human and the Technological*, between humanistic and scientific education, the approach to complexity and the urgency of a systemic view, the importance of inclusive innovation and the construction (also) of a digital culture so as *to inhabit the digital* (1996), the "new" ecosystemic "Nature". I could go on and on. It is yet another confirmation of the extent to which all the more or less enlightened talk and recent "narratives" regarding these topics were only, at best, good intentions that were never implemented; ones in which nobody ever seriously believed.

Fashionable topics and issues to champion, from time to time, to appear/declare oneself ready and prepared for anything and everything at all times, "experts" in everything and all things. We all saw, in the course of the last global and systemic emergency, how things fared, and we are still seeing how they are faring.

Until the next emergency, the next "black swan".

The consequences of this *systemic shortsightedness* and underestimation/non-consideration do not regard, evidently, those who think of or try to pose the question of the urgency of an "other" mode of "true" (pardon my simplification) thought and/or of a "thought system"—which, I repeat, must/should proceed, and not only from today, hand in hand with a radical reform of education and teaching/learning. The consequences do not regard those who study and conduct research on similar topics and issues who, and even more so, if unassociated with, autonomous and

[23] See Dominici P., (1995–1996) (2003), *op.cit.*, and following.

independent of political parties, they find no space within the so-called groups of experts in question. The consequences—it is important to be clear on this point-regard people and communities, the quality of educational processes and institutions, of democracy itself; even more so at moments like the past several years marked by extraordinary, continuous and systemic emergency. Despite knowing about and understanding the logic of emergencies, the discourse and narratives associated with "the grand illusions of the hypertechnological civilization"[24] return to impact with increasingly greater force and invasiveness. Everything is observable (?), measurable in quantitative terms—quantifiable, actually—and, consequently, *everything is predictable* (except for the "black swans") even controllable and predeterminable, using systems of artificial intelligence. Consequently, only 'facts'(?) are needed/count or, rather, what we recognize as "facts", on the basis of hypotheses and theories which, at a given moment, appear to be convincing or apparently effective at least. Only "data" presented as "factual "count, even if unsupported by variables, correlations or sure proof (as well as so-called *methodological notes*). For these reasons, too, no one was –or will be—able to predict with absolute certainty the anything but linear evolution of the last pandemic (or any future series of pandemics) or other systemic emergencies. For some time now, in spite of the awareness acquired regarding the characteristics/dimensions of scientific observation and the role of the "observer", within many institutional and organizational environments the famous "objectivity principle", which immediately recognizes and attributes to Science the fundamental "objective" (precisely that) of observing and investigating the "facts" in their pure and essential objectivity, without any contamination/interference on the part of the Subject, has not only returned but has become hegemonic. After all, it is the *constitutive principle of the epistemology of modern science*, subsequently demolished, however, by a good number of great scientists (Heisenberg, Einstein, and Prigogine, to name but a few). It is an epistemological principle, later translated into pedagogical and organizational cultures, which still has profound implications for all areas of organizational and social practice. What comes to mind is the hegemony of *technocracy* and the role/typology of the "experts" involved in decision-making processes. And so, for a hypertechnological and hyper-connected civilization, only certain types of "useful" fields of knowledge are helpful/count, only technical skills, hands-on knowledge, know-how, and *hyper-specialized technicians* are valid. In short what count are "know-how"/"knowing how to perform", problem solving, certain exclusive, specific "skills" and—I repeat—a rather technocratic and, by now, ideological view of skills, competence, knowledge and Society.[25]

All these are fundamental and essential dimensions, otherwise, we would never emerge from any quandary, all the more so during a critical phase like that of these past years with its numerous unpredictable implications. The unavoidable question

---

[24] For reasons of clarity, in this case too I wish to recall them: rationality, control, predictability, measurability, elimination of error.

[25] See Dominici P. (1995–1996), 2014, 2019, 2021, 2022, *op.cit.*

which continues to condemn us to inadequacy and cultural delay is the systematic underestimation and non-consideration of the other dimensions of complex systems, of the emergencies that characterize/constitute them and, broadening our gaze further, of the dimensions and variables that characterise our lives/existences, very often unobservable at the beginning. And so, if we persist in following this short-sighted and inadequate perspective, what happens is that we continue to think (?) that only those fields of knowledge and those skills should be involved that are, or at least claim to be, capable of simplifying, facilitating or, in any case, "showing"/representing/feeding imaginaries and imaginations regarding how, despite everything, any problem/emergency whatsoever can be easily dealt with, controlled, solved.

1.3 *Simplifying Everything,* Even What Cannot Be Simplified

Technocratic societies and democracies, through *automation*, digitalization and systems of artificial intelligence (despite being, as I always repeat, extraordinary opportunities), intend to move precisely in that direction which, all told, means designing and creating social organizations and systems, not as open, dynamic, complex and constantly evolving "organisms" (as they are), but as if they were perfect "machines" or, in any case, always perfectible, making it possible to plan and predetermine every action, every choice, every form of behaviour, every strategy. In the background, once again, the utopia/dystopia of building a *civilization without error* and a *Society-Mechanism* which does not foresee/does not accept/does not allow any form of unpredictability, a-normality and/or deviance from the norm. Even if the worst dystopia, undoubtedly the most disturbing, is that of striving to *make the human being into a machine*, an apparently "perfect" machine, incapable of making mistakes and, above all, less and less aware of its own fragility and vulnerability. Once again, what I have called "the ultimate error" (1995–1996) returns: the confusion between "complicated systems" and "complex systems" which has produced so many, to say the least, negative effects over time.

In this regard, after many years of study and research, as well as experience, I have the clear feeling that, often, a little at all levels of action of organizational and social practice, those who, in the presence of situations/dynamics that escape their control (illusion of control), insist on the issue/metaphor of the "black swan" (and I am obviously not referring to Taleb and his famous *The Black Swan*), of the unique and unpredictable or, in any case, highly improbable event, do not seek/do anything other than—I repeat—apply/build ("classical") "post-rationalizations", capable of calming and reassuring others and themselves that, despite some anomalous events, everything remains " under control" and predictable. Once again, we are unable to calculate and become aware of the "factors" and "variables" which, not only today, condemn us to lack of preparation and inadequacy. As I have continued to repeat for some time now, everyone has been talking about *critical thinking*, the paradigm shift (sic!), complexity, complex systems, and even *the centrality of the Person and/*

or the Citizen. Everyone, absolutely everyone, talks/writes about/discusses the contamination of knowledge, interdisciplinarity and multi-disciplinarity, cross-contamination between humanistic and scientific education, without even studying/preparing/documenting them a little, repeating propagandistic slogans and sensational inaccuracies and not only of a terminological nature. Speeches and texts of all kinds, beyond a fascinating and persuasive use of words, continue—to say the least—to underestimate, as they fail to consider, the dimensions of the hypercomplexity they speak about. Beyond words, videos, press and media appearances, accompanied by exciting and motivational discourse, when those splendid speeches or social marketing campaigns are applied and operationally translated, it appears immediately clear that they are mere narratives and/or cosmetic speeches (to put it mildly). One among the many possible examples is the statement that to 'innovate' (with all the nuances and ideological significance of the word) all we need are data, technologies and, above all, *technologists*. At most, the problem is regulatory and one of legal compliance: therefore, everything is focused on the exclusive involvement of jurists or legal experts, in addition to the inevitable involvement of computer scientists and, more generally, technical skills and knowledge. But which kind of social and cultural innovation can be promoted and/or activated in this way? Only one of an exclusive kind, imposed, as has always been the case, from above, by more or less enlightened elites.

For decades now, more or less all of us have been writing, debating (?) and talking about these issues, (then there are those among us who study and research these subjects with rigour and method). Yet, what point have we reached? How far have we come? Every time, we come up against empirical evidence, data and research (UNESCO, OECD, Istat, Censis, etc.) which promptly underline/highlight our structural and systemic tardiness, from every point of view. All (almost all) of those who wield power and make decisions have been falling, for decades, into the trap of this misleading, deceptive approach which, among the many consequences it generates, continues/will continue to reproduce that 'model' of technological innovation imposed *from above* and without a "culture of complexity" capable of supporting it in any way. This is an approach that will never permit complex organizations and ecosystems to *metabolise* the technological transformation which, as we have reiterated many times, produces/triggers/determines—even if no "causal link" exists a transformation which is, above all, anthropological. Thus, a far cry from "building from below" (another concept always evoked), from *systemic change*,[26] *social*

[26] On these questions, I refer in particular to two recent scholarly articles: (1) Dominici P., *Beyond the Darkness of our Age. For a Non-Mechanistic View of Complex Organization as Living Organisms*" in "Rivista Trimestrale di Scienze dell'Amministrazione", 2022, and later at http://rtsa.eu/RTSA_2_2022_Dominici.pdf?fs=e&s=cl; (2) Dominici P., "*From Below: Roots and Grassroots of Societal Transformation, The Social Construction of Change*", in CADMUS, World Academy of Art and Science, 2021. "That systemic change must begin from grassroots communities and single individuals and groups, and by definition can never be a top-down imposition, implicates a necessary rethinking of our educational institutions, which are still based on logics of separation and on "false dichotomies" (quote) and later at http://cadmusjournal.org/article/volume-4/issue-5/essay5-social-construction-change

innovation, citizenship (digital and otherwise), and participatory democracy, a far cry from long-term policies, what we have been doing for too many years is diametrically opposite and based on the traditional logic of control (short-term logic) and emergency.

Everything around us seems to move frenetically, due to the many narratives and the systematic stimulation caused by a phantasmagorical and well-constructed form of *storytelling*, while, on many issues, everything is essentially at a standstill. Recalling—and recontextualizing — a famous philosophical aphorism of earlier days, we are witnessing and re-witnessing *the eternal recurrence of the same*.

And it will not be long before the usual voices (and how many there are of them!) will be raised to say that they have always "thought" (!) and supported this, despite their lack of supporting "evidence" and the existence of previous positions that (often) affirm diametrically opposite theses and visions (here, I am referring to articles/written texts, interviews, scientific and/or popular publications, etc. etc.). I add and, in these cases, I always go back and question myself about the rampant conformism and the absence of critical thinking regarding innovation while wondering: if these ideas/thesis/critiques have always been expressed and practiced (?)—something that also concerns the issues of multi/inter/transdisciplinarity and "critical thinking" (obligatory, it seems, to quote this concept directly in the English language), not only within educational and training institutions—then why we are stuck where we were (over) twenty years ago and still continue, on many issues, to navigate by sight only to find ourselves completely unprepared and inadequate when it comes to the emergent, the unexpected and the unpredictable? An *emergent emergency* and a *predictable unpredictability* which—I repeat—are structural, ontological characteristics, founding aspects of complexity, or rather, of hypercomplexity. Of our being "living, human, social beings".

Once again, despite the extraordinary advances in the field of scientific research and technologies, we are experiencing/inhabiting an era of *maximum unpredictability*, uncertainty, and precariousness. It is, above all, an era of radical *obsolescence* of all the dimensions and processes that characterize social action and human life. Thinking, knowledge, and skills included.

1.4 *Complex Syntheses*. On Thinking and on Technological Transformation

The hypertechnological and hyperconnected civilization, in addition to taking the form of a civilization of rationality and total control, presents itself and self-represents itself as an increasingly advanced civilization which is, above all, capable of eliminating *Error and unpredictability* from processes, systems, ecosystems and life. This *civilization*, innervated by automation and processes of simulation (= efficiency/control) which, increasingly attempts to *delegate everything to technology* and to fuel certain, even illogical, illusions, can resort exclusively, solely, to that technical knowledge and those skills that appear best able to support them.

The prolonged systemic emergency which started ("officially") in February 2020 (?)—and has since been replaced by other forms of emergency—unleashed/triggered, within the media, the media ecosystem, the social networks, among journalists and sometimes, even among some scientists, besides the now traditional obsessive quest for presence/visibility at all costs, an irrepressible production of slogans, labels, incurable dichotomies and sterile aporias, of ideological and polarised/polarising readings, leaving no space for in-depth analysis or awareness of the urgent need for a *systemic approach to complexity*.

The rediscovery and acknowledgement of the fundamental and foundational, if not the absolute, value of science and scientific knowledge, of the strategic centrality of education, freedom, individual and collective responsibility—relational concepts of the well-known "cultural question" (which unfortunately is always recognized in alternating flows, dictated by convenience and not only, and always of a political nature)—are proof of the fundamental role played by social and cultural factors/variables. And along with these we also find the role played by alarm, fear and the *culture of indifference*, and the weight that they continue to possess, today more than ever. These feelings are nourished by an ecosystem of information and communication designed (above all) to be fast and viral, according to the logic of control and surveillance which, as we are well aware, is centred on processes of automation and hyper-accelerated simulation.

These are topics and issues of vital importance that, unfortunately, have always been underestimated and rediscovered only during situations of emergency. These are events that reveal all our vulnerabilities and uncertainties most clearly and which, more than any other kind of situation, make us understand how inadequate and unfounded our certainties, our certainties (illusory)[27] are and, with them, the educational, pedagogical, organisational, political, cultural and social paradigms which support/reinforce them, at least in appearance.

They are situations that, every time, among the many dimensions to be taken into consideration—the phase of transition, which is only in part an "economic crisis"; the now dated historic crisis of the agencies of socialization; the affirmation of educational polycentrism; individualism and the weakening of social bonds, the loss of credibility of the educational and political institutions, the complexity and unpredictability of the ongoing change; the absence of an adequate "thought system"—have in the past highlighted and continue to highlight the substantial inadequacy and incompleteness of the education and training systems, sometimes/often accompanied by superficiality and an inability to implement decisions effectively. In recent decades, the choices, strategies and tools adopted, with the precise intention of rethinking education and training, have proven to be completely fallacious and ineffective, as well as being based on reductionist and deterministic approaches. It has not sufficed to try to adapt/adjust/extend processes of education and training to technological change and the so-called *digital revolution*. Frankly, it has not taken

[27] See: Sloman S., Fernbach P. (2017), *L'illusione della conoscenza. Perché non pensiamo mai da soli,* Cortina, Milano 2018; see also: Hustvedt S. (2016), *The Delusions of Certainty,* Italian translation: *Le illusioni della Certezza*, Einaudi, Torino 2018.

long to realise this, even if we are still continuing to move in the same direction as before, following fashions and trends, in the name of a more traditional "*technological solutionism*" and of *a banal and acritical neophilia* that inevitably leads us to underestimate the ethical and epistemological implications of the ongoing technological, and above all, anthropological transformation.[28]

Choices, strategies and tools (together with a lack of resources and investments) incapable, even more so given the current state of things, of even defining/creating the minimum conditions (knowledge, skills, critical thinking, culture of error, empathy and responsibility, relational and communicative space, etc.) to help us understand, better still "inhabit", the complexity, unpredictability, ambivalence, variability and dynamism of educational and training processes, and, more generally, of Life and the Social. These are all dimensions– with some nuances –structural to/connotative of complex systems. At the same time, they are all dimensions that algorithms and systems of artificial intelligence cannot contemplate/predict, let alone allow or integrate.

1.5 The Systemic Error. Seeking to "Manage" the "Organism-Society" as if It Were a "Mechanism-Society"

Crises, disasters and emergencies have highlighted, dramatically, how decisive social bonds are, and, above all, how decisive processes of education and training are and how erroneously they are re-designed and rethought based on *logics and dialectics of confinement and dismantling* perfectly in line with what I have called the "Mechanism-Society". As a result, we continue to provide education and training (in the best case scenario...since there is also the prospect of absence of thought) in "disjunctive thinking". These are themes and questions which I have taken up several times over the years, and to which I shall return again later. As I was saying, educational and training processes are rethought in appearance only, applying the logic of labels, recourse to sophisticated marketing operations, with many "special effects" and a systematic endeavour to simplify everything, even what cannot be simplified. The goal, undeclared but all too evident, remains/has always remained the same: *to educate and train "mere executors of functions and rules"*[29] because, after all, it is those elements that are so desperately needed by technocratic and

[28] See: Dominici P. (1995–1996, 2005 and following) *op. cit.*

[29] *Ibidem.* On these issues, see also Ivan Illich's "classic" which constitutes an important, well-argued critique of mass education, but also of that ideal of false progress—also achieved by means of mass school and university—which produces, through the educational and training institutions themselves, conformism and uncritical obedience to institutions: Illich I. (1971), *Deschooling Society*, Italian translation, *Descolarizzare la società*, Mimesis, Milan 2019. These are themes and questions of vital importance, to which we have returned many times in studies and research conducted over the years.

dystopian "Mechanism-Societies". In recent decades, working in this direction, we have also contributed to reinforcing the (structural) social and cultural conditions of what I had called elsewhere the "society of irresponsibility"[30] and of "non-knowledge", which hinder, more or less unconsciously, the spread/affirmation—at all levels of organizational and social praxis—of a systemic vision and a "culture of complexity" which, in its multifarious dimensions, also plays the role of a "systemic culture of prevention".

From this point of view, the sharing economy and society are "objects of study" as well as complex phenomena, which bring to our attention issues and different levels of analysis that cannot be simplified or worse still, trivialized. In any case, they cannot be attributed simply to the economic and rational (?) *nature* of exchanges and certain social dynamics. Yet, in the face of the growing complexity of human ecosystems, we are witnessing the yet again reductionist and deterministic definition of responses and strategies.

The systemic, systematic, strategic and cultural error was/is/continues to be that of thinking/imagining/planning/figuring Society not as a living "organism" (which it is), but as a gigantic "mechanism", comprising entities, social actors, cogs and linear processes that are manageable, replaceable if they fail to be functional/efficient, observable, measurable, predictable, and predeterminable.

A "Mechanism-Society" in which we try to manage (=control) every dimension and device (?), down to its very core, simplifying what cannot be simplified and, in particular, *reducing* politics to "technocracy" and/or education to "technological-digital education"; in other words, trying to regulate/manage/channel/(almost) imprison the so-called sharing economy without thinking/imagining/planning/understanding the sharing *society*, with the complex relationships that characterize it and the potential opportunities/risks that distinguish it. These are risks and opportunities which—it should be made clear—concern only minimally the technological and digital dimension and practice.

An approach and perspective (almost a paradigm by now) which, besides continuing to confuse "complex systems" with "complicated systems", and communication (*the social process of knowledge sharing*) with (hyper)connection, have contributed to affirming an overall vision of innovation and change as phenomena related exclusively to technological and economic variables and dynamics (*hegemony of the economistic paradigm*); a paradigm and a model of development which, besides everything else, have the Person and the Citizen coincide exclusively—including their relative rights/duties and freedoms/responsibilities—with the "figure" of the Consumer, producing disquieting diversions and possible short-circuits for citizenship, inclusion, social bonds, and democracy itself which have not been in very sound conditions for some time now; (the crucial issues of *profiling*, transparency and, more generally, *surveillance* also come to mind). This approach, characterised by scarce awareness, does not take into adequate consideration the strategic importance of trust and cooperation, of the mechanisms and "devices" fundamental

[30] See Dominici P. (2009), *La società dell'irresponsabilità,* FrancoAngeli, Milan 2010.

for the very existence of social bonds, aside from being the real "factors" genuinely capable of fostering the sharing economy. These factors are the object of discussion and debate by many, even though only a few have truly grasped the importance of their implications. Furthermore, as stated several times in the past, *the sharing economy and society* oblige us to rethink our ideas/definitions/operational translations of the "social contract",[31] the space granted to fields of knowledge and skills, to communal life, to inhabiting relational, systemic and communicative spaces, as well as the local and global communities, whose openness and inclusivity can by no means be taken for granted. We should not forget, in this sense, that the so-called sharing economy also requires that *People and Citizens be taught the principles of freedom and responsibility*, which are capable of regenerating, in part at least, those *weakened social bonds* and, why not, that welfare state, which has been overwhelmed by the *debris* of a global financial crisis which, with grave tardiness, has been seen to be a crisis that is not purely economic. From this point of view, too, the multiplicity and plurality of the "new" forms of sharing and cooperation, as well as economic exchange (I am thinking, above all, of the gift economy), are difficult to circumscribe and frame within legal norms, however complex and detailed. The real problem lies, also in this sense, in building a *culture of complexity and sharing*, of cooperation and collaboration, as yet almost totally absent, despite the fact that narratives regarding citizenship and digital democracy and the great expectations generated by the digital revolution, also in terms of participation, which continue to be fueled by the media (but not only). It is almost as if time had passed without our realizing it; or, to put it better, without our being aware of the many strategic errors committed, also and above all, by the so-called "experts" entrusted with providing policymakers with support.

The underlying systemic and strategic error is—and continues to be—*that of trying and claiming to manage (=control) the sharing economy without having fully understood and imagined the sharing society.* In other words, referring back to the metaphor and the operational definitions I have proposed in the past, the underlying systemic and strategic error is that of wanting to continue building an (almost) perfect, error-free, unpredictability-free, conflict–/anomaly-free "Mechanism-Society".

Project and paradigm, the "Mechanism-Society", complementary and functional to this hypertechnological and hyper-connected civilization, based on the rationales of a global and transnational system and on its pedagogical, social, economic and political cultures, that had found (it would seem) in the three years of pandemic an

---

[31] See, in particular, Dominici P. (1995–1996), (2003), (2011) and following, *op.cit.* Concepts, topics and issues on which large transnational bodies/organizations such as UNESCO and the United Nations are working at present.

umpteenth global and systemic emergency—once again erroneously considered a "black swan"—capable of creating new inequalities and asymmetries. As I have also stated in the past, the progressive, as well as pervasive, diffusion of *connection technologies,* with its high rate of technological innovation, has for some time been reconfiguring social and political architectures, favouring the affirmation of new modes of economic production based entirely, rather than on ownership/possession, on their ability *to process and disseminate information and knowledge*, though these dynamics have not triggered equally decisive processes of greater emancipation and social inclusiveness, quite the opposite, actually(!).

The "Mechanism -Society" is a type, an *ideal type* (Weber) of society, a mega-structure and a *social model* we use to fool ourselves into believing that we are capable of controlling/managing, shaping, directing, even *pre-determining* (through systems of artificial intelligence). The "Mechanism" involved seems capable, in many ways, of countering/eliminating every "anomaly", every "abnormality", every "deviance", every conflict and every unpredictability (every "black swan"), by issuing, as in the past, scores of laws and decrees (a necessary but insufficient measure), with the intervention of technology, algorithms and artificial intelligence (despite the extraordinary opportunities the latter may/will turn out to provide, taking into account all its possible implications) while, more, in general, *delegating* everything to technology and, ever more frequently, today and in the future, to automation and processes of simulation.

We are still there, locked inside movement and dynamics based on posing and/or worse still, on simulating; stuck within a series of illusions, typical of the hypertechnological civilization,[32] that continues to convince us that everything is controllable. When, however, we realize—at all levels and in every area of social and organizational practice—that "things are not what they seem", we are left with fear and indifference, closing ourselves off from everyone and everything.

The Mechanism-Society lives and thrives—in addition to the economic-financial flows involved—on narratives and templates, subsequently embodied in pedagogical, educational, social and cultural models. For some time now, we have been finding ourselves obliged to deal with certain novel "narratives" regarding schooling (education) and, with a few, not overly accentuated nuances, universities as well. These idealized narratives and models, usually imported from abroad, are "new" and original only in appearance. They are rooted in important interests, not only economic ones, which present/recognize/define Schools and Universities as "places of innovation". An innovation that is idealized and crafted, and above all, *imposed from above*—thus exclusive—which redefines and fuels new inequalities and asymmetries.

---

[32] See Dominici P. (1995–1996), *op.cit.,* and following.

The pathway of the most classical tradition of recent decades, based on an idea/view/conception/paradigm of innovation and change, which, as I was able to affirm many years ago, continues to be imposed from above and fails to present any characteristic of the systemic change that has been evoked so many times. These processes of change, despite the continuous and apparently convincing narratives used to present them, continue to reveal their "nature" as a kind of exclusive innovation, of "innovation for the few", to the sole advantage of elites and limited groups.

This idea and this model of innovation—there would be much to say, even on the choice of the term and its *ideological implications*—are grounded in an overall idea and vision—in my opinion, totally inaccurate and inadequate—of School and University and, consequently, of Society itself, since, besides being based on the principle of *exclusion* and on asymmetries that we are now widely aware of, they do not permit us—and we have verified this in recent years—to fully grasp/metabolize the complex dimensions of the change taking place, or even of the full potential of the continually-evoked technologies themselves which, I continue to repeat—depending on the logics of the individual or systemic choices made, – may/will turn out to be either "technologies of control" or "technologies of cooperation". The so-called digital revolution that continues to be evoked is not producing the desired effects, owing, above all, to the lack of a systemic and overall vision relating to *complex human ecosystems*.

This model and this paradigm of education and educational processes aim, explicitly, openly, according to a given idea of Schools and Universities, at requiring (primarily and above all) to "*churn out*", produce *workers, technicians and specialists,* while involving young students in the creation of services and products deemed "innovative".[33] The lofty objective (?) is to counter the well-known and frequently evoked *misalignment* (*mismatching* as it is often called today) between school, university and the world of employment. In plainer terms, the crucial objective is to meet the "demands of the market", despite the fact that for some years now we have been experiencing a sensitive phase of transition and extremely rapid obsolescence of *the fields of knowledge and skills*. All told, even "the market" does not seem to know exactly where we are going, due partially but not solely to the speed of the ongoing changes and of the scientific and technological discoveries that are determining increasingly uncertain and unpredictable scenarios. In any case, one important objective is that of intercepting the expectations of the market which, unfortunately and paradoxically, is freezing our educational and training institutions in a state of constant cultural delay. As things stand at present, besides the preservation of an extremely radical form of mono-disciplinarity, pedagogical and

[33] There is often talk in this regard, of *engagement* → making it essential to use the English word in order to convey the originality and freshness of ideas and processes.

organizational cultures are also kept in a condition of immobility. Little room for manoeuvre is accorded to non-reductionist, non-deterministic, non-conservative approaches. The clear impression—here we are talking about "systemic logic" -is that there is an increasing tendency to disqualify and lower the educational level of our children with a view to instructing and training, as previously mentioned, "mere executors of functions and rules". According to this perspective/vision—perfectly coherent and in line with the paradigm of the hypertechnological civilization and with what I have called the grand illusions of the hypertechnological civilization (the Mechanism-Society), – School is assuming, more and more, the form of an institution of information, responsible for the transmission of notions and "know-how"—therefore no longer as an institution of education. Its task now is (only) to train "executors", "officials", "technicians", and, at best, "hyper-specialists". In other words, this view of School and, more in general, of education, combines/weds perfectly with that overall vision/perspective of a Society and of democracies guided and dominated increasingly by *technocracy* and by *experts in technical fields of knowledge* and technical skills.

The "Machine-Society", the "Mechanism-Society", and not the "Organism-Society": this is the final overall outcome of this general view/approach, which has been imported and accepted without critical analysis, as has happened in the past with other imported and hegemonically imposed models/paradigms. I am also referring here to what, for many years, I have called the STEM "doctrine"[34]: a hegemonic, reductionist, deterministic paradigm, completely inadequate in terms of the possibilities for decoding, understanding and, ultimately, inhabiting the structural and constitutive *hypercomplexity* of the living, of the social, of the Human.

---

[34] I have always returned to this fundamental point over the years, having been among the few scholars, even at the international level, openly critical of this "paradigm"(?), perfectly functional to the construction of the "Mechanism-Society-" and, in some way, to the protection of the many interests that feed not only the (global) ecosystem of education, training and research. Among the most recent scientific articles related to these issues, I wish to suggest: Dominici P., *The Digital Mockingbird: Anthropological Transformation and the "New Nature"*, in "World Futures. The Journal of New Paradigm", Routledge, Taylor & Francis, Feb. 2022; Dominici P., *Beyond the Darkness of our Age. For a Non-Mechanistic View of Complex Organization as Living Organisms*, in *Rivista Trimestrale di Scienze dell'Amministrazione. Studi di Teoria e ricerca sociale*, n°2, Anno 2022—http://rtsa.eu/RTSA_2_2022_Dominici.pdf?fs=e&s=cl

Educating (?)—maybe more appropriately, given the present state of things, means "instructing", "indoctrinating", "training"[35]—creating highly skilled *technicians* and super-efficient *executors*, because the civilization of automation and simulation (which under other assumptions and approaches, represents an extraordinary opportunity) requires "mere executors". After all, *intelligent machines* (?) and systems of artificial intelligence) — keeping in mind that they are programmed and created, (should this need to be spelt out) — by other "human beings", will think/be able to think of all the rest. It all adds up, from this point of view, I repeat, to the basic aim that was/is the hope, the illusory and deceptive certainty, regarding ourselves as well, of creating/building a *Mechanism-Society,* whose parts and procedures are perfectly efficient. This is tantamount, consequently, to a kind of society that is controllable in every aspect. No need to worry, this has been and continues to be the ever-recurring, "everything-under-control" message, even among our institutions of education and training.

[35] Perfectly in line with the dimensions analysed, in recent months I have noticed, for the umpteenth time, how strong certain myths/narratives are and how certain narrators/gurus/super experts are always ready to advance and formulate the well-known proposition of "simple solutions to complex problems" (1995–1996). When a problem pf any kind arises, there are "experts of everything" who always know what to do and what will happen, even in times of *pandemic* crises (!). These are the people who, in these years of ideological intoxication based on massive doses of "uncritical newism in a manner", of reductionism and technological determinism—I repeat, also due to the many interests at stake—have outlined scenarios and illusory and deceptive, sometimes bankruptcy, scenarios and prophecies, even more so due to the necessary change that should have radically changed the educational processes and training. Processes and dynamics that have made social and cultural inequalities and asymmetries even more evident. Those who have insisted, for years and years, on "only digital and media education are needed", "those who"... the only problem is the "digital divide" (there were many, really... except then realizing that there existed, in addition to the "cultural divide", even more complex and problematic dimensions); those who "only digital skills" and, precisely, "only STEM" (the acronym, then, for obvious reasons, was, as known, integrated with other letters), those who "enough useless knowledge and degrees"; those who continually repeated how they served "only and only technical and hyper-specialized profiles" (specialization vs interdisciplinarity/complexity are other "false dichotomies", as I defined them in the mid-nineties), perhaps while they evoked/evoked the "contaminations between knowledge" (sic!); the same ones who spoke, in fact, of "useless" knowledge, disciplines and studies, promoting and exalting the principle/value of the usefulness of knowledge and research which, as I have been repeating for years, is the deceptive principle on which they were counter-reformed our educational and training institutions (we have returned to them over the years); "those who", faithful disciples of the rhetoric/narratives of disintermediation, repeated several times "what is the use of studying communication"—and what is the use of humanistic training (we should go far back in time)—what is the use of preparing in the related sectors, so much, after all, to "know how to communicate" or "manage complexity" (I have always repeated, a real oxymoron) you don't need particular knowledge and skills (according to them): even in recent years, we are realizing it well how much (urgent) need there is for prepared, even methodologically, and responsible "communicators". A discourse that obviously also concerns all the professional profiles of the information ecosystem (not just journalists). On the contrary, after years of short-sighted perspectives and strategies, we should worry about giving them a truly multidisciplinary/interdisciplinary, more profound, rigorous and articulated preparation, and not just "technical" and knowledge of applications, software and hyper-connected environments.

As I have repeated several times in the past, despite everything, we also find ourselves on the brink of making *an irreversible quantum leap*—the so-called "technological transformation", which is, above all, an "anthropological transformation"—determined by extraordinary scientific discoveries which have triggered processes of complex synthesis between "worlds" and between systems[36] that, in the past, were/appeared to be clearly distinct, not to say separate. *Processes of complex synthesis* capable of recomposing fractures and gaps legitimized and strengthened, above all, by the *overall architecture(s) of the fields of knowledge* erected in the meantime. It is not from today only that technology (= culture) has become an overwhelming part of *the synthesis of new values and criteria of judgement*.[37] We find ourselves *inside* a hypertechnological civilization and one of *intelligent matter* (?) based entirely on programming/planning and on the complete and total (hyper) simulation of all processes, dynamics and actions. This civilization, made of

[36]Over the years, I have returned, repeatedly, to the issue of the complex syntheses existing between nature and culture and between the "natural" and the "artificial", as well as to the various related issues (anthropological transformation, paradigm shift, "false dichotomies", human-machine interaction, the deceptive and counterproductive separation between technology and culture, the significance of being human in the era of the information and digital revolution, etc.) and last but not least the urgency of a systemic approach to complexity. See, in particular, P. Dominici (1995–96), *Per un'etica dei new-media. Elementi per una discussione critica*, Firenze Libri Ed., Florence 1998; see also *La comunicazione nella società ipercomplessa: istanze per l'agire comunicativo e la condivisione della co + noscenza nella network society*, Aracne Ed., Rome 2005; Dominici P. (2008), "Sfera pubblica e società della conoscenza" in Morcellini M., Mazza B. (a cura di), *Oltre l'individualismo. Comunicazione, nuovi diritti e capitale sociale*, Franco Angeli, Milan: Franco Angeli 2008; *La società dell'irresponsabilità*, FrancoAngeli, Milan 2010; *La comunicazione nella società ipercomplessa. Condividere la conoscenza per governare il mutamento*, Milan, FrancoAngeli 2011; (2014), *La modernità complessa tra istanze di emancipazione e derive dell'individualismo*, in "Studi di Sociologia", n°3/2014, Vita & Pensiero, Milan. Again, on these themes and issues, among the recent scientific publications see: P. Dominici, *The Hypercomplex Society and the Development of a New Global Public Sphere: Elements for a Critical Analysis*, in, RAZÓN Y PALABRA, Vol. 21, No. 2_97, April–June 2017—pp. 380–405; P. Dominici, *For an Inclusive Innovation. Healing the fracture between the human and the technological*, in, European Journal of Future Research, Springer, 2017; *Controversies on hypercomplexity and on education in the hypertechnological era*, in, A. Fabris & G. Scarafile, *Controversies in the Contemporary World*, John Benjamins Publishing Company, 2019; "*La complessità della complessità e... l'errore degli errori", in* the TRECCANI encyclopaedia–"Italian Language" section– Istituto Enciclopedia Italiana "Treccani"—December 2018. See also C.J. Preston (2018), *The Synthetic Age*, Italian translation, *L'era sintetica. Evoluzione artificiale, resurrezione di specie estinte, riprogettazione del mondo*, Einaudi, Turin 2019; M. Tegmark (2017), *Life 3.0. Being Human in the Age of Artificial Intelligence*, Italian translation, *Vita 3.0. Essere umani nell'era dell'intelligenza artificiale*, Raffaello Cortina Ed, Milan 2018; also Éric Sadin (2018), *L'Intelligence artificielle on l'enjeu du siècle*, Italian translation, *Critica della ragione artificiale*, LUISS University Press, Rome 2019; H. Fry (2018), *Hello World. How to Be Human in the Age of the Machine*, Italian translation, *Hello World. Essere umani nell'era delle machine*, Bollati Boringhieri, Turin 2019. Of the classics see: H. Jonas (2001), *The Phenomenon of Life*, Italian translation, *La cibernetica e lo scopo: una critica*, ETS, Pisa 1999. And, finally, an extraordinary classic—in my opinion –: Koyré A. (1961), *Les philosophes et la machine. Du monde de l'"à-peu-près" à l'universede la précision*, Italian translation, *Dal mondo del pressappoco all'universo della precisione*, Einaudi, Turin 1967.

[37]*Ibidem.*

hyper-connected ecosystems, marked by a progressive, as well as exponential growth of the size of the *technologically controlled* which seems capable of definitively marginalizing the Human (= error and unpredictability) along with *the dimension of responsibility*. What therefore comes about is that the challenges of change and hypercomplexity are precisely those of rethinking/redefining the centrality of the Person and of the Human, within increasingly hyper-connected and interdependent environments and ecosystems, which are simultaneously marked by the increasing fragmentation and fragility of social bonds. These environments and ecosystems no longer contain any boundaries/limits between *the natural* and *the artificial*, nor, at least apparently, between *the real* and *the virtual.*

The process of globalization—the radical, omnipresent outcome of the contradictory and ambiguous processes of Modernity—has, among many other things emphasised further *the transformation of the modes of the practice of human action* and the proliferation of levels of interaction and interconnection, which redefine and encounter new forms of the logic of power and domination.

The fundamental role of politics -as has been previously observed[38]—is becoming increasingly marginal, within a global reality moulded completely by the liberalization of markets and information flows, which, obviously, make in-depth reflections even more urgent as far as the role and importance of truly functioning and operational "global institutions", capable of governing the processes under discussion, are concerned. It is urgent to find ways of countering/containing/resizing the worrying instances/dynamics of the imbalance of power which favours the IT giants, the major virtual platforms and, more generally, the corporations of the digital world. This means that the very issue of democracy "intersects here with the more general issue of the dominion of some people over others who are obliged to work for them. Indeed, during the course of human history there exists a constant axis along which, even after the formal disappearance of slavery, mastery continues to be exercised massively in an indirect and impersonal way. This happens also through machines (initially those designed to facilitate labour and subsequently those equipped with AI and the ability to learn)", which "usurp" first physical, then intellectual strength, to the advantage of those already in power, who wield more and more of this power the more it remains hidden. All told, technologies evolve, but relationships of domination remain firm because they become not only more complex, but also more invisible".[39]

The world system, marked by "open dialectics" and traversed by "centripetal" and "centrifugal" forces,[40] bears witness to *processes of complex synthesis,* destined

---

[38] *Ibidem.*

[39] See R. Bodei, *Dominio e sottomissione. Schiavi, animali, macchine, Intelligenza Artificiale*, Il Mulino, Bologna 2019, p. 19.

[40] See P. Dominici (1995–96), *op.cit.,* 1998; you can see also: *La comunicazione nella società ipercomplessa: istanze per l'agire comunicativo e la condivisione della conoscenza nella network society*, Aracne Ed., Rome 2005; Dominici P. (2008), "Sfera pubblica e società della conoscenza" in Morcellini M., Mazza B. (a cura di), *Oltre l'individualismo. Comunicazione, nuovi diritti e capitale sociale*, Franco Angeli, Milan; *La società dell'irresponsabilità*, FrancoAngeli, Milan 2010; *La comunicazione nella società ipercomplessa. Condividere la conoscenza per governare il mutamento*, Milan, FrancoAngeli 2011.

to further determine new epistemological fractures in the pathways of science and knowledge. At the same time, the *nature* of *interactive media* and the complex architectures of the new global ecosystem—besides the *hyper-velocity* of digital technology—have obliged social systems, together with the production, distribution, political and cultural processes that characterize them, to structure themselves according to the (inclusivity vs exclusivity) logic of networks, which, at least initially, seemed capable of undermining old hierarchies and decision-making processes. Moreover, they have highlighted above all else the inadequacies, anomalies and weaknesses of traditional interpretative and cultural models, previously capable of making social and organizational life relatively predictable. The *control* (?) of the complex dynamics of world capitalism and technological progress has always involved the management of knowledge (complexity), but also, evidently, the management of information and data, which requires "shared knowledge" (2003) and a (collectively and socially constructed) culture of knowledge. In other words, it requires a "culture of complexity", even more so at a time when *narratives*, at times even deceptive and misleading, regarding the so-called digital revolution[41] have erroneously associated the *digital* with what many deem the absolute value of *simplification*.

Technological innovation is a determining variable for the evolution of social systems and organizations, but acting on its own it does not suffice. Once more, culture, shared knowledge and education are necessary to enable the systems to metabolise change and effectively manage phases where the control associated with the accelerations dictated by technology is wanting—and which, let us not forget, is always a product of culture and not something "external". However, education, culture, shared knowledge and training also serve (and a great deal of them) to enlarge *the platform of rights and freedoms* (entailing responsibility as well), by increasing awareness of the fact that in order to be citizens, not "subjects", it is no longer sufficient to be connected and/or informed (which, in any case would already be a good start). The digital civilization of transparency and total surveillance aims at making us feel more and more secure and (perhaps) connected to others but, at the same time, it questions matters of identity, fundamental rights, not to mention our freedom (a relational concept). From this point of view, we have yet to understand how to *inhabit* the new global ecosystem, manage its dynamics, as social bonds appear increasingly weakened amid *the need for security* and the *solidarity of fear... if we wish to avoid being managed.*

The attempt to *inhabit* this global civilization (hypercomplexity)—*an ecosystem of ecosystems* (1996)—has to come to terms with its intrinsic, almost structural, vulnerability, which is associated precisely with the factors that have always made it appear (appear only!) perfectly functional, efficient, hypermodern, rational, super-accelerated and apparently controllable, predictable and measurable in all its aspects and dimensions. This is a "civilization without error(s) and anomalies", which aspires, at least, to be so and in the shortest time possible.

---

[41] These are not simply media narratives.

However, contrary to certain "narratives", the advent of the hypertechnological civilization determines an unprecedented broadening of the dimensions of the dynamism, variability, precariousness, uncertainty, insecurity and vulnerability existing within and outside of social systems, making the functions performed by communication even more strategic in terms of the *reduction of complexity* (Luhmann), the management of risk and uncertainty, and the mediation of conflict.[42] It is useful to reiterate the point that we find ourselves going through a very sensitive crossover phase—yet another hyperinflated era of transition– to global and multidimensional modernity, marked by a disquieting process of further accumulation of *anomalies*,[43] which have undermined the certainties associated with consolidated cognitive paradigms.

The hegemony of instrumental rationality and the (self-regulated) market economy has led to the triumph of a *logic of domination* which has been extended to the whole of social life. This process has also weakened those ties that transform individual choices into collective projects and action. At the level of social cohabitation, a highly individualized form of global society has been generated, which places much more responsibility on the shoulders of the individual social actor, which requires a certain form of *responsible freedom.*[44] In this respect, the development of forms of mediated communication,[45] besides the opportunities they provide, might also cool down the mechanisms involved in the *production of social capital*, by weakening social bonds even further. Moreover, the belief that technology, and networks, in particular, under the banner of a *banal technological and acritical solutionism,* are capable of solving problems of all kinds, including the rapprochement between Politics and citizens, may turn out to be yet another fatal mistake. Political and social practice, while finding/having found new virtual *arenas* for the construction and organization of consensus and/or public opinion, requires the crucial passage from theoretical to practical, concrete action if it is to affect the political decision-maker. To achieve this, we need informed, critically trained social actors in *flesh and blood*, as the active and cognizant recipients of social cooperation within their networks.

Globalization constitutes, not only as of today, the *empirical condition* of the modern world,[46] a condition associated with the concept of complex connectivity, intended as a constant process of increased interlacing of the interconnective and

---

[42] *Ibidem.*

[43] See T. Kuhn (1962), *The Structure of Scientific Revolution*, Italian translation *La struttura delle rivoluzioni scientifiche*, Einaudi, Turin 1969.

[44] See Dominici P., *op.cit.*, (1996), (2005), (2010) and following. See also Cesareo V., Vaccarini, 2006.

[45] See Thompson J.B. (1995), *The Media and Modernity. A Social Theory of the Media*, Italian translation, *Mezzi di comunicazione e modernità*, Il Mulino, Bologna 1998.

[46] See, in particular, Sassen S., (1998). *Globalization and its Discontents*; Italian translation, *Globalizzati e scontenti. Il destino delle minoranze nel nuovo ordine mondiale*, Il Saggiatore, Milano 2002.

interdependent networks that characterize modern social life.[47] This is a process that may be interpreted not only as the triumph of Western *subjectivist and instrumental rationality*, but also of that of an all-encompassing and totalizing ideology that surrounds, encompasses, moulds all spheres of praxis and real life. And the criticism of globalization,[48] which produces a disruptive kind of individualism,[49] is, in reality, a critique of the global capitalist system, guilty of having destroyed the ancient alliance between capitalism and democracy and of having focused exclusively on economic and technological development without considering the implications for society and individuals.[50] It is worth our while to reiterate the way the world economy has progressively weakened the mechanisms and *devices* of democratic regimes and produced profound repercussions, not only on the structures and hierarchies of the global production system, but also, and above all, *on the overall architecture of the rights and protections* of people, of workers, in particular. It portrays the way transition from a society based on labour to a *risk society,*[51] takes place and causes the definitive affirmation of *a political economy of insecurity*.

The world economy is progressively weakening the characteristics typical of democratic regimes. All this has profound repercussions not only on the structures and hierarchies of the system of global production but also, and above all, *on the overall architecture of the rights and protections* of people, citizens and workers. In this sense, I am fond of recalling the notion that we are striving to manage the *sharing economy* without having understood the *sharing society* at all. Yet, despite their

---

[47] Tomlinson J. (1999), *Globalization and Culture*, Italian translation *Sentirsi a casa nel mondo. La cultura come bene globale*, Feltrinelli, Milan 2001, p. 14.

[48] See Gallino L. (2000), *Globalizzazione e disuguaglianze*, Laterza, Rome—Bari; Gallino L. (2001), *Il costo umano della flessibilità*, Laterza, Rome—Bari; Stiglitz J.E. (2001), *In un mondo imperfetto. Mercato e democrazia nell'era della globalizzazione*, Donzelli, Rome 2001; Stiglitz J.E. (2002), *Globalization and Its Discontents*, Italian translation, *La globalizzazione e i suoi oppositori*, Einaudi, Turin 2002.

[49] See Touraine A. (1997), *Eguaglianza e diversità. I nuovi compiti della democrazia,* Laterza, Rome -Bari 1997; Touraine A. (2004), *Un nouveau paradigme. Pour comprendre le monde aujourd'hui*, Italian translation, *La globalizzazione e la fine del sociale. Per comprendere il mondo contemporaneo*, Il Saggiatore, Milan 2008.

[50] See Magatti M., (2009), *Libertà immaginaria. Le illusioni del capitalismo tecno-nichilista*, Feltrinelli, Milan; Magatti M., Giaccardi C., *Generativi di tutto il mondo unitevi. Manifesto per la società dei liberi*, Feltrinelli, Milan 2014; Magatti M., Giaccardi C., *Supersocietà. Ha ancora senso scommettere sulla libertà?* il Mulino, Bologna, 2022.

[51] See, in particular Beck U. (1986), *Risikogesellschaft. Auf dem weg in eine andere Moderne*, Italian translation, *La società del rischio. Verso una seconda modernità*, Carocci, Rome 2000; Beck U. (1993), *Die Erfindung des Politischen*, Italian translation, *L'era dell'E*, Asterios, Trieste 2001; Beck U. (1994–1997), *I rischi della libertà. L'individuo nell'epoca della globalizzazione*, Il Mulino, Bologna 2000; Beck U. (1999), *Schöne neue Arbeitswelt. Vision: Weltbürgergesellschaft*, Italian translation, *Il lavoro nell'epoca della fine del lavoro. Tramonto delle sicurezze e nuovo impegno civile*, Einaudi, Turin 2000; Beck U. (2007), *Weltrisikogesellschaft. Auf der Suche nach der verlorenen Sicherheit*, Italian translation, *Conditio Humana. Il rischio nell'età globale*, Laterza, Rome -Bari 2008; Beck U., Giddens A., Lash S. (1994), *Reflexive Modernization*, Italian translation, *Modernizzazione riflessiva. Politica, tradizione ed estetica nell'ordine sociale della modernità*, Asterios, Trieste 1999.

current strategic function, communications and the social production of knowledge can only accompany this explosion of the *world of artificiality and technology,* no longer subordinated (it would seem) to "nature" (configured, to all intents and purposes, as a phenomenon difficult to manage) and make it the "visible" subject of public discussion. In other words, in our *hypercomplex society*, the dimension of what is technically controlled has become hypertrophic compared to what is non-technically controlled. Despite this, however, technological practice has rendered our "life worlds" extremely more uncertain, fragmented and unsafe for the individual subject.

1.6 The "*Persistent Paradox*": Continuing to Teach Concepts of Control and (Total) Rationality in Times of Unpredictability

As mentioned previously, we live in an era that is still founded on illusion and, at the same time, on the axiom that *rationality* and *predictability* govern social action and systemic and organizational strategies. Ours is an era that idolizes the very concepts of (total) "control", "management", "measurability" and "predictability", illusions based on an outdated organizational and social paradigm which, however, still deeply characterises social organizations, as well as educational and training institutions. These illusions have been translated into and practically embodied in visions and utopias, linked to the possible advent of a *perfect* civilization, *devoid of errors and/or deviations*, where *everything* is plannable and planned, pre-determinable and pre-determined. Illusions, visions, utopias/dystopias, that find new *lymph* in technological devices, in systems of artificial intelligence, in the extraordinary availability of data and in hyperconnected environments/ecosystems which seem capable of providing the necessary (artificial) technological, logistical and environmental support to organizations, social systems, political regimes (democratic and non), increasingly oriented towards forms of technocracy that leave very little room for complexity, mediation and negotiation processes, politics and the traditional social mechanisms of trust and cooperation. It is no coincidence that we continue to *think*, imagine, define and design organizations and social systems resorting to mechanistic metaphors and even go so far as to compare and confuse complex with complicated systems. Neither is it a coincidence that we continue to try to create and then manage those very organizations and social systems as if they were, in fact, cogs or *machines*, driven by *engines*, veritable *mechanisms* made of replaceable or modifiable "parts" that are individually isolable and "repairable", i.e. *complicated systems.* Behind this vision/these visions of organizations and, more generally, of the Social, lies the false belief that *technological* and *legal factors* are *conditions* not only *necessary* (as is the case), but also sufficient (as is not the case) to generate cohesion, cooperation, trust, innovation, but also efficiency and change. In other words, we continue to believe that in order to trigger processes of innovation and/or manage complexity (I have always referred to this as a "contradiction in terms", if not actually an "oxymoron"), a regulatory framework constantly updated by *insertions* of technological innovation, suffices. Just like that, *simply through top-down intervention(s).* A far cry from systemic and cultural change! In this way, the human, social and relational factor continues to be overlooked completely, or seen as something that can arrive as a consequence, but this is not how things stand.

The gigantic, historical/atavistic misunderstanding between "complex" and "complicated", between "complex systems" and "complicated systems"—as I have been repeating for many years now– is the basis of many of the problems that have always afflicted, first and foremost, educational and training institutions and complex work organizations and, secondly, political institutions and nation-states.

In times of unpredictability, maximum uncertainty and obsolescence, *we continue to provide an education that emphasizes control, predictability and (economic) rationality*, scarcely *aware of our limits*, of the complexity and ambivalence of organizational and social processes, under even greater pressure from technological innovation and the so-called digital revolution.

The problem, once more, is that we are tied to the idea that these processes are thoroughly *governable* and predictable/pre-determinable. From this point of view, too, contrary to what has been done in recent decades, it is not a question of adapting and extending educational and training processes to encounter technological progress. Here, by way of clear contrast, it is necessary to eradicate them from their very foundations and overall approach, grounded in *false dichotomies* and the logics of separation/reclusion, to genuinely rebuild—not through superficial cosmesis (marketing)—*the entire architecture of the fields of knowledge*[52] *and skills*. We are already experiencing and inhabiting a hypertechnological civilization, based on systems of automation and simulation, which marginalize the human and the sphere of responsibility, *under the illusion of being able to eliminate/expel error* and unpredictability from systems. We are dealing with a cultural paradigm that is strongly hegemonic, but based on *fragile* foundations, which envisions that eliminating error and unpredictability will give us the possibility of becoming more similar to the machines themselves, to their (apparent) perfection, a perfection that, in reality, will prevent them from becoming truly intelligent. Because it is precisely error that characterize human beings and their freedom, which first of all, means the freedom to think they may be wrong, but also the freedom to err consciously, *tout court*. In fact, human beings have not yet become aware that becoming *similar* to machines makes us/will make us lose our uniqueness, our diversity and variability, our peculiarity and specificity. Yet what we are even less aware of is that the power to direct and/or pre-determine our lives/existences does not represent/constitute the achievement of a presumed omnipotence, but, rather, the achievement of a negative, counter-productive objective, which is, to some extent, devastating: the end of our being (free) "human beings".

---

[52] See Dominici P., (1996–2019), *op.cit.*

Indeed, the complex process of changing paradigms finds our educational and training institutions completely unprepared. New inequalities and asymmetries are emerging precisely in the institutions that should be contrasting/should have contrasted them.[53] Therefore, a radical rethinking of education and training would have fundamental repercussions also, and above all, in terms of citizenship,[54] of a "new social contract"[55] and the possibility of inhabiting *(hyper)complexity* (1996).

On the contrary, despite the fact that the educational and training paradigm of the hypertechnological and hyperconnected civilization has revealed all its weaknesses and inadequacies, *we continue to insist on training mere executors of functions and rules*, as I have said before, who know only how to "isolate" and "separate", who only know how to use "linear thinking" (and causal), without finding the *connections* and correlations between the parts, without the ability to look at the whole, availing themselves of a systemic vision of things. Instead, we should educate and train people *to see objects as systems and not systems as objects.*[56] Treating *complex systems* as if they were *complicated systems* means starting over from the illusion of being able, in any case, to manage/control/predict/predetermine phenomena/processes, continuing to perpetuate what, before the term became popular, I had called "the ultimate error". I am thinking, in particular, of Artificial Intelligence,[57] of

---

[53] In one of its recently published studies, *Rapporto sulla conoscenza*, [Report on knowledge], the Italian institution of statistics, ISTAT, (2018) confirmed that schools and universities have become once more "agencies of selection", not of emancipation. Matters of geographical, social, cultural and economic origins play a decisive role in defining the future/destiny of our young people. The lack of investment in education, training and research completes the picture. In this regard, Italy—according to the OECD report "*Education at a Glance*" (2019)—invests little (and badly) in public education: 3.6% of its GDP, 1.4% less than the OECD average (of about 5%); the same is true of research, since my country invests less than 1% of GDP, compared to an average of 1.5% of the other countries taken into consideration.

[54] Citizenship is a complex issue which, for some time now, has ceased to be a simply legal issue. On these topics see, in particular: the classic study by T.H. Marshall, *Citizenship and Social Class* del 1950; S. Veca, *Cittadinanza. Riflessioni filosofiche sull'idea di emancipazione*, Feltrinelli, Milan 1990; Zolo, 1994 e 2000; R. A. Dahl (1998), *On Democracy*, Italian translation, *Sulla democrazia*, Laterza, Rome -Bari, 2000; R. Dahrendorf *Dopo la democrazia*, Laterza, Rome-Bari 2001; R. Bellamy, *Citizenship. A Very Short Introduction*, Oxford University Press, Oxford 2008; Dominici, *op.cit.*, 2008, 2014–2018; Norris P., *Democratic Deficits: Critical Citizens Revisited*, Cambridge University Press, Cambridge 2011; Balibar É., *Cittadinanza*, Bollati Boringhieri, Turin 2012.

[55] See Dominici P., (1995, 1998 and following), *op.cit.*

[56] *Ibidem.*

[57] See Boden M.A. (2018), *Artificial Intelligence. A Very Short Introduction*, Italian translation, *L'Intelligenza Artificiale*, Il Mulino, Bologna 2019. Tra i grandi classici, See Turing A.M. (1992), *Collected Works of A.M. Turing*, Italian translation, *Intelligenza Meccanica*, Bollati Boringhieri, Turin 1994. Among recent works, see: Crawford K. (2021), *Atlas of AI. Power, Politics and the Planetary Costs of Artificial Intelligence*, Italian translation, *Né intelligente né artificiale. Il lato oscuro dell'IA*, Il Mulino, Bologna 2021.

expectations for new automation systems, of information management, claims about the invulnerability of new sophisticated technologies.[58]

From this point of view, *putting the person* (not the individual) *back at the centre* means recovering the "complex dimensions of educational complexity" to use to teach and train people and not just individuals, nurture flexible, creative minds,[59] "hybrid figures"[60]—with respect to educational and training courses—that must be educated and qualified to *inhabit* conflict, borders, contradictions, variety, the emergent, recognizing the real benefits that can be gained through dialogues among the fields of knowledge. The alternative: persisting in a short-sighted perspective that looks solely at the market as the reference goal of education and training, an error we risk paying dearly for, especially if we consider the rapid *obsolescence* of knowledge and skills that is already characterizing today's social and working context and is destined to do so more and more.

We need to be aware of how (hyper)complexity is intimately connoted *by*—and at the same time, feeds *on* constant and dynamic instability, *variability*[61] and *diversity*, all of which are structural, ontological and not reducible, much less referable, to mathematical formulae and/or sequences, albeit infinite, of data. "Chance" and unpredictability play a fundamental role in all the non-linear evolutionary dynamics of living systems.

So, once again, we need to be careful. Here we are talking about *complex,* not complicated, *systems*. It is important not to go on repeating and perpetuating the above-mentioned "ultimate error". I repeat, and with insistence, that this is a systemic and strategic error, one of vital importance, a lack of understanding at the basis of many of the issue for which we are unable to provide/find an answer. *Under the delusion that we are capable of simplifying and facilitating everything*, from education to learning, from communication to democracy, we continue to navigate around the definition of "simple solutions to complex problems",[62] in the name of a kind of *solutionism* (often of juridical and technological origins) that blends well with the objectives of what I have called the "Mechanism- Society", a mega-system acknowledged and managed as a "complicated system", comprising *apparently innumerable* sub-systems that must/should be rendered increasingly transparent and

[58] I confess that I have always been very critical of the very concept of "artificial intelligence", since it makes intelligence coincide exclusively with the ability to calculate and find solutions to problems.

[59] See also Illich I., op.cit. (1971).

[60] See Dominici P., 1995–1996 and following., op.cit.

[61] As regards "variability" as a primary property of the living, as well as the *urgency* of a new paradigm and a reaction to reductionist visions/interpretations, but also on the very idea of "order" as a real obstacle to knowledge and interpretation of the world, see the "anarchist conception of the living": Kupiec J-J. (2021), *Et si le vivant était anarchique*, Italian translation, *La concezione anarchica del vivente*, Milan: Eleuthera 2021. On these issues, see also: Benasayag M., Cany B. (2021), *Les Nouvelles Figures de l'Agir*, Italian translation, *Corpi viventi. Pensare e agire contro la catastrofe*, Feltrinelli, Milan 2022.

[62] See Dominici P. (1995–1996), (2003–2019), *op.cit.*

observable to the public sphere and to single citizens, in order to be thoroughly—and genuinely — manageable and predictable.

Going back to the "ultimate error," it is well worth our while to reiterate, with clarity and determination, an itemized, fundamental concept which holds that *not all phenomenal reality can be explained using "causal" and linear thinking, not all is observable* ("emergent properties") and, as a result, *not all is measurable and foreseeable.*[63] *As a result, not everything can be described/represented using the scientific languages of certainty* (which produce reductionism and determinism); not everything can be explained in terms of simple, universal scientific laws, already inscribed/present in Nature and in the world of phenomena. *Not everything can be simplified and/or reduced* to the application of mathematical formulae and algorithms. Above all, as I have said before, not everything is *manageable* (on the contrary!) even though the technologies of connection and the seemingly inexhaustible availability of data are, from this perspective, bringing us back exceptionally *uncertain* certainties and securities, sometimes illusory, such as the atavistic aspiration/ambition to control and to exercise dominion over Nature itself. We were able to verify this during the 2020–2022 global, systemic emergency, where the cause of "everything" was a virus which, supported/facilitated by the *organized, scientific disorganization* of the nation-states, threw the entire *hyperconnected and technologically hypercontrolled/supervised* world system into a crisis of epic proportions.

On the contrary, within the illusion and the deceptive prospect of being able to *simplify* and facilitate everything, to eliminate error and unpredictability from our lives and our existences, to manage and direct, even measure, complexity/hypercomplexity, we continue to circumnavigate the definition of "simple solutions to complex problems", scarcely aware, moreover, of how extremely slim the *border between being a citizen and a subject* is; a borderline that has become even slimmer within this hypertechnological and hyper-connected civilization of ours– based, let us not forget, on stringent logics of control and surveillance, which increasingly resize the social spaces of autonomy and freedom/responsibility. A civilization that, not only at the level of narratives/*storytelling*, harbours the more and more prevalent illusion, even without taking into consideration other variables and contributing factors, that it is precisely the digital dimension and the technologies of connectivity (1996) that are capable of creating the conditions of a "true" (direct) democracy and of *a (finally) active and not hetero-directed participation*. These are dangerous *dystopias*, linked yet again to reductionist and deterministic analyzes of reality.

Because—as stated several times in the past—instruments, procedures, language, data and their visualization can be simplified, but "democracy is complexity" and cannot be simplified, exactly like educational processes, life itself and the forms it uses to manifests itself! Within what we have called the "civilization

[63] On these issues see in particular: Dominici P., op.cit., 1996–2020; Blastland M., op.cit. (2019).

without error", the borderline between citizenship and subservience[64] is even more subtle than it used to be and, for the moment, at least, we are stuck within the illusion of a less asymmetrical relationship with power and with a participation that is only "simulated",[65] a kind of participation fueled by a certain rhetoric of *simplification* and *disintermediation*. As I insist on repeating, *Democracy is complexity*[66]*: relational, social and human complexity.*

---

[64] On these questions, and on the intimate correlation that citizenship and subservience have with the "quality" (a complex concept, to be untangled), of educational and training institutions and with access to a good level of education/training, see once again the important work of: Illich I., (2019), *op. cit.*

[65] Regarding the concept of the "simulation of participation", see: Dominici P., (1996–2018), op.cit.

[66] The expression *Democracy is complexity* was proposed long before it became popular (Dominici, 1995–1996 and following). Democracy is, to all intents and purposes, a complex process involving numerous links and levels of connection involved, the reduction/simplification of which can always involve risks. It is a form of (hyper)complexity which, in this case, too, can be seen as including dialectics of conflict; a passage from linearity to complexity; a transition from simplicity to complexity; a reciprocity of sets and multiplicities; an epistemology of interdependence; the organization of experiences and knowledge; a continuous dialectic rapport between freedom and equality; the recognition of error as a producer of knowledge; a systemic and interdisciplinary/transdisciplinary approach to problems; recognition of conflict and mediation; the simultaneous presence of order and chaos; a multiplicity of identities and subjectivities; the enhancement of multiplicity and diversity; a pluralism of principles, values and views; the enhancement of heterogeneity and recognition of the value of deviance; the unpredictability and vulnerability of people and systems. On these issues, see K. Minogue's critical analysis (2010), *The Servile Mind. How Democracy Erodes the Moral Life*, Italian translation, La mente servile. *La vita morale nell'era della democrazia*, IBL Libri, Turin 2012. As regards this approach and "way" of studying and investigating the (hyper)complex system of "democracy", see also the interesting article: "La democrazia funziona come un sistema naturale", at https://greenreport.it/news/economia-ecologica/la-democrazia-funziona-come-un-sistema-naturale-se-la-polarizzazione-riduce-la-diversita-rischia-il-collasso/

Chapter 2
What About Social Bonds? The Grand Illusions of the Hyper-Technological Civilization and the Complex Interaction Between *the Human* with Technique and Machines

Society can only be understood through a study of the messages and the communication facilities which belong to it; and that in the future development of these messages and communication facilities, messages between man and machines, between machines and man, and between machine and machine, are destined to play an ever-increasing part…
Norbert Wiener
Technology tells us how to carry out certain operations. It is silent in the face of why.

Franco Ferrarotti

Abstract In this increasingly interdependent and interconnected era, the hegemonic paradigm of the hyper-technological civilization brings in its wake a series of grand illusions intimately linked to the possibility of marginalizing the Human, delegating our choices, responsibilities and even our educational processes to systems of artificial intelligence and interconnected technological devices, which seem capable of restoring certainties that are instead purely illusory: grand illusions of rationality, total control, predictability, measurability, and the elimination of error. In the meantime, the human dimension is dwarfed by the logics of a market in which dignity, in Weberian terms, is known only of the *thing* and not of the Person, a global market left at the mercy of its "self-normativity", which highlights even further the weakening of social bonds we are experiencing, along with the alarming ambition that human beings can or should resemble machines, in order to enhance our capacities, eliminate error and even worse, neutralize the unpredictability of our actions and decisions.

P. Dominici, *Beyond Black Swans: Inhabiting Indeterminacy*, Philosophical Studies Series 160, https://doi.org/10.1007/978-3-032-09029-4_2

Keywords Hegemonic paradigm · Hyper-technological civilization · Grand illusions · Eliminating error · Interdependent era · Interconnected era · Rationality · Total control · Measurability · Predictability

The world system is taking shape increasingly as a *vast web of global networks* or, to formulate it better, an *ecosystem of ecosystems*, made of interconnected hyper-technological metropolises, at whose borders masses of the new poor, comprising those who have no access to and are not included within the new global ecosystem, press. It is necessary to develop international and transnational political strategies capable of dealing with human history's umpteenth great industrial revolution: a transition which, due to the speed and complexity of the changes triggered and, unlike those that went before, has no precedent upon the evolutionary pathway of social systems. A global transition, which, in addition to having exponentially increased cognitive and communicative possibilities by eliminating the space-time barrier, has also called into question our systems of values, identity and belongingness. This means that new dynamics and new networks of social interaction are emerging that completely redesign the boundaries between public and private spaces,[1] between relationships with power and aesthetic enjoyment.

The technological revolution, from which the advent of the interconnected/hyperconnected society arose, has created the conditions—for the moment, we hasten to add, especially within the so-called "advanced societies"—for increasing integration between intelligence and machines, and, above all, for a growing degree

[1] See J. Meyrowitz (1986), *No Sense of Place: The Impact of Electronic Media on Social Behavior*, Italian translation *Oltre il senso del luogo. L'impatto dei media elettronici sul comportamento sociale*, Baskerville, Bologna 1993. In this essay the author, acknowledging the power to influence social behaviour that the media and, in particular, the electronic media, possess, argues that this power does not derive from the nature of the messages they convey, but from the new configuration and organization of the "social environments" determined and within which networks of social interaction develop. The relationship between "physical place" and "social place" has greatly weakened, as a result of the fact that the new means of communication have made individuals lose their former "sense of place", by eliminating boundaries between the spheres of the "public" and "private", even altering the very concepts of "identity" and "social behaviour". Meyrowitz, combining the theories of Goffman and McLuhan, speaks of an "overlapping of spheres of socialization" caused by the process of restructuring the social stages upon which we play our multiple roles. In other words, the electronic media—by transforming the meaning of time and space in social interaction—have profoundly changed the "situational geography of social life", configuring themselves as real "cultural environments". The media system tends, increasingly, to erase the existing differences between the "worlds of experience" of individual social actors, even prefiguring a new social order. On these topics, see the important works by E. Goffman (1959), *The Presentation of Self in Everyday Life*, Italian translation *La vita quotidiana come rappresentazione*, Il Mulino, Bologna 1969 e di M. McLuhan (1964), *Understanding Media: The Extensions of Man*, Italian translation *The tools del comunicare*, Il Saggiatore, Milan 1967.

of interdependence between the parts (sub-systems) that comprise the world-system. At the same time, the nature of the interactive media and the new ecosystem obliges social systems, along with the production, distribution, political and cultural processes that characterize them, to structure themselves according to a logic of networking that undermines old hierarchies and decision-making processes. It highlights, above all, the anomalies and weaknesses of traditional interpretative models, previously capable of making social and organizational life relatively controllable/predictable.

The problem is that, at least for now, we have not defined a form of "thinking", a "thought system", or alternative models (the urgency of the paradigm shift is plain for all to see), though we have refined our methods of data collection and processing considerably. Hence, while the numerous critical issues (control, total surveillance, asymmetries, functional illiteracy, etc.) and resistance to change continue and are destined to remain, along with the failure to focus decisively on (and invest in) education, training and research, the *connective tissue* of civil, political and human rights and, more specifically of (global) citizenship, that has been seriously damaged in recent years, will not be taken care of. It is precisely the interactive logic inherent in the model/paradigm itself of the Big Network, based on communication/hyper-connection and on the *sharing* of all resources of information and knowledge which could drive/could have driven the evolution of modern social systems towards increasingly open, inclusive and democratic models of society and provided the basis for a strong (transnational) project of ethical globalisation. Up to this moment, at any rate, we have been moving in the opposite direction. The (illusory) control of the complex dynamics of world capitalism and technological progress remains, in any case, a (complex) question of knowledge sharing and management, but also, patently, of the management of information and data, and requires, not only as and from today, "shared knowledge" (2003) and a culture of knowledge and complexity (collectively and socially constructed): an invaluable and inexhaustible *collective good* that all of us, within the limits of our possibilities and the roles we play, are/should be duty bound to nourish and disseminate, especially in an era which is putting to the (extreme) test the fundamental perspective of a *new Humanism*, of a *Planetary Humanism* (Morin), not to mention the very concept of human dignity itself. The sharing of cognitive resources and skills—combined with adequate (and complex) literacy, *schooling* and training policies at various levels—represents a pathway we cannot fail to take. Technology, and, in particular, digital technology, have, in an unprecedented, profound and capillary way, compared to the past, obliged us to transform reality, not simply to adapt to it.

We need to deal with the systemic and global emergence (of emergencies) that has thrown the entire world system into a state of crisis, and with it, *homo sapiens* himself—I repeat—convinced that he is *the undisputed lord and ruler* of Nature and the ecosystems he inhabits. From this point on, it is a question of reformulating hypotheses and questions –in a phase, moreover, when everyone is proposing exclusively simple answers and solutions to problems that are obviously complex—also with respect to the possibility of creating a New Humanism for this hyper-technological civilization of ours, so rife with risk (Beck). This New Humanism– as stated several times in the past—needs to start, precisely, from an overall reconsideration of thinking, of knowledge (as *shared knowledge*), of the *space between fields of knowledge* (and, on a second level, between skills) and, above all, of relational space (freedom is responsibility- the centrality of educational processes). Concretely, this New Humanism needs to place at its centre not only the Person/the human being but also, and above all—and this is the crucial point—(a newly) inhabited "Nature", within the framework of a renewed *systemic and symbiotic relationship*.

The New Humanism envisaged here needs to place the Person and *Nature*, not just Technique/Technology or the Law, at the centre of the complex process of change that is taking place; otherwise, we shall go on making the same age-old mistakes (see, in particular, environmental issues and climate change), in continuing to consider environments/ecosystems in terms of resources to be exploited only and/or systems over which to exercise dominion. A New Humanism that does not consist simply in the all-too-frequently accepted, anti-historical reaffirmation of certain (albeit fundamental) values imposed, probably, from above, in a completely different global historical context, marked by disturbing features, such as the "ethical void" (Jonas) and "anomie" (Durkheim), by indifference and moral torpor, by uncertainty and precariousness that have, at this point, become existential conditions. In a world system dominated by the logics of a market in which dignity, in Weberian terms, is known only of the *thing* and not of the Person. A global market left at the mercy of its "self-normativity", which highlights even further the triumph of individualism, of new asymmetries and the consequent weakening of social bonds we are experiencing. A New Humanism that will have to redefine certain categories (humanity, identity, dignity, Person, value, rights, etc.) in order to rethink the position of the human being in the world, in a renewed, as well as complex, relationship with ecosystems (we shall later return to the issue of "complex adaptive systems") and with the revolutionary, and in many cases invasive, technological innovations that characterise our reality.

It is a matter of *open dialectics,* as I stated years ago. This is precisely because, in the era of increasingly interdependent and interconnected/hyperconnected globalization and systems, the world system appears to be characterised more and more by fragmentation as well as by social and cultural resistance, which, in turn, are interrelated with new forms of inequality and asymmetry. A (complex) world system—as mentioned before- rendered increasingly unstable by centrifugal and centripetal "forces" (emergent properties) which are simultaneously and irreversibly dynamic and chaotic. It is precisely during this phase of global change, that technology, technological solutions, the possibility of manipulating life in all its forms, the so-called "digital revolution", the infinite availability of data, processes of automation and simulation, and AI systems, seem capable of restoring—with all their potency, even in terms of the imaginary—certainties that are purely illusory (rationality, control, predictability, measurability, elimination of error). Indeed, it is precisely during this phase that we must come to terms with a global ecosystem which, in addition to being chaotic and complex from every point of view, has become irreversibly polycentric. At the same time, globalization has led to a progressive downsizing of the sphere of politics, creating a void that economic and technocratic powers have hastened to fill,[2] within which the self-destructive potential of this technological civilization has advanced as well. A hypercomplex kind of civilization which is urgently demanding the utopia of a planetary society (transnational actions and policies)—narratives and representations do not suffice—which might, however, run the risk of basing itself on *the solidarity of fear* and *the need for security.*[3] It is necessary, therefore, to insist on one point: the risks and dangers inherent in the global technological civilization can only be faced by using global strategies drawn up on the basis of knowledge and the communication of knowledge, fundamental for their management. What Beck calls *the solidarity of fear* is destined to create social systems that bar their doors against change and confrontation (2003) and erect walls and fences (ideological and/or cultural), despite the extension of networks of interdependence and interconnection. Interdependence vs fragmentation—control vs cooperation. With its ultra-powerful technological tools at the service of the control and transformation of reality, the society of instantaneous communication has by now acquired the essential structure of a "surveillance society",[4] where all social actors are willing to give up part of their liberty for the sake of greater security. This is a question as "old" as the origin of human societies,

[2] See Habermas J. (1991), *Vergangenheit als Zukunft*, Italian translation, *Dopo l'utopia. Il pensiero critico e il mondo d'oggi*, Marsilio, Venice 1992; Habermas J. (2013), *Im Sog der Technokratie. Kleine Politische Schriften XII*, Italian translation, *Nella spirale tecnocratica. Un'arringa per la solidarietà europea*, Laterza, Rome-Bari 2014; Habermas J., Luhmann N. (1971), *Theorie der Gesellschaft oder Sozialtechnologie. Was leistet die Systemforschung*, Italian translation, *Teoria della società o tecnologia sociale*, Etas Kompass, Milan 1973; Berlin I., *Two Concepts of Liberty*, Clarendon, Oxford 1958; Habermas J., "The Scientization of Politics and Public Opinion", In *Toward a Rational Society: Student Protest, Science and Politics*, Beacon, Boston 1970; Fischer F., *Technocracy and the politics of expertise*, Sage, Newbury Park [etc.] 1990; Friedman J., *Power Without Knowledge: A Critique of Technocracy*, Oxford University Press, Oxford 2019.

[3] See Castel R. (2003), *L'insécurité sociale. Qu' est-ce qu' être protégé?*, Italian translation, *L'insicurezza sociale. Che significa essere protetti?*, Einaudi, Turin 2004; Ceri P., *La società vulnerabile. Quale sicurezza, quale libertà*, Laterza, Rome -Bari 2003.

[4] See, in particular: Foucault, 1975; Lyon D. (1994), *The Electronic Eye. The Rise of Surveillance Society*, Italian translation, *L'occhio elettronico. Privacy e filosofia della sorveglianza*, Feltrinelli, Milan 1997; Lyon D. (2001), *Surveillance Society. Monitoring Everyday Life*, Italian translation, *La società sorvegliata. Tecnologie di controllo della vita quotidiana*, Feltrinelli, Milan 2002; Lupton D. (1999), *Risk*, Italian translation, *Il rischio. Percezione, simboli, culture*, Bologna, Il Mulino, 2003; Lupton D. (ed.), *Risk and Sociocultural Theory: New Directions and Perspectives.* Cambridge: Cambridge University Press, 1999b; Lupton D. (2001), *Surveillance Society. Monitoring Everyday Life*, Italian translation, *La società sorvegliata. Tecnologie di controllo della vita quotidiana*, Milan, Feltrinelli, 2002; Lupton D., Tulloch J. (2003), *Risk and Everyday Life*, Sage, London; Bauman Z., Lyon D. (2013), *Liquid Surveillance. A conversation*, Italian translation *Sesto potere. La sorveglianza nella modernità liquida*, Editori Laterza, Bari 2014; Byung-Chul H. (2012), *Transparenzgesellschaft*, Italian translation, *La società della trasparenza*, nottetempo, Rome 2014; Byung-Chul H. (2013), *Im Schwarm. Ansichten des Digitalien*, Italian translation, *Nello Sciame. Visioni del digitale*, Nottetempo, Rome 2015.

a question that we should never stop investigating as a complex "object" (system), which harks back to an apparently incurable dialectic.

Through—as always—reasoning and technology, Subjects/Social Actors in the Hypercomplex Society, having designed and built instruments that have brought about an unlimited transformation of the world system, have reached a position from which it is possible to direct and *pre-determine* biological evolution itself. At the same time, what has been rolled out is a virgin territory of collective praxis that is still revealing itself to be, not only for ethical reasons, a no-man's land.[5]

The extraordinarily innovative modalities of power and technological practice, in addition to the characteristics of the globalization process, radically undermine the dominant paradigms and, with them, traditional ethics, confirming all the (hyper) complexity and implications of this umpteenth industrial revolution, which puts the principle of precaution itself to the test.[6] The German philosopher Hans Jonas, as early as the end of the 1970s,[7] intuited and denounced the existence of a disturbing ethical "void"—due also to today's relativism of values—which does not provide adequate resistance to the process of the colonization of values forged increasingly by high technology. Jonas himself came to formulate the concept of the "heuristics of fear",[8] starting from awareness of the planetary expansion and lasting consequences for humanity of this new form of power and its technological prostheses. In other words, technological practice, as it evolves, tends to gain more and more autonomy, acquiring, above all, a propulsive function that transcends the will and strategies of individual actors, determining a compulsion to repeat that becomes normativity, potentially capable of turning free beings into slaves. These are topics and issues of vital importance concerning which, even today, there seems to be little awareness.

The global and hyperconnected civilization is characterized by an *intrinsic,* almost structural, *vulnerability*, related, precisely, to the factors that make it appear perfectly efficient, hypermodern and super-accelerated. This global civilization is

---

[5] See Dominici P. (1995–1996), (2005), (2014), (2019) and following, *op.cit.*; on these themes and related issues, see the perspective of "biocentrism", proposed by Robert Lanza in Lanza R., Pavšič M., Berman B. (2020), *The Grand Biocentric Design*, Italian translation, *Il grande disegno biocentrico. Come la vita crea la realtà*, Il Saggiatore, Milan 2022. See alsoil nostro recente articolo scientifico: Dominici P., *The Digital Mockingbird: Anthropological Transformation and the "New Nature"*, in "World Futures. The Journal of New Paradigm", Routledge, Taylor & Francis, Feb. 2022.

[6] See Sunstein C. (2001), *Republic.com, Italian translation, Republic.com. Cittadini informati o consumatori di informazioni?*, Il Mulino, Bologna 2003; Sunstein C. (2005), *Laws of Fear. Beyond the Precautionary Principle*, Italian translation, *Il diritto della paura. Oltre il principio di precauzione*, Il Mulino, Bologna 2010.

[7] See Jonas H. (1979), *Das Prinzip Verantwortung,* Insel Verlag, Frankfurt am Main, Italian translation, *Il principio responsabilità. Un'etica per la civiltà tecnologica*, Einaudi, Turin 1990.

[8] See Sofsky W. (2005), *Das Prinzip Sicherheit*, Italian translation, *Rischio e sicurezza*, Einaudi, Turin 2005; Sofsky W. (2007), *Verteidigung des Privaten*, Italian translation, *In difesa del privato*, Einaudi, Turin 2010.

marked, however, almost paradoxically, by a widespread perception of the insecurity and unpredictability of phenomena, which turns out to be even more lacerating because of the illusions generated by technology and the expectations of greater freedom, rights and security it induces but betrays too often. This civilization, even where possible and desirable, finds it very difficult to transform risks into opportunities and is obliged to deal with the renewed needs of an ethics based on prudence and, above all, on respect for the Other and his/her destiny (Jonas). In other words, the advent of radical modernity determines an unprecedented enlargement of the dimensions of uncertainty, insecurity and vulnerability within social systems, making the functions performed by communication even more strategic when it comes to the management of complexity (which is not manageable), risk, uncertainty and *conflict mediation* (1996).

Ulrich Beck brings us back to one of the essential premises of this work, which has been addressed and debated previously in greater depth and from numerous perspectives, regarding a decisive factor which we see as a pivotal aspect of the twentieth century: namely, the crisis of scientific rationality and the consequent acknowledgement of the non-absolute, probabilistic value of knowledge. And since the moment that the monopoly of scientific rationality in processes for reducing complexity and defining (global) risk began fading, "chasms" have been thrown open, which in cognitive terms and in relation to choices of resolutory strategies can only be filled, at least in part, through cooperation between scientific and social rationality. Further causes are the hypertrophy of the sphere of risk and insecurity in terms of the world-system[9] and the conflictual pluralization of civilizational risks, which have led to a proliferation of collective definitions and representations of risk and, to quote Weber, of "the meaningless infinity of the world process".[10]

The close correlation between risks and values (and interests) makes the discussion even more difficult and uncertain, precisely because of the multiplicity of value

---

[9] See: Eibl-Eibesfeldt I. (1988), *Der Mensch—das riskierte Wesen*, Italian translation, *L'uomo a rischio*, Bollati Boringhieri, Turin 1992; Beck U. (1986), *Risikogesellschaft. Auf dem weg in eine andere Moderne*, Italian translation, *La società del rischio. Verso una seconda modernità*, Carocci, Rome 2000; Beck U. (2007), *Weltrisikogesellschaft. Auf der Suche nach der verlorenen Sicherheit*, Italian translation, *Conditio Humana. Il rischio nell'età globale*, Laterza, Rome-Bari 2008; Douglas M. (1985), *Risk Acceptability According to the Social Sciences*, Routledge, New York, Italian translation, *Come percepiamo il pericolo. Antropologia del rischio*, Feltrinelli, Milan 1991; Douglas M. (1992), *Risk and Blame: Essays in Cultural Theory*. London: Routledge, Italian translation, *Rischio e colpa*, Il Mulino, Bologna 1996; Luhmann N. (1990). *The Autopoiesis of social Systems,* in Luhmann N., *Essays on Self-Reference*, New York: Colombia University Press; Luhmann N. (1991), *Soziologie des Risikos*, Italian translation, *Sociologia del rischio*, Milan, Bruno Mondadori, 1996; Lupton D. (1999), *Risk*, Italian translation, *Il rischio. Percezione, simboli, culture*, Bologna, Il Mulino, 2003; Lupton D. (ed.), *Risk and Sociocultural Theory: New Directions and Perspectives*. Cambridge: Cambridge University Press, 1999b; Lupton D. (2001), *Surveillance Society. Monitoring Everyday Life*, Italian translation, *La società sorvegliata. Tecnologie di controllo della vita quotidiana*, Milan, Feltrinelli, 2002; Lupton D., Tulloch J. (2003), *Risk and Everyday Life*, Sage, London.

[10] See, Weber M., (1922).

systems brought into play. It is worth our while to reiterate the fact that we are going through an extremely sensitive phase—yet another, dilated period of transition—leading to global, multidimensional modernity, marked by a disturbing process of further *accumulation of anomalies,*[11] which undermines the certainties associated with consolidated cognitive paradigms.

2.1 The Illusory Certainties of the Hypertechnological and Hypercomplex Civilization

The knowledge society, on the one hand, seems capable of defining the conditions and interpretative models to use to deal with risk, uncertainty, complexity, and the unpredictability of systems and organizations, thus giving us the illusion of total control over the environment and ecosystems. On the other hand, it has not as yet implemented fully the *social mechanisms*[12] of (self) protection from the unpredictability of individual and/or collective behaviour, a veritable "weak link" in the systemic chain of security (and not only), dictated often—at all levels of the chain of decision making and responsibility—by a kind of rationality, total only in appearance, a "limited" rationality,[13] in actual fact, which is obliged to deal, more and more, with the strategic dimensions of individual and collective perception. In the era of the global market and hyperconnection, the *social production* of capital goes hand in hand with the social production of risks. The balance of power between the two forms of logic has been inverted precisely by virtue of reflexivity, which denies productive forces the opportunity of concealing their most latent side effects. The complex and manifold mechanisms associated with the social production of knowledge, supported by the new ecosystem of communication, unveil this dimension by relocating it within the public sphere, which provides, despite myriad difficulties, the opportunity to thematize questions and requests (also from below) initially excluded from the public debate.

The hegemonic paradigm of hyper-technological civilization, with all its paradoxical features, brings in its wake, as mentioned previously, a series of grand illusions intimately linked to the possibility of marginalizing the Human, delegating (carte blanche) our choices and our responsibilities to systems of artificial intelligence (?) and interconnected technological devices. In other words, the dimension of the technologically controlled, which has already become hypertrophic, will continue to increase and, with it, the illusory certainty/belief/ideological view that human, social and relational factors are increasingly less important and determining within productive, social and cultural spheres. In recent decades, these illusions

[11] Kuhn T., (1962), *op.cit.*

[12] See Coleman J.S. (1990), *Foundations of Social Theory*, Italian translation, *Fondamenti di teoria sociale*, Il Mulino, Bologna 2005.

[13] Simon H.A., (1947, 1959, 1997), *op.cit.*

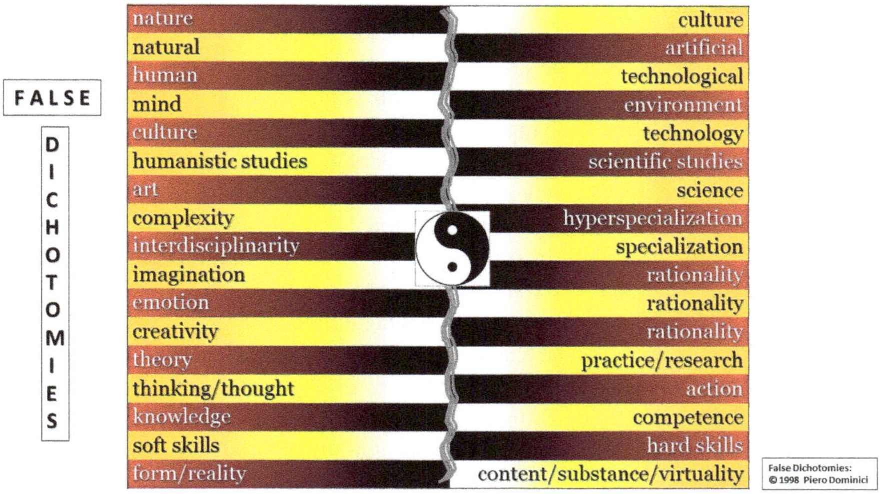

Fig. 2.1 *Healing the Fracture:* "False Dichotomies" and the urgency of *Rethinking Education* (Dominici, 1995–2018)

have led us to make grave mistakes, above all regarding the vital importance of rethinking education and training, the processes of education and training, and the organizational culture of educational and training institutions.

As we become increasingly convinced that digital technologies are capable of solving problems of all kinds, of protecting us from all dangers, we continue to ignore/underestimate "the" decisive variable: the human factor, which "underlies" all dynamics, processes, mechanisms, algorithms, as well as the emergencies typical of complex systems. What is staring us in the face is the urgent need to make long-term, strategic, even courageous choices, that regard not only the possibility of adapting and/or managing change but also the very opportunity of choosing between liberty and security, between freedom and control/surveillance, between participation and the freedom to be vassals[14]! The utopia of going beyond the freedom of being vassals. Beyond the freedom of an apparently safer, though undoubtedly less free life. These are ancient dilemmas and challenges which—I repeat with all my might—are primarily of an educational and cognitive nature.

The fundamental work that needs to be done—and which continues not to be done—brings directly into play the importance of rethinking/reinventing education, continuous training and scientific research itself, all of which remain marked and, in many cases, bound to the logics of control and separation, of false dichotomies (Fig. 2.1), of forms of reductionism and, above all, of *mono-disciplinarity*, confused more or less deliberately with (albeit necessary) scientific and methodological rigour. Although the ongoing series of global and systemic emergencies amply highlight all our vulnerabilities and inadequacies, a proper type of awareness is still

[14] See Dominici P., *op.cit.*, 2005, 2017.

wanting. Technical knowledge, (hyper)specialism, technology, digital solutions and complex connectivity alone cannot fully guarantee our "well-being", "freedom", "protection", "security" and "democracy".

In the *era* of *artificial intelligence,* of total control and programming/(hyper) simulation, the challenges of change are, precisely and, almost paradoxically, those of rethinking/redefining the centrality of the Person and the Human factor, within environments and ecosystems where borders/limits between natural and artificial no longer exist. It requires us to rethink the relational and communicative space in our training and learning/teaching institutions so that we may relaunch education from a systemic perspective that can only be socio-emotional. And yes, indeed... Order and Chaos. It is no longer sufficient, after having defined them, to try to distinguish between them in order to restore the balance and control that has been lost, because order and chaos coexist and cohabit; they are ever-present, providing feedback within the systemic, complex framework of the living,[15] and even more so, of the social, which forever continues to defy full comprehensibility and intelligibility. Within *open dialectics,*[16] whose moment of synthesis remains unfound, and where change presents itself, not only as *evolution of social complexity*, but also, and above all, as a challenge to education and the culture of communication. A complex challenge that asks us for a new kind of imagination, rethinking educational and training processes thoroughly in order to teach empathy and communication, freedom[17] and responsibility—not fear- but unpredictability (*ibidem*), complexity, the "scientific method"—with full awareness of the relative criticalities involved- and a systemic view of problems and phenomena.

Furthermore, the hyper-technological, hyper-connected civilization of intelligent (?) machines continues to feed another age-old and counterproductive illusion: that of being able to eradicate error, doubt and unpredictability, not only from complex organizations, but also from the processes of education and training themselves. The current global ecosystem of communication, although nourished by incessant flows of data and information, appears, paradoxically, to be increasingly

[15] See, in particular, the work of the famous biologist François Jacob (1970), *La logique du vivant*, Italian translation, *La logica del vivente*, Einaudi, Turin 1971, who, in one of the most significant passages of this work, stated that "Nothing says that a movement of the conscience, a feeling, a decision, a memory will ever be described in terms of physics and chemistry. Not only because of their complexity, but also because we know after Gödel, that a logical system does not suffice to describe itself" (p. 337).

[16] *Ibidem.*

[17] On the matter of educating as a "practice of freedom", see also the interesting work by the American scholar Gloria Jean Watkins, published under the pseudonym of *Bell Hooks*: Hooks B., *Teaching to Transgress: Education as A Practice of Freedom*, Italian translation, *Insegnare a trasgredire. L'educazione come pratica della libertà*, Meltemi, Milan 2020. See also Gardner H. (1991), *The Unschooled Mind. How Children think and How Schools should Teach*, Italian translation, *Educare al comprendere. Stereotipi infantile e apprendimento scolastico*, Feltrinelli, Milan 1993.

characterized by "limited rationality"[18] which cannot but increase—well beyond our incompleteness- how we perceive confrontation, as (active and participant) *observers* and as the objects of observation, with a hyper complexity that is not only augmented but difficult to control/manage, and above all, impossible to observe, let alone measure or replicate and which remains, therefore, beyond the scope of prediction. And perhaps it was inevitable that precisely this hypertechnological civilization, apparently "rational" in all its aspects, should be marked by trajectories, discontinuities, and dialectics lacking moments of synthesis, as well as by paradoxes which make its evolution ever more ambiguous and unpredictable. Among the many paradoxes of the era that we are going through and experiencing, perhaps the most obvious, but the one we are least aware of, concerns education, training and the relative processes that accompany them. It is exactly in this hypertechnological civilization, innervated by complex systems (interdependent and interconnected), dominated by (increasingly invasive) techniques and technologies and animated by new utopias and dystopias, that we continue to repeat and perpetuate the gravest error, one that is, to say the least, a strategic error. I have, in the past, defined it as "the Great Mistake": believing that education and training should simply be adapted to technological change by continuing to keep culture and technology separate, by progressively reducing the space available to the humanistic and more creative disciplines (including both their artistic and aesthetic forms), above all, by continuing to fuel "false dichotomies" (1995). As I have had the opportunity to say in the past, "This kind of separation, with all its dramatically negative repercussions—as mentioned—is reinforced and reproduced in educational, training and refresher courses, not only those held for figures of fundamental importance such as teachers, lecturers, researchers, trainers, and so on, but also during the education and training (and updating) of professional figures who are, in fact, totally immersed in and responsible for producing narratives and storytelling regarding culture and technologies, (journalists, popular opinion-makers, communicators, etc.). Furthermore, on the issue of the fracture between culture and technology, I would like to add that it is one of the many cultural fractures that concretely hinder the transition to hypercomplexity (hypercomplexity at last understood and inhabited) and the definition of the structural conditions of truly "inclusive innovation". Keeping "culture" and "technology" separate means continuing to believe—and act on the basis of this belief—that studies in the humanities and the sciences can go on—at least as regards educational and training institutions—surviving and evolving in ways completely autonomous from each other. Unlike certain hegemonic

---

[18] See Simon H.A. (1947), *Administrative Behavior*, Italian translation, *Il comportamento amministrativo*, Il Mulino, Bologna 1958; Simon H.A. (1959), *Theories of Decision-making in Economics and Behavioral Science*, in "American Economic Review", 49, pp. 253–83; Simon H.A. (1962), *The Architecture of Complexity*, in "Proceedings of the American Philosophical Society", 106, pp. 467–82; Simon H.A. (1979), *Models of Thought*, Yale University Press, Connecticut, New Haven; Simon H.A. (1997), *Models of Bounded Rationality*, Volume 3, *Empirically Grounded Economic Reason*, Italian translation, *Scienza economica e comportamento umano*, Edizioni di Comunità, Turin 2000.

narratives and storytelling, technique and technologies actually dwell *within* culture, inside the cultures and socio-cultural contexts that produce and distribute them, facilitating or hindering their metabolization and virality. Moreover, they are "complex cultural products" which already represent an (almost) perfect synthesis between the so-called "two cultures" (Snow)."[19]

2.2 The Human and Its *Complex Interaction* with Technology and Machines

This progressive appropriation of the *levers* of their own evolution by human beings calls traditional models and categories into serious question, forcing us (?) to revise/reformulate even the very definition of the concept of *Person*. It obliges us to rethink the *Human* and its—to some extent ambiguous—interaction with *technology*,[20] with the *technological* and, in particular, with *intelligent machines*[21] (?) and *robots*.[22] This complex human-machine/human-technology interaction is a factor from which *complex synthesis* cannot but stem (1996) although we are as yet unable to assess its perspectives, developments and implications. A "complex synthesis",[23] as I wrote long before it became a buzzword, which does not concern me so much regarding a possible, probable, *resemblance* between machines or robots and human beings; on the contrary, I look upon this kind of dynamics favourably, since it will facilitate the synthesis and interaction referred to here. What greatly worries me is the mere idea/aspiration/vision/narration that human beings can or should resemble—or desire more and more to resemble—machines, for the purpose of enhancing their abilities/capacities boundlessly, but above all, of eliminating error or the possibility of making different (even mistaken) choices and, even worse, of neutralizing the

---

[19] See Dominici P. (2003), *op.cit.*, pp. 61–62.

[20] See, in particular, Mumford, L. (1934), *Tecnica e cultura,* Il Saggiatore, Milan 1961; Mumford L. (1967). *The Myth of Machine*, Italian translation, *Il mito della macchina*, Milan: Il Saggiatore 1969; see also Foucault M., *Tecnologie del Sé. Un seminario con Michel Foucault*, Bollati Boringhieri, Turin 1992.

[21] On these topics, see also the interesting works by Giorgio Israel: Israel G., *La macchina vivente. Contro le visioni meccanicistiche dell'uomo*, Bollati Boringhieri, Turin 2004; Israel G., *The Science of Complexity. Epistemological Problems and Perspectives*, in "Science in Context", 18, Year 2005, pp. 1–31. See also Mumford L. (1921), *The Story of Utopias*, Italian translation, *Storia dell'utopia*, Donzelli, Rome 1997 (with a preface by F. Crespi), (the first Italian edition of the work dates back to 1969); Mumford L. (1934), *Technics and Civilization,* Italian translation, *Tecnica e cultura,* Il Saggiatore, Milan 1961 and, by the same author (1967), *The Myth of Machine*, Italian translation, *Il mito della macchina*, Il Saggiatore, Milan 1969. See the classic by Koyré A. (1961), *Les philosophes et la machine. Du monde de l'"à-peu-près" à l'universede la précision*, Italian translation, *Dal mondo del pressappoco all'universo della precisione*, Einaudi, Turin 1967.

[22] See Nourbakhsh I.R. (2013), *Robot Futures*, Italian translation, *Robot fra noi*, Bollati Boringhieri, Turin 2014.

[23] See Dominici P. (1995–96, 2003–2005, 2017) and following.

unpredictability of our actions and decisions. In other words, of eliminating, amid "new" utopias and *dystopias,* the very factors that makes us "human beings".

Furthermore, it is necessary to become definitively aware of the fact that the true strategic "factor" of change and processes of innovation is cultural: a complex variable capable, in the long run, of triggering and accompanying economic, political and social processes. This "cultural factor" that, precisely as a consequence of that process of *overturning* the *complex interaction between 'biological evolution' and 'cultural evolution',* is proving to be increasingly decisive. Today, it is cultural evolution that is determining biological evolution.[24] What do I mean by this? As already mentioned, I intend to say that human beings are gradually taking possession of the 'Archimedean levers' of their own evolutionary pathways and that the extraordinary scientific discoveries and technological innovations of our era are increasingly putting us in a position to control evolutionary mechanisms.[25] Beware, however, of the belief that technology is/may be an independent, neutral factor, even a factor "external" to culture. That it has never been, nor will it ever be! Continuing to entertain this idea leads/would lead us to perpetuate an enormous error of perspective which then translates/would then translate promptly into strategies and policies that are completely inadequate and off-track with respect to the complexity of the *digital transformation.* Yet, although this seems obvious, it is still an extremely widespread belief/view among scholars as well as among the so-called *experts.* Let us remember that the division between technology and culture is one of the falsest of the false dichotomies. "Inside" and "outside" belong to yet another of the "false dichotomies" which—not only when it comes to our relationship with technology—enchain cognitive and existential pathways and trajectories, preventing us from seeing/reaching beyond, from imagining life and knowledge beyond our incompleteness, past our incompleteness, starting from our very incompleteness. "False dichotomies" that prevent us from grasping/recognizing the whole, the complex, the Human, life: it is time to inhabit the boundaries and tensions that this hypercomplexity entails.

With this radical, definitive overturn, at least in appearance, the *cultural factors* prove themselves to be absolutely decisive—and, in the future, they are likely to be increasingly so—when determining the evolution of social systems and the manifestation of life itself. In other words, I am positing that it is the cultural factors that will determine the pathways and trajectories of technological and social progress; to the point of re-defining the very concept of 'life', re-defining what 'BEING HUMAN(S)' means today, in the civilization of hypertechnology, of automation, of "intelligent matter" and of complex syntheses. It will be the cultural factors to establish 'what is possible and what is not', at a time when—I repeat—the traditional boundaries between *the natural and the artificial* have been completely done away with!

---

[24] *Ibidem.*

[25] On the possible ethical and political implications, see Habermas J. (2001), *Die Zukunft der menschlichen Natur. Auf dem Weg zu einer liberalen Eugenik?*Italian translation, *Il future della natura umana. I rischi di una genetica liberale*, Turin, Einaudi 2002.

A hypertechnological civilization that intends to achieve perfection, balance, total control and the absence of error and deviation, drifting ominously toward control, surveillance, freedom and the power of the few over the multitudes. Like a perfect *artificial mega-mechanism*, synchronized and hetero-directed in every one of its dimensions and processes.

2.3 The Human Dimension, Education and Social Bonds

Rethinking education, therefore, to remain human first and foremost, and to attempt to *inhabit hypercomplexity* where order and chaos coexist and feed on each other. On the other hand, the complex synthesis between the natural and the artificial can only determine an evolution that is already translating into an increase/growth of complexity, with the emergence of new levels of interconnection and interdependence.

The change and complex anthropological transformation[26] we are going through present themselves, therefore, not only as an *evolution of social complexity*, but also, and above all, as a challenge to education and the culture of communication. It is time to facilitate the creation of bridges between fields of knowledge and skills, between (the dissolved borders of) the natural and the artificial, between knowledge and life, between the human and the technological.

Inhabiting hypercomplexity (1995–96), inhabiting the future, starting today to define and build the social and cultural conditions of inclusive innovation and of a less asymmetrical and unequal society. *Education* is *"the" question,* the question of questions actually (along with training and research, which need to be rethought along the same lines). Technology/technologies and the digital are extraordinary opportunities, whose value and implications we have not yet grasped and not only in terms of cultural change. Furthermore, within *complex adaptive systems*—as mentioned previously, marked by *non-linearity, unpredictability, the ability to self-organize and adapt*[27]—the problem is not/will not be a matter of simply knowing

[26] *Ibidem.*

[27] See, in particular Bocchi G., Ceruti M. (Eds.) (1985), *La sfida della complessità*, (a cura di), Bruno Mondadori, Milano 2007; Ashby, W.R., (1956), *op.cit.*; Barabási A.L. (2002), *Link. La scienza delle reti*, Einaudi, Turin 2004; Bateson, G. (1972), *op.cit.*; Bertalanffy von, L. (1968), *op.cit.*; Capra, F. (1975), *op.cit.*; Capra, F. (1996), *op.cit.*; Emery, F.E. (ed.)(2001), *La teoria dei sistemi,* FrancoAngeli, Milan; Gallino, L. (1992), *L'incerta alleanza,* Einaudi, Turin; Gandolfi, A. (2008), *op.cit.*; Foerster von, H. (1981), *op.cit.*; Kauffman, S.A. (1993), *op.cit.*; Maturana H., Varela F. (1972), *Autopoiesi e cognizione. La realizzazione del vivente*, Venezia, Marsilio, 1985; Maturana H.R., Varela F.J. (1985), *L'albero della conoscenza*, Garzanti, Milan 1987; Mead, G.H. (1934), *Mente, Sé e Società*, Barbera, Florence 1966; Morin E. (1977–2004), *op.cit.*; Prigogine I. (1996), *op.cit.*; See Dominici P., op.cit. (1996–2017); Dominici, P. (2018), *op.cit.*

how to manage/control technologies, making the most of their potential. What we need to increase, to begin with in terms of educational processes, is the awareness of our limits, inadequacies, and incompleteness in order to *build and educate toward a culture of doubt and error,*[28] educate people on the complexity of life from *a perspective that can only be critical and systemic* (where technology and the digital can prove undoubtedly decisive if we wish to *go beyond*) and place complex variables at the basis, not only of scientific research, but of life, of our existence, of "cohabitation", of meeting the OTHER who differs from Us. Rethinking education and, with it, a culture of communication which permits us to inhabit the future, draw up a new social contract and create a new kind of democracy; aware that only the "*end of certainties*"[29] *can generate knowledge and creativity and, at the same time, awareness that knowledge is rooted in life's errors.*[30] We need to deal (and, above all, "live") with a *hypercomplexity* which, as mentioned before, has demolished former borders and separations, questioning our "being human" and placing on the agenda—it is worth repeating—the urgency of a "new epistemology". We are still too unaware that the (hyper)complexity of which each of us is an active and integral part, cannot be *managed* in any way. We can only learn and try to *inhabit it*. In the midst of systemic, non-linear, chaotic and inexorable dynamics, we need to commit ourselves, in this perspective, all levels of action (individual and systemic) and

[28] Dominici P., (1995–1996), (2005), (2015–2022), op.cit., and following.. See also Perkinson H.J., *Didattica dell'errore. Aspetti pedagogici del pensiero di K.R. Popper*, Armando, Rome 1983; Mollo G., *Il valore dell'errore nella dinamica dell'apprendimento*, in "Cultura e Scuola"—Aprile-giugno 1986, n°98, pp128–145; Baldini M., *Epistemologia e pedagogia dell'errore*, La Scuola, Brescia 1986; Benes R. et al., *Per una pedagogia dell'errore*, Asterios, Trieste 2017. For a more in-depth look at these themes and issues, among the "classics" in the area of the so-called "hard sciences", I limit myself to referring to: a Oppenheimer J.R (1954), *Science and the Common Understanding*, Italian translation, *Scienza e Pensiero comune*, Bollati Boringhieri, Turin 1965; Feynman R. P. (1963), *Six easy pieces*, Italian translation, *Sei pezzi facili*, Adelphi, Milan 2000; Feynman R. P. (1985), *QED. The Strange Theory of Light and Matter*, Italian translation, *QED. La strana teoria della luce e della materia*, Adelphi, Milan 1989. Among the classics of Human and Social Sciences, I refer once again to: Canguilhem G. (1966), *Le normal et le pathologique*, Italian translation, *Il normale e il patologico*, Einaudi, Turin 1998; Montessori M., *L'errore e il suo controllo*, in *La mente del bambino*, Garzanti, Milan 1970; Barth R.S., *Open Education and The American School*, Agathon Press, New York, 1973; Postman N., *Ecologia dei media. La Scuola come contropotere*, Armando, Rome 1981; Popper K.R., *Scienza e Filosofia*, Einaudi, Turin 1969; Perkinson H.J., *Didattica dell'errore*, Armando, Rome 1983, Mach E. (1982) *Erkenntinis und Irrtum*, Italian translation *Conoscenza ed errore*, Memisis, Milan 2017. *Read* also and with care Schulz K. (2010), *Being Wrong. Adventures in the Margin of Error*, Italian translation, *L'arte di sbagliare. Avventure nel margine dell'errore*, Bompiani, Milan 2019.

[29] See Prigogine I., (1996), *La fine delle certezze. Il tempo, il caos e le leggi della natura*, Bollati Boringhieri, Turin 1997.

[30] See Canguilhem G., (1966), *Il normale e il patologico*, Einaudi, Turin 1998.

practice, in order to build a "culture of error"[31] that trains for and teaches unpredictability[32] by rethinking the very concepts of freedom and responsibility in a relational key.

2.4 Amid Limits and Inadequacies. For a "Culture of Error and Complexity"

Our socially and culturally constructed limits and inadequacies, our incompleteness make our vulnerabilities evident in the face of the vitality, richness and dynamism of complexity. Technologies are and are destined to become increasingly strategic, at all levels, precisely if, in addition to supporting human beings in all their actions, they will help them transform their limitations into opportunities (questions of knowledge and knowledge management), into knowing how to inhabit the tensions, hybrid zones and conflicts of the hypercomplex society by healing *the fracture between the human and the technological.* This will happen only *if we stop keeping culture and technology separate.* Technology and Culture → Technology vs Culture → Technology is Culture.

A hyper-technological civilization, which in addition to aspiring towards total efficiency, something that is impossible to achieve, not only fails to recognize the strategic relevance of error and unpredictability and refuses to admit error, conflict or *deviance* but chooses to condemn and stigmatize them, is a civilization destined, inevitably, to dwell within a condition of *perennial cultural delay and limited freedom.* On the contrary, as there is a desperate need for creativity and imagination if

[31] See Dominici P., *op.cit.*, (1995–2022). See also: Perkinson H.J., *Didattica dell'errore. Aspetti pedagogici del pensiero di K.R. Popper*, Armando, Rome 1983; Mollo G., *Il valore dell'errore nella dinamica dell'apprendimento*, in "Cultura e Scuola"—Aprile-giugno 1986, n°98, pp128–145; Baldini M., *Epistemologia e pedagogia dell'errore*, La Scuola, Brescia 1986; Benes R. et al., *Per una pedagogia dell'errore*, Asterios, Trieste 2017. To further explore these themes and issues, among the "classics" in the area of the so-called "hard sciences", I limit myself to referring to Oppenheimer J.R (1954), *Science and the Common Understanding*, Italian translation, *Scienza e Pensiero comune*, Bollati Boringhieri, Turin 1965; Feynman R. P. (1963), *Six easy pieces*, Italian translation, *Sei pezzi facili*, Adelphi, Milan 2000; Feynman R. P. (1985), *QED. The Strange Theory of Light and Matter*, Italian translation, *QED. La strana teoria della luce e della materia*, Adelphi, Milan 1989. Among the classics of the Human and Social Sciences, I refer, once again, to Canguilhem G. (1966), *Le normal et le pathologique*, Italian translation, *Il normale e il patologico*, Einaudi, Turin 1998; Montessori M., *L'errore e il suo controllo*, in *La mente del bambino*, Garzanti, Milan 1970; Barth R.S., *Open Education and The American School*, Agathon Press, New York, 1973; Postman N., *Ecologia dei media. La Scuola come contropotere*, Armando, Rome 1981; Popper K.R., *Scienza e Filosofia*, Einaudi, Turin 1969; Perkinson H.J., *Didattica dell'errore*, Armando, Rome 1983, Mach E. (1982) *Erkenntinis und Irrtum*, Italian translation *Conoscenza ed errore*, Memisis, Milan 2017. *Read* also and with care Schulz K. (2010), *Being Wrong. Adventures in the Margin of Error*, Italian translation, *L'arte di sbagliare. Avventure nel margine dell'errore*, Bompiani, Milan 2019.

[32] *Ibidem.*

we are to inhabit hypercomplexity, it is necessary to bring error, doubt, conflict, and unpredictability back to the core processes of education, training and research, *grafting them* onto a logical, methodological, epistemological, transdisciplinary learning/teaching and, later, onto a system of continuous, lifelong learning. These are destined, more and more, to be the variables capable of making the difference, above all in the era of intelligent machines, above all in the era of *automation* and *simulation*.

Chapter 3
Beyond Linearity. The Complexity of Complexity and an Epistemology of Error

What we observe is not nature itself, but nature exposed to our method of questioning.

Werner Heisenberg, 1958

Abstract In total opposition to complicated systems, a complex system (whose whole is much greater than the sum of its parts) is not observable from outside, and its evolution cannot be predicted or controlled, nor can it be broken down into separate parts, as they are all intersecting, interdependent, interconnected and constantly interacting. Such was the revolutionary paradigm that was beginning to be discovered at the turn of the last century, regarding the dynamic, non-linear, self-organizing, and emergent features of living systems, made by a number of brilliant scientists from a variety of fields, going well beyond the artificial fracture between disciplines—veritable "false dichotomies"—persisting stubbornly even today. In any case, the ongoing *anthropological transformation of* our civilization is being fueled by a numerosity, a velocity and a virality of information and communication never before experienced, leading to a condition of *hyper*complexity which we must learn to inhabit, breaking free of our cognitive and epistemological cages to realize that everything is interdependent and relational, that objects are systems rather than vice-versa, and even more, that what we think of as objects are themselves relations: *everything* is relationship.

Keywords Complicated systems · Complex systems · Non-observable · Artificial fracture · False dichotomies · Anthropological transformation · Hyper-complexity · Everything's related · Self-organizing features · Non-linear dynamics

P. Dominici, *Beyond Black Swans: Inhabiting Indeterminacy*, Philosophical Studies Series 160, https://doi.org/10.1007/978-3-032-09029-4_3

An important premise: adopting a systemic approach (or view) to complexity, and not only to complexity, talking about complexity and/or complex systems or, even simply evoking/talking about *systems thinking*, implies and requires, beyond the type of *problem* observed, a (profound) awareness of a series of implications:

1. Awareness that the hypotheses, perspectives and mechanistic formulae of classical science are wholly inadequate when seeking to describe and understand *chaotic systems*, even more so, *complex systems* (although, in my opinion, the distinction is, in this case, only terminological and not substantial; at least when it comes to the social and human sciences). In the past, classical science was always seeking order, reduction and simplification, under the banner of what a great philosopher (Nietzsche) would have called *forms of the undifferentiated.*[1]
2. Awareness of the inadequacy, in the broadest sense of the word, of reductionist and deterministic explanations and models which are, evidently, unable to consider structural instability, evolution through differentiation, the infinite potential of action and the irreversibility of complex systems.
3. Awareness that *analytical thinking* (although fundamental, in any case) and the *practices of the breakdown of complexity* make the description and recognition of complexity itself, with its infinite, irreversible and dynamic systemic interactions, practically and concretely impossible.
4. Awareness that the principles, explanatory models and traditional hypotheses of the "exact sciences" (among these, the principle of causality and various forms of determinism) are, once again, inapplicable and inadequate when seeking to describe and observe the *quality/qualities* of complex systems, also because they are founded on and attributable to variables defined and recognized, from time to time, as independent and essentially "stable".
5. The adoption of the (complex) concept of "system" itself implies awareness that the set, the *whole*, never coincides with *the sum of the parts* that constitute the system itself (more so, if complex) and that the parts/constituent elements of the system(s) are all interconnected and interdependent. In this sense, among the classics that deal with this issue, not solely regarding cybernetics, W. Ross Ashby—while referring to essentially "closed" (mechanical) systems—defined a *system* as "a list of variables" endowed with infinite potential and possible realizations, holding that "*every material object contains no less than an infinity of vari*ables and therefore of possible systems".[2] We also need to mention the quest for a unified view of systems and, more

[1] See Nietzsche F. (1887), *Genealogia della morale*, Adelphi, Milan 1984, d.III, cap.XXV, pp. 147–154.

[2] See Ashby W.R., *An Introduction to Cybernetics*, Chapman & Hall, London 1956. See also Emery F.E. (ed.) (1969), *Systems Thinking*, Italian translation, *La teoria dei sistemi. Presupposti, caratteristiche e sviluppi del pensiero sistemico*, FrancoAngeli, Milan 2001.

generally, of reality, carried out by von Bertalanffy[3] who, having studied and worked on *the theory of organic systems* and *the theory of open systems*, characterized by continuous exchanges of energy with the reference environment (see also Prigogine's studies and research), proposed a *general theory of systems* in an effort to define/recognize the universal elements valid for any type of system. In any case, systems are and can be defined as sets of mutually interacting components/units, marked by diversification and, at the same time, *structural isomorphism*.

6. Awareness that the systemic models defined and produced by the physical sciences and mathematics are not applicable to the complex systems belonging to the world of the organic, the vital. They are even less applicable to the Human and Social.
7. Awareness of the importance of taking into account fundamental characteristics such as *hierarchy* (presence of interdependent sub-systems) and *evolution through progressive differentiation* within the framework of a systemic and overwhelmingly dynamic form of interaction capable of eluding any type of dimensional parameter.
8. Awareness that a systemic approach, systems thinking and the concept itself of "complex systems", bring with them the need to consider the infinite *potential* and possibilities of an evolution that is complex, irreversible, immeasurable and "non-linear".
9. Awareness that the complex relationship between chance and chaos, between order and disorder—the crucial theme/question of "singularity" comes to mind—cannot be traced back/reduced to any kind of definitive/absolute law/rule.
10. Awareness of the extremely strategic and decisive role of the *observer* who, because he/she is never "neutral" or "external" with regard to the object/system observed, is always the *bearer* of a limited, partial point of view, which, when referred to linguistic and symbolic systems, is conventional and (more or less) arbitrary. Again, from this standpoint, the meanings and models proposed and constructed by an observer are a constituent element of the (hyper) complexity that she/he seeks to investigate.
11. Awareness of the "non-isolability" of the object—a matter of mere abstraction—with all the relevant implications and consequences.

[3] It is worth our while to underline that for von Bertalanffy, despite the fact that systems, with their functional organization, tend towards a "selective order", they are also characterized by irregularities and elements of disharmony. Therefore, biological models are not sufficient to describe forms of life exhaustively, and even less the ways in which they organize themselves. The study of individual and collective social action, and, even more so, analyses of society, require and need adaptability and flexibility. See Bertalanffy von L. (1968), *General System Theory: Foundations, Development, Applications*, Italian translation, *Teoria generale dei sistemi*, Isedi, Milan 1975, p. 332.

12. Awareness that actions and modes of behaviour, related and attributable to complex systems, while presenting infinite possibilities of (non-linear) evolution, *are not predeterminable.*

Having established this premise, there is, from this point of view, a further distinction to be made between "complicated systems" and "complex systems", to which we shall return later in more detail. In my opinion, this is a fundamental, unavoidable distinction, even if, as I claimed when such things were not as yet trendy (1996), it is destined to be swept away by what I have called a "new epistemological fracture", meaning the advent of robotics, artificial intelligence and AI systems.

A *complicated* system, unlike a complex system, is *observable* in its dimensions and, consequently, *manageable* and predictable in its developments. We can break down/separate its parts when seeking to observe, study, understand and, indeed, predict their behaviour. As we know, predictability (= measurability) means control, the possibility of managing the temporary absence of equilibrium and/or the possible accelerations of a system. In the case of a complex system, on the other hand, if we wish to describe it and, even more so, if we seek to understand its functioning, we need to observe and study it as a whole, focusing not on the parts that comprise it—trying to trace its "linear" relationships (which we would not find)—but on the numerous connections and the equally numerous levels of interconnection between the parts. In this case, moreover, we are dealing with types of systems which, also given the characteristics of the parts that constitute them (cells, animals, human beings, social systems/sub-systems, etc.), can never be controlled fully, much less simplified.

It is well worth our while to point out, to further clarify our theses, developed over years of study and research, that, as I often repeat, *the opposite of complexity is not simplification, but reductionism.*

In the analysis and (attempt at) management of complex systems, it is truly difficult even to speak of predictability and/or control because the sum of the parts never coincides with the whole, the totality of a complex entity, precisely because of the presence of *emergent properties*, of the ability of the "parts" to self-organize and proliferate. The quality and quantity of connections, often quite remote in space/time, always makes the difference because a system does not limit itself to oscillating between order and disorder, between order and chaos,[4] between balance and dynamism/entropy.

[4] On these topics and issues, on the relationship between "order" and "chaos", in particular, see: Strogatz H.S., Sync: *How Order Emerges from Chaos in the Universe, Nature, and Daily Life*, Hachette Book Group, 2004; Waldrop W. M. (1992), *Complexity. The Emerging Science at the Edge of Order and Chaos,* Italian translation, *Complessità. Men and ideas on the border between order and chaos,* Instar Libri, Turin 1995; Prigogine I. (1992), *Le leggi del caos*, Laterza, Bari-Rome 1993 (see also ed. 2003). See the excellent publication of: Cramer F. (1988), *Chaos und Ordnung. Die Complexe Struktur des Lebendigen,* Italian translation, *Caos e Ordine. La complessa struttura del vivente*, Bollati Boringhieri, Turin 1994.

3.1 A Step Back: A Brief History of Complexity

Along the cognitive and evolutionary pathways of the exact sciences there has always been a rooted conviction that one can/could, in some way, come to terms with, manage and control the complexity of phenomena. This is an important as well as multidisciplinary debate to which the great American mathematician and scientist Warren Weaver gave an immediate and significant boost with his article *Science and Complexity* (1948),[5] in which, among the various theses developed and the issues addressed, besides asking himself what Science and its boundaries might be (*The Boundaries of Science*), he proposed three macro-categories, to which all observed phenomena can be conducted. In order of temporal sequence and evolution they were: simplicity (*The Problems of Simplicity*—a description of systems with few variables); disorganized complexity (*Problems of Disorganized Complexity*—thanks to the adoption of probabilistic and statistical tools, a description of systems with large quantities of variables); and, finally, organized complexity (*Problems of Organized Complexity*—the study of [complex], internally organized systems, comprising strictly interdependent and interconnected parts, which cannot be described by the mathematical, probabilistic and statistical tools used to describe *disorganized complexity*).

The first systemic models of complexity were provided by the biological sciences during the first decades of the last century, in an attempt to explain the behaviour of complex organisms by going beyond the reductionist schemes and mathematical formulae of classical science or, to put it even better, of the traditional sciences. As the instruments of observation, analysis and detection were perfected and refined, the *anomalies* (Kuhn) and the substantial inadequacies of the observations and descriptions of the traditional sciences (Heisenberg) emerged in an increasingly clear manner. In the field of physics, a fundamental contribution was made by one of the great precursors of the very concept of complexity, the 1932 Nobel Prize laureate Werner Heisenberg, one of the first to formulate the principles of quantum mechanics. Heisenberg, highlighting the fallacy of one of the *dogmas* of classical science, that of the "objective observer", stated clearly that "*What we observe is not nature itself, but nature exposed to our method of questioning*".[6] Interesting reconstructions and analyses are available[7] to help us trace the history of

---

[5] See Weaver W., (Rockefeller Foundation, New York City), "Science and Complexity", *American Scientist*, 36: 536 (1948).

[6] See Heisenberg W., *Physics and Philosophy: the Revolution in Modern Science*, Prometheus Books, Buffalo N.Y., 1958; see also other works of this great scientist (and, allow me to add, philosopher): Heisenberg W., *I principi fisici della teoria dei quanti*, Bollati Boringhieri, Turin 1976; Heisenberg W. et al., *Discussione sulla fisica moderna*, Bollati Boringhieri, Turin 1980; Heisenberg W., *Lo sfondo filosofico della fisica moderna*, Eds. Gembillo G. and Giannetto E.A., Sellerio, Palermo 1999; Heisenberg W., *Indeterminazione e realtà*, Eds. Gembillo G. and Gregorio G., Guida, Napoli 2002.

[7] See, in particular: Capra F.—Luisi P.L. (2014), *The Systems View of Life*, Cambridge University Press, Cambridge; Gentili P. L. 2018, *Untangling Complex Systems: A Grand Challenge for*

the initial, but also of the subsequent, pathways of thought and research that led to a systemic view of complexity.

At the same time an urgent, inexorable need to attempt a unification of disciplines and knowledge emerged again to follow, with great determination and without the least hesitation, the pathway of multidisciplinarity, interdisciplinarity and of what was to be called *transdisciplinarity.* Later, a number of prestigious scientists[8] began speaking of a "new science" that they claimed was capable of going beyond classical science. This complex and well-structured project has regained strength and energy in recent decades, both within public debate and scientific communities. Within the context of an extremely multifarious scientific debate, which has involved numerous disciplinary fields and attempted, having moved along research pathways for several decades,[9] to define a theory/a science of complexity, Herbert Alexander

Science, Boca Raton (FL, USA): CRC Press, Taylor & Francis Group; Dominici P., *For an Inclusive Innovation. Healing the fracture between the human and the technological*, in, European Journal of Future Research, Springer, 2017—DOI: 10.1007/s40309-017-0126-4; Dominici P. (2017), "The hypertechnological civilization and the urgency of a systemic approach to complexity. A New Humanism for the Hypercomplex Society" in Neto A.B.F., Nieto E.C., Chiodi G., Petroccia S., (Eds.)., *Governing Turbolence. Risk and Opportunities in the Complexity Age*, Cambridge Scholars Publishing, Cambridge; Dominici P., *Dentro la Società Interconnessa. La cultura della complessità per abitare i confini e le tensioni della civiltà ipertecnologica,* FrancoAngeli, Milan 2019; Dominici P. *Controversies on hypercomplexity and on education in the hypertechnological era*, in, A. Fabris & G. Scarafile, Eds, *Controversies in the Contemporary World*, Amsterdam-Philadelphia: John Benjamins Publishing Company, 2019; Dominici P., *La complessità della complessità e l'ultimate error*, in Enciclopedia Italiana "Treccani", Treccani, Anno 2019; Dominici P., "*Educating for the Future in the Age of Obsolescence*", in CADMUS, Volume 4—Issue 3, November 2020, pp. 93–109.

[8] See, in particular Ashby W.R., *An Introduction to Cybernetics*, Chapman & Hall, London 1956; Bertalanffy von L. (1968), *General System Theory: Foundations, Development, Applications*, Italian translation, *Teoria generale dei sistemi*, Isedi, Milan 1975; Kauffman S. A. (1971), *Gene Regulation Networks. A Theory for Their Global Structure and Behaviours*, in "Current Topics in Developmental Biology", 6, pp. 145–182; Kauffman, S. A., *Origins of Order: Self-Organization and the Nature of History*, Oxford Univ. Press, NY 1993; Bateson G. (1972), *Steps to an ecology of mind*, Italian translation, *Verso un'ecologia della mente*, Adelphi, Milan 1976; Bateson G. (1979), *Mind and Nature. A Necessary Unity*, New York: Dutton, Italian translation, *Mente e natura. Un'unità necessaria*, Adelphi, Milan 1984; Anderson, P., *More is Different,* Science, 04 Aug 1972, Vol. 177, Issue 4047, pp. 393–396; Anderson P., Meyer A., Eisenhardt K., Carley K, Pettigrew A., *Introduction to the Special Issue: Applications of Complexity Theory to Organization Science*, "Organization Science" 1999, Vol. 10(3); Capra F. (1975), *The Tao of Physics*, Italian translation, *Il Tao della fisica*, Adelphi, Milan 1982; Capra F. (1996), *The Web of Life*, Italian translation, *La rete della vita. Una nuova visione della natura e della scienza*, Rizzoli, Milan 2001.

[9] Among the "classics" see: Schrödinger E. (1944), *What is Life? The Physical Aspect of the Living Cell*, Italian translation, *Che cos'è la vita?,* Adelphi, Milan 1995 (See also his concept of "negentropy" or negative entropy, or syntropy, capable of bringing systems back to order. For Schrödinger, "matter feeds on negative entropy"... and again "living matter escapes decay towards states of equilibrium"...); Morin E. (1973), *Le paradigme perdu: la nature humaine*, Italian translation, *Il paradigma perduto. Che cos'è la natura umana?*, Feltrinelli, Milan 1974; Bertalanffy von L. (1968), *General System Theory: Foundations, Development, Applications*, Italian translation, *Teoria generale dei sistemi*, Isedi, Milan 1975; Monod J. (1970), Le hazard et la nécessité, Italian translation, *Il caso e la necessità. Saggio sulla filosofia naturale della biologia contemporanea,*

Simon, winner of the Nobel Prize for Economics (1978), needs to be recalled. In 1962 he published an article, *The Architecture of Complexity,* which was, subsequently, to influence the entire scientific debate.

The topic of complexity and interdisciplinarity was and is a multidimensional and systemic project which, cyclically, also draws attention to the urgency of defining and constructing a common language—as well as a "method"—for use in the various fields of knowledge and disciplines. It was precisely research in the fields of biology and physics that definitively contributed to reinforcing the conviction and point of view, never sufficiently widespread, that complex systems and phenomena could not, in any way whatsoever, be reducible/attributable to relatively simple mathematical formulae, or to systems of elementary rules which, moreover, in the past, had proven capable of providing convincing representations of the complexity of reality. It is not just a question of a crisis of the traditional linear models and of the reductionist approaches that characterized classical science. It is a radical crisis affecting the ancient classifications and traditional taxonomies. These issues and questions have currently turned out to be even more urgent and unavoidable in the face of the paradigm shift and the ongoing process of *anthropological transformation.*

Even the traditional models, linked to *analytical thinking*, find themselves obliged to deal with phenomena, systems, organisms, organizations, realities and, above all, systemic interactions/relations, which elude any idea or view of complete balance and stasis/conservation and, at the same time, with possibilities of decomposition and re-composition of the "parts". These "objects of study"

Arnoldo Mondadori Ed., Milan 1997; Morin E. (1977), *La Méthode I. La Nature de la Nature*, Italian translation, *Il metodo 1. La natura della natura,* Raffaello Cortina Editore, Milan 2001; Morin E. (1980), *La Méthode II. La Vie de la Vie*, Italian translation, *Il metodo 2. La vita della vita*, Raffaello Cortina Editore, Milan 2004; Prigogine I. (1980), *From Being to Becoming: Time and Complexity in the Physical Sciences*, Italian translation di Bocchi G. e Ceruti M., *Dall'essere al divenire,* Einaudi, Turin 1986; Miller J.G. (1971), *Living systems*, Italian translation, *Teoria generale dei sistemi viventi*, FrancoAngeli, Milan 1986[3]; Cramer F. (1988), *Chaos und Ordnung. Die Complexe Struktur des Lebendigen*, Italian translation, *Caos e Ordine. La complessa struttura del vivente*, Bollati Boringhieri, Turin 1994; Prigogine I. (1996), *La fin des certitudes. Temps, chaos et les lois de la nature*, Italian translation, *La fine delle certezze. Il tempo, il caos e le leggi della natura*, Bollati Boringhieri, Turin 1997; Maturana H., Varela F. (1972), *Autopoiesis and Cognition. The Realization of the Living*, Italian translation, *Autopoiesi e cognizione. La realizzazione del vivente*, Venezia, Marsilio, 1985. On these topics and on the concept of "autopoietic organization", see Maturana H.R., Varela F.J. (1985), *The Tree of Knowledge*, Italian translation, *L'albero della conoscenza*, Garzanti, Milan 1987; Prigogine I.—Stengers I. (1984), *Order out of Chaos*, New York: Bentham Books; Heisenberg W., *Physics and Philosophy: the Revolution in Modern Science*, Prometheus Books, Buffalo N.Y., 1958; Margulis L., *Symbiosis in Cell Evolution in the Archean and Proterozoic Eons*, Freeman, New York 1993; Nicolis G.—Nicolis C. (2007), *Foundations of Complex Systems*, Singapore: World Scientific; Goodwin B., *How the Leopard Changed Its Spots. The evolution of Complexity,* Princeton Science Library, Princeton 1994; Israel G., *La macchina vivente. Contro le visioni meccanicistiche dell'uomo*, Bollati Boringhieri, Turin 2004; Bertuglia C.S., Vaio F., *Non linearità, caos, complessità*, Bollati Boringhieri, Turin 2003; Israel G., *The Science of Complexity. Epistemological Problems and Perspectives*, in "Science in Context", 18, 2005, pp. 1–31.

(systems) cannot (and already could not, at the time) be traced back to—much less be structured within—a framework of older operational concepts and definitions, which are no longer capable of realizing or describing what is real, the incessant and unstable dynamism of what is real, nor of the interdependent interconnectedness of every dimension of reality. Following the biological sciences and physics, the social sciences, cybernetics and, subsequently, philosophy, semiotics and linguistics as well, began to draw up systemic hypotheses or, in any case, approaches that allowed a glimpse into possible evolutions towards complexity and a systemic approach to complexity itself. The confrontation with *growing complexity*, the *capacity for differentiation and self-organization of systems*, already overdue in earlier times, is all the more non-deferrable[10] today.

The concept of *system* and the systemic approach have proved themselves, from the beginning, to be powerful explanatory and cognitive "devices" capable of recognizing and enhancing, apart from the complexity of processes and phenomena, the by now eminently probabilistic (non-absolute) and statistical character of knowledge. The concept of system[11],with all its implications, is also capable of bringing

[10] On these and similar topics, in particular, within the literature of multidisciplinarity, which is, to say the least, endless, see: Wiener N., 1948, 1950; Ashby, 1956; Neumann von, 1958, 1966; Hajek von, 1964; Bertalanffy von L., 1968; Piaget, 1970; Kauffman, 1971, 1993; Bateson G., 1972, 1979; Anderson, 1972; Holland, 1975; Le Moigne, 1977; Haken, 1977; Lovelock, 1979; Foerster von, 1981; Maturana- Varela, 1980, 1985; Luhmann N., 1984, 1990; Gleick, 1987; Gell-Mann, 1994, 1995; Morin E., 1973, 1977–2004, 1990–2019; Krugman, 1996; Prigogine I.-Stengers I., 1979, 1984, 1997; Nicolis G.- Nicolis C., 2007; Israel G., 2005; Dominici P., 1996–2020.

[11] See, in particular Hall, A.D., Fagen, R.E., *Definition of system*, in "General systems", 1956, n. 1, pp. 18–28.; Bertalanffy von L. (1968), *General System Theory: Foundations, Development, Applications*, Italian translation, *Teoria generale dei sistemi*, Isedi, Milan 1975; Le Moigne J.-L. (1977), *La théorie du système général. Théorie de la modelisation*, Paris: Presses Universitaires; Emery F.E. (ed.) (1969), *Systems Thinking*, Italian translation, *La teoria dei sistemi. Presupposti, caratteristiche e sviluppi del pensiero sistemico*, FrancoAngeli, Milan 2001; Kauffman S.A. (1971), *Gene Regulation Networks. A Theory for Their Global Structure and Behaviours*, in "Current Topics in Developmental Biology", 6, pp. 145–182. Regarding the approach(es) of Sociology and, more in general, of the Social sciences: Durkheim É. (1893), *De la division du travail social*, Italian translation, *La divisione del lavoro sociale*, Milan 1971; Pizzorno A. (1973), "L'incompletezza dei sistemi", in Rositi F. (Eds.), *Razionalità sociale e tecnologie dell'informazione. Descrizione e critica dell'utopia tecnocratica*, Comunità, Milan; Habermas J.- Luhmann N. (1973), *Teoria della società o tecnologia sociale*, Etas, Milan 1983; Luhmann N. (1984), *Soziale Systeme*, Suhrkamp, Frankfurt 1984, Italian translation, *Sistemi sociali. Fondamenti di una teoria generale*, Il Mulino, Bologna 1990; Luhmann N. (1986), *Okologische Kommunikation. Kann die moderne Gesellschaft sich auf ökologische Gefährdungen einstellen?*, Italian translation, *Comunicazione ecologica, Può la società moderna adattarsi alle minacce ecologiche?*, FrancoAngeli, Milan 1989; Luhmann N. (1990). *The Autopoiesis of Social Systems*, in N. Luhmann, *Essays on Self-Reference*, New York: Colombia University Press; Bayley, K.D., *Sociology and the new systems theory, toward a theoretical synthesis*, SUNY Press, Albany, N.Y., 1994; Mead G.H. (1934). *Mind, Self and Society*, Italian translation, *Mente, Sè e Società*, Barbera, Florence 1966; Mead G.H., *La voce della coscienza*, Jaca Book, Milan 1996; Morin E. (1980), *La Méthode II. La Vie de la Vie*, Italian translation, *Il metodo 2. La vita della vita*, Raffaello Cortina Editore, Milan 2004; Morin E. (1986), *La Méthode III. La Connaissance de la Connaissance*, Italian translation, *Il Metodo 3. La cono-*

closer together the dimensions of the natural and the artificial present in all the phenomena and objects of study. From this point of view, *a systemic approach to complexity* seems to be able to overcome the traditional logic of separation between fields of knowledge and disciplines, by providing the cognitive/epistemological elements that enable us to go beyond any kind of deterministic and linear perspective.[12]

The complexity of the *living,* not to mention that of the social, is never thoroughly comprehensible or intelligible. It is never reducible/conductible back to mathematical formulae and/or to more or less infinite sequences of data. In other words, the "quality" of living, of social and human complexity, cannot be translated into quantitative data. This feature, evidently, has profound implications in terms of representation, visualization, modelling and the communication of complexity itself. Indeed, precisely at this level of analysis and practice, representations, models and communication regarding complex systems become an integral (fundamental) part of them, both in terms of (individual and collective) observation and perception. We seem, as yet, to be unaware of this.

3.2 For an Epistemology of Error. The Strategic Relevance, at All Levels, of a "Culture of Error"

The problem is (and has always been) that of defining and distinguishing between *order and chaos*. The problem is knowing how to *recognize* this complexity, within which, order and chaos (order and disorder) coexist and cohabit, are both present and engaged in mutual feedback. Furthermore, it is important to be particularly careful—at every level of thought and analysis as well as of

scenza della conoscenza, Raffaello Cortina Editore, Milan 2007; Parsons T. (1951), *The Social System*, Italian translation, *Il sistema sociale*, (intro. by L.Gallino), Comunità, Milan 1965; Simon H.A.(1962), *The Architecture of Complexity*, in "Proceedings of the American Philosophical Society", 106, pp. 467–82. See also, with particular attention, studies and research by the physicist Giorgio Parisi, with particular reference to the theory of "disordered systems" Parisi G., *Field Theory, Disorder and Simulations*, World Scientific Pub. Co. Inc., Singapore 1992; Parisi G., *La complessità da un punto di vista di un fisico*, in Rendiconti Accademia Nazionale delle Scienze—*Memorie di Scienze Fisiche e Naturali* 119° (2001), Vol. XXV, pp. 7–14; Parisi G., Urbani P., Zamponi F., *Theory of Simple Glasses*, Cambridge University Press 2020. Among recent publications see Rainie L., Wellman B. (2012). *Networked: The New Social Operating System*, Italian translation, *Networked. Il nuovo sistema operativo sociale*, Guerini, Milan 2012; Simonigh C. (ed.) *Pensare la complessità. Per un Umanesimo Planetario. Con interventi di Edgar Morin, Gianni Vattimo e Gustavo Zagrebelsky*, Mimesis, Milan-Udine 2012.

[12] See: Wiener N., op.cit., 1948, 1950; Heisenberg W., op.cit., 1958; Lorenz E.N., *The Essence of Chaos*, Univ. of Wash Press, Seattle, 1963.; Bertalanffy von, op.cit., 1968; Watzlavick P. et al., op.cit., 1967; Emery F.E., op.cit., 1969; Morin E., op.cit., 1973; Holland J.C., op.cit., 1975.

organizational and social practice—not to place ourselves within a perspective of a quest for/knowledge/acquisition of a single existing, "objective" and "self-evident" "Truth",[13] which simply needs to be *revealed*. This perspective/view/approach is still dominant, even within the sphere of scientific research, as well as in educational/training processes and teaching, which bears with it an explicit condemnation of *error* and the well-known *stigma* attached to those who commit it.

We are very far from that process of a social and cultural definition and construction of a "culture of error" which, as I have not tired of repeating for over twenty years, should be taught, shared and constantly *re-negotiated* from the earliest years of school. Even in terms of macro-systems, what should be clear to us, but unfortunately, isn't (quite the opposite!)[14] is that "systems", even those of super-sophisticated artificial intelligence—and not only People—that do not commit *errors* are not/cannot be "intelligent". At the same time, I hasten to add that we ought to come to an agreement, once and for all, on the operational definition of "intelligence" which, in my opinion, does not concern so-called *problem solving* alone—an ability often erroneously associated and made to coincide with education/training in complexity and/or with the ability/skill to "do maths".

On the contrary, we continue to be taught, we continue to teach/learn—even at the *management* level of complex organizations—that the most important thing, that of vital importance, is "not to make mistakes", that is, "not to commit errors".[15] In this regard, the analysis (and studies) conducted recently by Gerd Gigerenzer, cognitive psychologist and former director of the *Centre for Adaptive Behaviour and Cognition* (ABC) at the Max Planck Institute for Human Development, are of great interest, as they concern decision-making processes related to the (complex) dimensions of risk and safety, starting from the distinction between "good errors" and "bad errors" as well as between "positive and negative cultures of errors", while also underlining how important it is during processes of innovation and change, *not to prevent people from committing errors*. Gigerenzer holds that every

[13] On the fundamentally important issue of "Truth" and, more specifically, of so-called "scientific objectivism", there is an endless, as well as dated, reflection, beginning with the doctrine of *docta ignorantia* by Nicola Cusano, who proposed a "dualistic view", aware of the inescapable division between the "sensible world" and the "intelligible world": See Cusano N. (1440), *De Docta Ignorantia*. Rusconi, Milan 1988. Among recent works see: Lynch M.P. (2004), *True to Life. Why Truth Matters*, Italian translation, *La verità e i suoi nemici*, Raffaello Cortina, Milan 2007; Godfrey-Smith P. (2021), *Theory and Reality. An Introduction to the Philosophy of Science*, Italian translation, *Teoria e realtà. Introduzione alla filosofia della scienza*, Raffaello Cortina Ed., Milan 2022.

[14] In this sense, we need to mention some great "classics" in the fields of the cognitive sciences and educational and training processes, as well as studies and research related (not only) to the issues of robotics and the so-called "intelligent machines". For an introduction to these topics see Castelfranchi Y., Stock O. (2000), *Macchine come noi. La scommessa dell'Intelligenza Artificiale*, Laterza, Rome-Bari.

[15] See Dominici P. (1996–2020), *op.cit.*

profession, firm or group has an error-making culture of its own. At one end of the spectrum stand the negative error cultures. The people who live in one of these cultures are terrified of making any kind of mistake, good or bad, and, if an error actually occurs, they go out of their way to hide it. A culture of this kind has little chance of learning from its errors and discovering new opportunities. At the other end of the spectrum stand positive error cultures, which make their errors transparent, encouraging good errors while learning from bad ones with a view to creating a safer environment..[16]

It is precisely from this perspective, however, that we can easily see how the extreme, radical division/subdivision/parcelization of work—which is still being structured according to a dated/obsolete view of *scientific organization*—along with the failure to create a culture of communication as a *social process of knowledge sharing* (1995), within and without complex organizations, whose interrelated dynamics and processes have been radicalized by automation and (hyper) accelerated simulation, which are considered (apparently) infallible, cannot avoid determining and/or reinforcing "linear", cause-effect and/or *input/output* relationships. Evidently, if not only work but also every other kind of social and human activity is divided and parcelled out, nothing more is needed other than "mere executors of functions and rules", who become the *ideal figures* for any type of organization as well as for any sort of political regime, not only democracies. At the same time, error (or deviance) itself is configured, once again, not only as something risky, to be eliminated, but also, and, above all, as "something" "avoidable" and "predictable". Of something that is *always predictable.* Yet we all know how things have always gone/ended up along the complex and non-linear evolutionary pathway of social systems and human societies; coming to the present day, we all know how things went during the last famous global, systemic emergency, whose aftermath is still, to this day, undergoing continuous and irreversible evolution.

Furthermore, besides the *hypercomplexity* typical of *complex adaptive systems*, humans and social systems have to deal with their own *fragility*, as well as with that of objects, sub-systems and environments. There are also systems and organizations that are capable of resisting better and longer against external *perturbations*. By the same token, there are realities, life forms, systems and organizations that turn out to be *anti-fragile*, capable, that is, of adapting to change and transformation, by constantly redefining actions, behaviour, strategies, models and paradigms.[17]

The most *fragile* realities and systems are precisely those that see error and unpredictability itself (the so-called *black swan*; intended also as *improbability*) as

---

[16] See Gigerenzer G. (2014), *Risk Savvy. How to make good decisions*, Italian translation, *Imparare a rischiare. Come prendere decisioni giuste*, Raffaello Cortina Editore, Milan 2015 pp. 58–59 and following.

[17] On these topics and, in particular, on fragility and antifragility, see: Taleb N.N. (2012), *Antifragile*, Italian translation, *Antifragile. Prosperare nel disordine*, il Saggiatore, Milan 2013.

variables/elements, at times deemed even dangerous and to be avoided at all costs, so, in the illusion of being able to eliminate error and unpredictability, experimentation and, with it, adaptation to change and/or new environmental conditions are actually hindered in a more or less (un)conscious manner. Proceeding in this direction, systems become increasingly fragile, contrary to what is generally believed. To avoid error and unpredictability, at any level, people and systems (which are organisms, not machines) are taught, educated, trained and shaped to behave and act, always, in the same way, using the same strategies and the same tools, because "that's how it has always worked". This means that human beings and systems are taught, educated (indoctrinated), formed, trained and moulded to always pursue the "same pathways", use the same reasoning, the same formulae as well as the same solutions and types of solutions for what "appear to be" the same problems and types of problems.[18]

This is an overall approach based also on a view of *learning* intended and acknowledged as a process of mere *accumulation*, not one of *systematization* and organization of knowledge. As I have never tired of repeating for years now, *errors* and the possibility of committing them, even knowingly, have always been decisive, strategic elements of vital importance for both *common-sense* and *scientific* knowledge based, founded, hinged on and grounded in *experimentation*, but also on recognition of completely unexpected and unforeseen signs, evidence, effects, consequences which, in many cases, risk not even being grasped in their *emergence*, in their essential and *multiform* manifestation. These are topics and issues—I repeat—that have enormous, non-evaluable repercussions, also, and above all, within processes of education and training.

The possibility and opportunity of committing mistakes, even with the intention, the determination to err, to abandon the "norm" and "normality", the pathway taken by everyone, is, in my opinion, the feature that best connotes us as "human beings", even more so, as "free human beings". What makes us unique and non-replicable, non-reproducible; what makes it hard to simulate our complex activities, least of all our thinking and intelligence; a far cry from replacing humans with robots and artificial intelligence. Prophecies with an agenda, whose functions are to lead to specific objectives.

Therefore, *an epistemology and a culture of error* (1995–1996) turn out, to say the least, to be necessary and decisive, in the first place, because they stimulate/trigger complex learning processes,[19] in the second, to try to prepare people for

---

[18] In this regard, I'd like to recall the definition proposed years ago, "Innovating means destabilizing". See Dominici P., op.cit., 1996–2020. On these topics, see also: Dennet D., *Strumenti per pensare*, Raffaello Cortina Ed., Milan 2014; Latour B. (2005), *Reassembling the Social*, Italian translation, *Riassemblare il Sociale*, Meltemi, Milan 2022.

[19] We have worked at length over the years on concepts and operational definitions of "epistemology and a culture of error", both as a teaching and a research activity; see Dominici P., *op.cit.*, 1996–2020. See also Perkinson H.J., *Didattica dell'errore. Aspetti pedagogici del pensiero di*

unpredictability, for the famous *black swans* and for *acknowledgement and recognition* of this complexity, in a way characterized by a *greater adaptive capacity* to respond to *ecosystemic* stresses and *perturbations*, while always paying attention not to fall—not this time, at least—into the usual errors: (1) reductionist and deterministic interpretations; (2) dialogical reasoning which always identifies apparently irreconcilable and incurable polarities and dichotomies; (3) linear and causal models that close one's view/views of complexity instead of opening it/them.

Life, the Social, the Human and the relational have always been (hyper)complex, and the problem that they have always constituted, complexity, is a fact of life (not an option). Unfortunately, we continue to face this issue, even in terms of scientific research, with very scarce awareness of the (not only epistemological) uncertainty and unpredictability that govern relationships and interactions, which are themselves, once again, complex.

According to Herbert Simon, if we consider complexity as architecture (or as if we were architects), one of its fundamental, basic structures is a *hierarchy*. A hierarchy in complex systems describes/depicts the *pathways* along which the *emergent properties* that characterize them manifest themselves. Most hierarchies, those of a physical and biological type, for example, are described as visible *substructures* located spatially within larger structures; therefore, a higher degree of *emergence* generally implies an undoubtedly *denser* spatial distribution.

When it comes to social and human hierarchies, these are best described by observing the *intensity* and quality of their interactions/relationships, rather than the physical proximity of people (who, in the well-known metaphor of the *web/network*, constitute their "nodes"). They can, therefore, be better defined in terms of the intensity of the interaction.

K.R. Popper, Armando, Rome 1983; Mollo G., *Il valore dell'errore nella dinamica dell'apprendimento*, in "Cultura e Scuola"—Aprile-giugno 1986, n°98, pp128–145; Baldini M., *Epistemologia e pedagogia dell'errore*, La Scuola, Brescia 1986; Benes R. et al., *Per una pedagogia dell'errore*, Asterios, Trieste 2017. To further explore these topics and issues, among the "classics" in the area of the so-called "hard sciences", I shall refer only to Oppenheimer J.R (1954), *Science and the Common Understanding*, Italian translation, *Scienza e Pensiero comune*, Bollati Boringhieri, Turin 1965; Feynman R. P. (1963), *Six easy pieces*, Italian translation, *Sei pezzi facili*, Adelphi, Milan 2000; Feynman R. P. (1985), *QED. The Strange Theory of Light and Matter*, Italian translation, *QED. La strana teoria della luce e della materia*, Adelphi, Milan 1989. Among the classics in the Human and Social Sciences, I refer again to Canguilhem G. (1966), *Le normal et le pathologique*, Italian translation, *Il normale e il patologico*, Einaudi, Turin 1998; Montessori M., *L'errore e il suo controllo*, in *La mente del bambino*, Garzanti, Milan 1970; Barth R.S., *Open Education and The American School*, Agathon Press, New York, 1973; Postman N., *Ecologia dei media. La Scuola come contropotere*, Armando, Rome 1981; Popper K.R., *Scienza e Filosofia*, Einaudi, Turin 1969; Perkinson H.J., *Didattica dell'errore*, Armando, Rome 1983. See also, with particular attention: Schulz K. (2010), *Being Wrong. Adventures in the Margin of Error*, Italian translation, *L'arte di sbagliare. Avventure nel margine dell'errore*, Bompiani, Milan 2019.

In social systems, as in all complex systems, this *hierarchy* includes a moment of ascension/increase where emergence in a sub-unit or a lower level of organization, triggers emergence in higher units. This is not the only way emergence works/performs within social systems, where conflict is a crucial element that guarantees evolution, change and development. It is quite clear how attempts made to eradicate or eliminate conflict from social systems or communities, similar to naive attempts made to eliminate error, are not only doomed to failure but constitute an unnecessarily repressive strategy, which also hinders critical thinking, healthy debate and the diffusion of the culture and *epistemology of error*, that is so decisive.[20] I might add that the very nature of complex systems makes it impossible to achieve the level of precision necessary to analyse or predict (much less control) its evolution(s). In actual fact, an infinite level of precision would need to be reached, something that neither humans nor their systems of artificial intelligence can hope to achieve, despite the narratives, ambitions and arrogance that accompany their (our) efforts in this direction.

The self-organization manifested by social systems, typical of *complex adaptive systems* (open, dynamic, nonlinear and nondeterministic), also referred to as *anti-chaos,*[21] is a spontaneous (dis)order that arises over time, without any need for external tuning. Social systems, as defined by Ilya Prigogine,[22] are, by definition, "far from thermodynamic equilibrium", meaning that a fixed or pre-established state or condition of behaviour will never occur, nor, in any case, can one be taken as *given once and for all.*

Every social process is deeply characterised by ambiguity and ambivalence (characteristics, among other things, incompatible with algorithms), despite the continued belief, particularly on the part of the *hard sciences*, that self-organization, deriving from chaos, can, in any case, restore, a *previously altered* state of *order*. In

[20] See Dominici P. (1996–2020), *op.cit.*

[21] See in particular Kauffman, Stuart A., *Origins of Order: Self-Organization and the Nature of History*, Oxford Univ. Press, NY 1993.

[22] See Prigogine I. (1996), *La fin des certitudes. Temps, chaos et les lois de la nature*, Italian translation, *La fine delle certezze. Il tempo, il caos e le leggi della natura*, Bollati Boringhieri, Turin 1997; Prigogine I.—Stengers I. (1984), *Order out of Chaos*, Bentham Books, N.Y. On these themes and questions, see also the interesting series of lectures held by Ilya Prigogine at the State University of Milan (Prof. Giulio Giorello), collected in the volume: Prigogine I. (1992), *Le leggi del caos*, Bari-Rome, Laterza 1993.

other words, one continues to believe/claim that chaos can be, in any case, the necessary precursor/pre-requisite of a higher level of order.[23]

On the contrary, I believe that self-organization in social systems can determine and/or create, just as easily, the conditions for a considerably higher degree of *disorder*, to which the system will tend to remain tied, failing to return to its original state of equilibrium or order.

When unforeseen or improbable events occur, the so-called "black swans", which we should always expect, of course, considering that they—along with what I have defined as *emergent emergencies*—are structural elements of complex systems, the systems will call on *memory*, which will be used in an endeavour to recover the original organization that has been disrupted. However, unpredictable evolutionary turns and accelerations in social systems, correlated to "nodes", which are not simply "points" of connection or links, being endowed, above all, with *consciousness* (that of human societies, for example), can bring about consequences that lead, in turn, to the definition of different strategies. Far from being a return to "normality" replaceable with a "*new normality*" after a phase of instability and acceleration, the previous, original condition, in my opinion, will never be fully recovered or re-established.

We need to emphasize the fact that, in everyday speech, the most common concept of "order" does not correspond to the order that emerges from chaos to become a self-organizing system. The concept of "order" is generally intended as a combination of organization, respect for hierarchies (authority) and balance. This concept/definition of *order*, however artificial, as well as the conventional concept that has dominated science for centuries, bringing scientists to focus exclusively on the *causa efficiens*, confuses the movement that produces an event with the cause that determines it; whereas, each event is determined by a series of co-causes together with their interactions, intersections, adaptations, self-organization and *feedback*. Similarly, in ordinary speech, the concepts of "chaos" and "chaotic" are associated with a state of total anarchy, disorder and randomness; while, contrarywise, as chaos scientists[24] have shown us, underlying *patterns of order* can be detected in chaotic systems.

---

[23] See Kiel L. D., *Managing chaos and complexity in government*. San Francisco: JoseyBass, 1994.

[24] On chaos and chaos scientists, see: Bangone G., Carlini F., Cini M. et al., *Gli ordini del caos*, Manifesto Libri, Rome 2000.

Along this line of thought, we cannot fail to note that currently, the word "complexity"[25] is widely used along with the concepts-definitions of "complexity

[25] Complexity, as I have been repeating for many years, has now become yet another "fashionable" concept/topic/problem/word-label, which is spoken of—by literally everyone—and I am referring not only to politics, media and public discourse—often by trivializing it and emptying it of true meaning, even among scholars and researchers, because evidently, there is no such thing as an 'expert in all fields'. Another point that deserves thorough reflection is that we continue to witness an authentic anthology, not only of definitions and theories (not all based on study or research), but also of the use of the *verbs* that accompany the concept, which are somehow intended to certify the originality of the analytic approaches and tools of intervention, another element that is extremely indicative of the above trends. Thus, some wish to "manage" complexity, declaring themselves "certain" that they can "control/manage" it (an aspiration that we humans have always held on to and are unlikely to ever relinquish); some want/would like to "measure" it in all its dimensions, through the application, no less, of rigorous methods. There are also those who, on the other hand, would like to "embrace" or "navigate" it, others who want/would like to "practice/ train" in complexity, still others who try to define a typology, within which it is even possible to "hide from complexity". I am referring, in particular, but not exclusively, to an article penned by the philosopher Luciano Floridi, entitled "*Quattro sfumature di complessità dalle quali è inutile scappare*", published in the *Corriere Innovazione* magazine of the Corriere della Sera, il 30/07/2021. Again, there are even those who believe they can "eliminate" it (the use of this verb, "eliminate", is truly emblematic and significant), declaring, what is more, that they have also identified the "tools" to do so. *Proposing the elimination of complexity*...this means not only failing to consider its properties and characteristics, but also, and above all, being unaware of its not being something "external", "provisional", or "temporary". "Eliminating" complexity is equivalent to eliminating life, the vital, the living, ecosystems, etc. Above all, it means not taking into consideration theories, studies, research, evidence and acquisitions from—I repeat –innumerable volumes of multidisciplinary scientific literature. Likewise, we should also look askance at those who proffer themselves as "external observers". For those who carry out scientific research, in any disciplinary field, this means not having understood the dimensions and intervening variables. Many more examples might be provided. The choice of these verbs, at times, is made at random, dictated only by the desire to produce/evoke suggestions, images and metaphors. In other cases, it is indicative of the view/conception/disciplinary field/approach referred to, more or less intentionally. In still others, more simply, it is dictated above all by the desire to avoid citing studies and research authored by other scholars, according to the dominant logics of unbridled competition and rivalry, or owing as well to various forms of envy, which are never lacking. In any case, the word "complexity" is often used as a synonym of "complication", "difficulty" and "more difficult" or, more precisely, "the most difficult": in any case, of a level of action and/or thought not (easily) feasible. Evoking my good friend Edgar Morin, the word "complexity" has passed from being a "problem-word" to becoming a "solution word". And yet, as we are well aware, especially those who study and carry out research (not only in these sectors or related to these topics), the scientific literature on complexity and complex systems is boundless, to say the least, dating back considerably in time as well. An enormous amount of work has been done over the years, in the attempt to formulate one or more operational definitions, by virtue and as a consequence of the approaches to/from the diverse disciplinary fields. As I was saying, there is an immense volume of multidisciplinary literature as we have seen, containing fundamental contributions, provided over time, from all disciplinary and research fields: from biology to physics, from cybernetics to philosophy, from economics to sociology, from chemistry to mathematics, from educational sciences to psychology, from medicine to organizational studies as well as the social, human and life sciences. And I am leaving out the utmost importance of the arts and other extremely significant disciplines in and for this complex evolution. Nonetheless—as already mentioned—we persist, even among the so-called experts (?) and scholars, in a (banal) confusion between "complex" and "what is difficult, more difficult",

science" and "complex systems", which are often equated, both by "laymen" and, at times, unfortunately, even by some scientists, with *complicated systems*. On the other hand, many automated and artificial systems are somewhat (hyper)complicated and characterized by multiple levels of connection and feedback. The empirical condition of hyperconnection tends to further strengthen the *current obsession with automation*—and with processes of hyper-accelerated simulation—often failing to take into due consideration whether this very automation can coexist with autonomy; a crucial question, to say the least!

This is possible only if we avoid confusing complicated, controllable and predictable systems with complex systems, which are not and never will be.

Speaking of *order*, the belief (or claim) that human beings can voluntarily create order out of chaos, or *introduce* stability into a dynamic system, is based, in my view, on unfounded assumptions and purely illusory certainties. Historically, whenever humanity has sought to establish order, render complex systems *static*, or create balance by means of planning, policies and strategies, all it has succeeded in

---

the complex with the "complicated", perhaps even "greater" in size and extension. Above all, "complex systems" continue to be confused with "complicated systems" (as I have already said, I have been speaking of this since the mid-1990s, proposing the definition of a "fatal error"), in both micro- and macro-system terms. One non-irrelevant problem is that we continue to oppose simplification to complexity. But, as I have been repeating for over twenty years, "the opposite of complexity is reductionism, not simplification" (1996). Similarly, "simplification" continues to be considered an absolute value, which, as I reiterate, it is not. In fact, as we have also been able to verify in recent years, as well as during the global pandemic, not everything can be simplified (or broken down into its constituent elements): education, communication, risk, emergency, politics, democracy, life itself—just to give a few examples—cannot be simplified, much less rendered reducible and/or traceable to statistical models, mathematical formulae and/or infinite sequences of data, however useful and strategic. At this point—I repeat—it is just another "trendy topic", a keyword, a label to be inserted into any text and/or discourse. In this sense, I also refer to the article by my colleague Mauro Magatti, published in the *Corriere della Sera* of the 20th of August 2021, published under the title: "*Vaccini. Dobbiamo accettare la complessità per vincerla*" *(Vaccines. We must accept complexity to overcome it),* (suggesting, as always, to read beyond the title or the accompanying image). I have linked the article here: https://www.corriere.it/opinioni/21_agosto_20/vaccini-dobbiamo-accettare-complessita-vincerla-186d9502-01da-11ec-9259-e06a1abb2d03.shtml. I do agree with the author's reflection on "complexity" (in which he correctly evokes Edgar Morin), above all on his view of the complex dimension of the issues, on the necessary requirement for thought, for a "gaze" on reality/the real, for an approach, and also regarding the (strategic) dimension of responsibility, which is closely related. Speaking of "responsibility", my own 'proposal' in this sense was published in an interview with the Huffington Post in 2017: "*La cultura della complessità come cultura della responsabilità*": https://www.huffingtonpost.it/2017/05/04/al-festival-della-complexita-la-lesso-di-piero-dominici-il_a_22069135/. Honestly, however, I am hardly convinced—for the numerous reasons already expressed, not only in this monography but also in studies carried out and publications issued over the years, of the idea/view that complexity can be "overcome" (again, as for many other widely used action-verbs, for many reasons); even though I believe this term may have been used in the title for journalistic purposes, chosen/proposed perhaps by the editorial staff. However—let it be clear—that, on these topics and questions, it is possible to work and carry out (there are those who *do* work and carry out!) scientific research in a serious and rigorous, as well as a (necessarily) multi/inter/trans-disciplinary manner.

producing/creating have been totalitarian regimes, whose survival has inevitably depended on the limitation of civil liberties and human rights.

The present-day dream of the so-called *digital revolution* to control chaotic and complex systems through engineering, algorithms[26] and artificial intelligence, will, I fear, come up against the self-same limitations and criticalities and end up behaving the same way, until it once again reaches a more or less irreversible *breaking point*. The age-old, historical aspiration to control and *predetermine* human choices and behaviour has returned to haunt us once more, with a new, extremely fascinating and seductive utopia called "digital democracy" (*direct,* without mediation) amid the rhetoric of disintermediation and of the horizontalization of processes, which can only exist as posturing and simulation. The complexity, adaptability and ambiguity of social systems imply that they are, to all intents and purposes, *dissipative systems far removed from* equilibrium, where time is an essential element whose arrow points in a sole direction.[27]

Every system has its own history and, above all, it has memory of its history. Moreover, each system has both an external memory—the history relating to the environment—and an internal memory, concerning the interactions and connections that took place and were generated within it. Yet, the history and the memory

[26] See, in particular, Domingos P. (2015), *The Master Algorithm. How the Quest for the Ultimate Learning Machine Will Remake Our World*, Italian translation, *L'Algoritmo definitivo. La macchina che impara da sola e il futuro del nostro mondo,* Bollati Boringhieri, Turin 2016.

[27] See, in particular, the works of Prigogine: Prigogine I. (1980), *From Being to Becoming: Time and Complexity in the Physical Sciences*, Italian translation di Bocchi G. e Ceruti M., *Dall'essere al divenire,* Einaudi, Turin 1986; Prigogine I. (1996), *La fin des certitudes. Temps, chaos et les lois de la nature*, Italian translation, *La fine delle certezze. Il tempo, il caos e le leggi della natura*, Bollati Boringhieri, Turin 1997. The scope of the revolution introduced by Prigogine was enormous: going beyond the acquisitions of "linear thermodynamics", marked and characterized by "predictability" (systems evolve, in any case, towards an increase in entropy up to the stage of thermal equilibrium), Prigogine observed how most natural and vital phenomena are correlated to another, far from static type of thermodynamics that is non-linear and *lacking in equilibrium.* Everything, including matter, becomes creative, generating multidimensional and, above all, unpredictable: "complex" structures and systems, precisely defined as dissipative, which behave in an unpredictable manner, unpredictable and impossible to explain based on the laws of traditional physics. The challenge of (hyper)complexity manifests itself even more clearly here: from now on, science itself must deal concretely with the concept of "probability" and, above all, with phenomena and processes that must necessarily be studied starting from the *correlations* and *connective bonds* that characterize them and are increasingly being seen to be *sensitive to external perturbations*. Consequently, it is no longer sufficient to analyze/study its constituent elements. Irreversibility and indeterminacy become/are acknowledged as structural conditions (ontological, to put it philosophically) of complex dissipative systems. As regards the question of time, see also the work of an important philosopher, to whom, moreover, Prigogine himself refers to in several interviews (and he is not the only scientist to have done so): Bergson H. (1907), *L'évolution créatrice*, Italian translation *L'evoluzione creatrice*, Laterza, Bari 1957. See also: Prigogine I.(1992), *Le leggi del caos*, Bari- Rome, Laterza 1993. On the subject of "time" and on the related debate between scientists and philosophers, how can we forget the famous exchange between Bergson and Einstein, which took place in Paris on the 6th of April 1922, at the *Sociéte française de philosophie*, where they discussed the concept/definition of "time" *in the aftermath of* the theory of relativity. See also another great classic of the social sciences: Elias N. (1984), Über die Zeit, Italian translation, *Saggio sul tempo*, Il Mulino, Bologna, 1986.

themselves of the system cannot tell us what will happen tomorrow. In particular, they cannot guarantee that what worked yesterday will work today or tomorrow. The so-called *chains of events* (a mechanistic metaphor), which follow every moment, are fluid and uncontrollable like running water while, even more decisively, what happens over time is *irreversible*. We can assign mathematical names such as "non-Markovian"[28] systems to complex systems, but in the unlikely event that, one day, it becomes possible/feasible to conceive/imagine complex "Markovian" systems *devoid of memory*, we will, in any case, have to find another category/definition for social systems.

As is known, the need to measure time, and to do so ever more accurately, is related to the need to coordinate and undertake more and more group and collective activities, which became specifically imperative after the advent of collective employment. It is precisely this urgency to coordinate the movements and actions of many individuals/people/groups that urges the human mind to design and create more and more precise, accurate instruments capable of measuring time, and not only time. Nothing must escape our control; every thought and action needs to be perfectly coordinated and replicable under every aspect.

Yet, *time* has never been an "object" external to social actors and life.[29] Time is, after all, the *individual structure* of all phenomena, the building block of our experience as *human beings*. Time is not visible, much less observable. It cannot be drawn on a graph like the fascinating fractal visualizations of chaotic systems. Sundials, hourglasses, wristwatches and atomic clocks can give us *the illusion of measuring*, and even of controlling time, but time is neither controllable nor measurable, nor is it possible to visualize it, in fact, except by resorting to *conventional* (arbitrary) *measurements*, the outcome of agreements and processes of social and cultural negotiation and mediation. Similarly, not only time but also complexity itself, as already argued, is impossible to visualize and/or represent graphically or physically, in all its dimensions and systemic relationships and –something that must be kept in mind—any attempt to do so becomes/will become (constituent and constitutive) part and parcel of the complexity we are trying to "portray", visualize and/or represent, without obtaining any kind of reduction and/or simplification. Once again, the fundamental principle of science, the "observability" of phenomena, is wanting. So, while the evolution of a *subunit/sub-system* can be observed and portrayed –such as, for instance, the evolution of a diseased cell, which can be visualized in order to analyse the pathology affecting it—on the contrary, when it comes to complex systems, we can only attempt to illustrate and represent their nodes, *relationships* and *interconnections*. All this is made possible more than by science/the sciences, by the world of the arts and the more creative disciplines. These are the only forms of

---

[28] See Rickles, D., Hawe P., Shiell A., *A Simple Guide to Chaos and Complexity*, Journal of Epidemiology & Community Health, (BMJ) 2007 Nov; 61 (11): 993–937. https://www.ncbi.nlm.nih.gov/pmc/articles/PMC2465602/

[29] See Elias N. (1984), *Über die Zeit. Arbeiten zur Wissenssoziologie* II, Frankfurt am Main, Suhrkamp, Italian translation, *Saggio sul tempo*, Il Mulino, Bologna 1986; Elias N. (1987), *Die Gesellschaft der Individuen*, Frankfurt, Suhrkamp, Italian translation, *La società degli individui*, Il Mulino, Bologna 1990.

knowledge and symbolic mediation capable of representing complexity, for example, through visionary projections.

In other words, the ability to visualize, represent, reproduce, portray/model complex systems, in all their (hyper)complexity, ambivalence and ambiguity, unpredictability and irreversibility, is associated not so much with science as with languages, techniques, the symbolic universes created by art, by the arts. From this point of view, art is the only *form of knowledge and interrogation of the world,* the only form of *symbolic mediation* capable of representing complexity, time, and the intangible and unobservable dimensions of the existing and the real.

It is the visionary projections of art and artists that render the forms and dynamics of complex systems accessible and visible; even though, as I have been repeating for quite some time, digital signalling and the technologies of connectivity—for the first time compared to the past, even the very recent past– offer/ seem to offer us an extraordinary opportunity to observe/identify/recognise/valorise, in a clear-cut, well-defined and effective manner, the radical interdependence and interconnection of phenomenal reality and, at the same time, of all variables and factors that characterize it. All in all, redefining our relationship with Technology and Machines, above all in terms of "epistemological possibilities", is of vital importance when seeking to comprehend the *systemic and relational dimension of Life, of existence — of our existences*. For which we are, unfortunately, unprepared, and who knows for how long we shall go on being so, both in terms of approaches (not only to technology) and of an educational and training paradigm.

The hegemony of "disjunctive thinking" and the consequent goal of separating and breaking down "everything"—any process and/or dynamic whatsoever- to find a solution (?) to problems, has always concerned a discourse, a view, an overall approach which, in many ways, also calls into question the tendency to consider technology as being *separate* from culture: a deeply rooted vision that has its roots in that type of culture and way of thinking itself, both absolutely fallacious, which value *quantity* (exclusively) over *quality,* even when observing, analysing and attempting to understand relationships and the (infinity of) human and social connections. This has always been one of the "fatal errors" of the hypertechnological civilization, which continues to keep us inside a global, systemic condition of substantial inadequacy and unpreparedness in the face not only of the so-called *black swans* but also of the irreversible and unstoppable manifestation and display of life.

3.3 "Objects" Are "Relations". Self-Organization and (Dis)order

As to the principle of "observability", it turns out to be ineffective, *almost useless*, as well as difficult to apply when studying and analysing social systems. The *human* and the social cannot be observed and studied in "ideal conditions" by an external *observer*. Just as our *perception* cannot be limited to one sense only (*perceiving* always involves all senses and all sensory levels), external causes and/or external views do not in effect exist. Seeing, knowing and acting are correlated complex actions; every action is a systemic form of behaviour with systemic implications.

As in all complex systems, but even more so in social and human systems, the observer is a constituent *part* of what is observed: there is no "inside" or "outside"—another "false dichotomy" invented to "unlearn us" of our limits, which are, indeed, objective. The act of *observing* changes the object/system that is observed, and the observer him/herself is, in turn, modified by what he/she is observing. In sociology, the terms that are often used are "participant observer", "participant and participated observation". Referring again to the great thinker Heisenberg and his *Uncertainty Principle,* we can say that just as there is a *limit* to what can be known about a quantum system, the observation of a social system, developed from a certain perspective, binds and limits the observer/participant him/herself, often precluding him/her from "other" observations and, above all, places him/her in the condition of being unable to adopt and/or understand other points of view. As Heisenberg made it clear, *we cannot separate physics from philosophy.*[30]

Every observation is, in fact, a construct, a *projection*. Similarly, every "object" (= system) is also a *mental construct*, our way of identifying/recognizing and perceiving processes and systems, which also consist of energy and, precisely, of *non-observable dimensions*. The "objects" are simply *concentrated landscapes* of interactions and relationships.

The philosopher/physicist Meinard Kuhlmann, in his 2013 article "*What is Real*?" proposes the intriguing hypothesis that the basic "building blocks" of matter may not be represented by particles or fields, but rather by pre-existing "bundles" of mere *relationships and/or properties.*[31] This is an extremely fascinating hypothesis—also and, above all, as regards our theoretical, epistemological perspective—which was explored previously, in another area, by Italo Calvino in a story entitled *The Spiral.*[32] It follows, evidently, as we should note, that the very idea of "objectivity" is also a "construct" or, at least, can and must be questioned or, at least, can and must be re-defined. In other words, *everything that exists* can be described as a *historical*, complex and, in some ways, subjective event, located in a context, generated by some systemic organization and relationship. Again, with regard to the principle of "objectivity" and the mathematisation of reality, the great scientific historian

---

[30] See Heisenberg W., *Physics and Philosophy: the Revolution in Modern Science*, Prometheus Books, Buffalo N.Y., 1958; by the great scientist (I venture to add) "philosopher", see also: Heisenberg W., *I principi fisici della teoria dei quanti*, Bollati Boringhieri, Turin 1976; Heisenberg W. et al., *Discussione sulla fisica moderna*, Bollati Boringhieri, Turin 1980; Heisenberg W., *Lo sfondo filosofico della fisica moderna*, Eds: Gembillo G. and Giannetto E.A., Sellerio, Palermo 1999; Heisenberg W., *Indeterminazione e realtà*, Eds: Gembillo G. and Gregorio, G., Guida, Napoli 2002. On these questions, see also the interesting work by Lanza R., Pavšič M., Berman B. (2020), *The Grand Biocentric Design*, Italian translation, *Il grande disegno biocentrico. Come la vita crea la realtà*, Il Saggiatore, Milan 2022.

[31] See Kuhlmann Meinard, *What is Real?*, Sci Am. 2013 Aug; 209 (2): 40.7) (PubMed).

[32] See Calvino I., "The Spiral", in *Cosmicomics*, English translation by Weaver W., Harcourt Brace, 1968.

Alexandre Koyré underlined the fact that Science, in addition to building a unified image of reality, has replaced "our world of qualities and sensorial perceptions" with a "*world of quantity*, of reified geometry, a world where, although there is room for everything, there is none for the human being. So, the world of science—the *real* world—moved away and separated itself completely from the world of life, which science has been unable to explain – even using a separative explanation such as to bestow a "*subjective*" appearance on it.[33] The world of science and that of life, while united at the level of practice, turn out to be profoundly separate from a theoretical point of view[34]: "two worlds – meaning two truths, that is, no truth. It is in this that the tragedy of the modern spirit that 'solved the riddle of the universe', consists, but only to replace it with another: the riddle of itself".[35] As we know, Science does not accept, cannot accept the *dimension of mystery*, least of all that relating to the Subject and, more generally, to the Human. It is clear that this perspective/vision/approach cannot acknowledge any value to knowledge that is not based on the "principle of objectivity" and various other kinds of reductionism.

One of the essential characteristics of complex systems is self-organization: life, besides always showing a tendency opposite to balance and stasis, is an example of self-organization and *irrepressible dissipative emergence*. As regards the approach and the analytical perspective drawn up, one element needs to be clarified: self-organization, inherent to complex systems, is generally considered as being a producer/*creator* of order stemming from a previous condition of chaos. On the contrary, since complex systems are "dissipative systems", where everything that occurs is irreversible, it appears evident that the achievement of a higher degree of *order* requires energy which is then diffused as heat,[36] leading to a new, further increase in entropy. This has led me to hypothesize, even in the past, that self-organization, in any case, the self-organization found in (complex) social and Vital systems, might, create a state of greater *disorder* in the long run.

At this point, it is useful to reiterate that in social systems, the action of every single individual/entity/part is always capable of displaying the *potential* necessary to destabilize the entire system. A far cry from (total) control and predictability!

Before examining, albeit briefly, an example of the ability of a very small (*micro*) action or event to produce an extremely large effect on complex (and chaotic) systems, and its implications regarding the use of algorithms, in particular, and artificial intelligence, in general, I would like to consider the issue of sensitivity to perturbations and initial conditions, something which unites both of the kinds of *nonlinear dynamic systems* we are considering here: complex and chaotic.

---

[33] See the important work by A. Koyré (1968), *Ėtudes newtoniennes*, Italian translation, *Studi newtoniani*, Einaudi, Turin 1983, p. 43. See also Godfrey-Smith P. (2021), *Theory and Reality. An Introduction to the Philosophy of Science*, Italian translation, *Teoria e realtà. Introduzione alla filosofia della scienza*, Raffaello Cortina Ed., Milan 2022.

[34] *Ibidem*.

[35] *Ibidem*, p. 43.

[36] On the crucial issue of "heat", see this fundamental work by Fourier J. (1822), *Théorie analytique de la chaleur*, Eng. Trans., *Analytical theory of heat*, Encyclopedia Britannica, Chicago 1952.

Sensitivity to initial conditions can effectively refer to any given moment when a process begins and the unforeseen effects it may have on the system itself. In complex, *non-linear* systems, sensitive dependence on perturbations and initial conditions is often referred to as the "butterfly effect": a complex and systemic effect illustrated for the first time by Lorenz[37] and taken up by I. Stewart[38] to describe the enormous impact that very small (micro) changes can have on complex systems, which make it always impossible to predict and *measure* their future developments. Attempts to carry out measurements of/in complex systems, according to Chaos-Theory concepts, initially introduced by the mathematician Benoit Mandelbrot, are largely a function of the perspective adopted. As a result, quantitative methods cannot achieve the necessary precision.[39]

In chaotic processes, which can be enumerated among complex systems, very often *self-similarity* can allow us to perceive irregular patterns that will repeat themselves within different scales of time and space.[40] Mandelbrot coined the term "fractals" to describe qualitative approaches to methods for "measuring" complex systems. However, in most highly complex systems in general, and in social systems in particular, it is impossible to measure the initial conditions with the necessary level of precision. These are conditions that are/ought to be infinite. Moreover, if these were combined with even the tiniest variations or the slightest "background noise", this would make/makes predictability totally impossible.[41]

When considering the irreplaceable role and importance of every single social Actor—more in general the *human factor*—within fully automated systems/ecosystems, a historical episode that springs to mind immediately and raises further highly significant questions regarding the possibility (or desirability) of trying to control/pre-determine social behaviour by recourse to digital automation and AI systems, occurred in the Soviet Union, in 1983. Back then the computerized defence system of the USSR–for those who may remember– detected a multiple missile attack by the United States. The officer on duty at the time, Stanislav Petrov, was well aware that the protocol to be followed, in the event of such an eventuality, was immediate nuclear retaliation. The Soviet automatic missile system was in place and ready to launch the counterattack, but Petrov hesitated, wondering why on earth the Americans would choose such a heinous and irrational move. Accordingly, instead of reporting a missile attack, he decided to report a malfunction of the systems, thus effectively saving the whole of humanity.[42] What might have happened if, at that precise moment, there had been someone else on duty, perhaps a soldier used to/trained only to carry out orders without questioning them and questioning himself?

---

[37] See Lorenz E.N., *The Essence of Chaos*, Univ. of Wash Press, Seattle, 1963.

[38] See Stewart, I. *Does God Play Dice? The Mathematics of Chaos*, Blackwell Pub. Oxford 1989.

[39] Mandelbrot B.B., *Fractals: Forms, Chance and Dimensions*, San Francisco WH Freeman 1977.

[40] See Gleick, J., *Chaos: Making a New Science*, NY, Viking Press, 1987.

[41] Mathews, KM, White, MC, Long, RG, *Why Study the Complexity Sciences in the Social Sciences?* In Human Relations 25, April 1, 1999, (Sage Journals), p. 446.

[42] Askenov P., *Stanislav Petrov: l'uomo che potrebbe aver salvato il mondo*; BBC world news, September 2013, https://www.bbc.com/news/world-europe-24280831

How might the world have changed that day if Petrov had not decided to ignore protocol? A scenario akin to the apocalyptic ending imagined by Stanley Kubrick in his film "Doctor Strangelove?" And what might happen, even today, within sophisticated defence systems if events were analysed and monitored exclusively by algorithms? We should never lose awareness of how ambiguous and ambivalent and how numerous complex human situations, processes and systems are and how they cannot be regulated, in any way, by algorithms, whose unequivocal sequences—which do not admit/allow/tolerate any ambiguity and/or ambivalence—can only lead to the *caging* and internment of human life.

3.4 Everything Is Interdependent. Everything Is "Relationship" and in Relation

The utopia of the "Mechanism-Society" and the overall idea/view that digital technology, the infinite availability of data (with the support of mathematical, statistical and probabilistic models) and AI systems—perhaps, alongside the domain/hegemony of technocracy and expert technical knowledge—allow us/will allow us to control, predict, pre-determine and direct human thoughts, emotions, styles of behaviour,[43] as well as (hyper-complex) social ecosystems, is based mainly on the idea/conviction/belief that emotions and human behaviour can, on the whole, be pre-determined and mapped and that social, economic and political systems/ecosystems evolve, in any case, through *cycles* comparable, to some extent, to *chaotic deterministic models*. Assuming, and not granting, that all this should turn out to be probable/possible and, above all, become possible to observe and verify empirically, the price to pay would be, in any case, the simplification, perhaps excessive, of the dynamics and the processes themselves involved and observed.

Historically, just to provide one example, even great empires can rise and fall, over and over again. Societies may tend to divide themselves into privileged elitist minorities and more or less disadvantaged majorities. As is known, inequalities and oppression can undoubtedly cause violent revolutions, which tend to end up, in one way or another, by evolving in the direction of rigidly controlled systems which, over time, see/will witness yet another increase in inequality and oppression, until a *new critical breaking point* is reached. Yet, all this still does not tell us anything about the relationships and daily vicissitudes of the people who live/go through these difficult/dramatic periods. Neither does it tell us anything of the journey/of

---

[43] Among recent scientific publications, I would like to indicate the special issue of the Journal "Nature", entitled "*Computational social science*": among the articles, see *The powers and perils of using digital data to understand human behaviour*, in "Nature" vol. 595, 149–150 (2021); Big-data studies of human behaviour need a common language, in "Nature" Vol. 595, July 2021. On these topics, particularly on "computational thinking" used to solve "gradually more complex problems", see also Ferragina P., Luccio F., *Il Pensiero computazionale. Dagli algoritmi al coding*, Il Mulino, Bologna 2017.

what is lived/experienced by each individual over *time* which, actually, constitutes the *basic* (ontological and structural) *makeup* of a single life. It tells us nothing about further new personal, social, systemic, existential *trajectories*, which may/might be triggered precisely by the action of every single individual and/or social actor. Furthermore, it tells us nothing about the possible decisions that might lead, not only to *exponential deviations* common to *dynamic chaotic systems*, but also to completely unexplored and unexpected interactions, adaptations and dynamics, involving other people and the environment itself, with extremely significant consequences, extended to the ecosystem(s) of reference. These are dynamics and processes capable of triggering, in turn, yet another process of *self-organization*, which does not require any further external stimuli/interventions, since a complex system is already able to "attune itself"—in a non-linear manner—and adapt continually to new environmental/eco-systemic conditions.

Social systems are, from this point of view, in continuous, inexorable and, above all, irreversible transformation.[44] From this perspective, looking at the history and the complex—non-linear—evolution of human societies, we can certainly verify how the periods of relative harmony and stability (equilibrium) have been far less frequent than the periods/phases of *critical dynamics* and sudden acceleration. On the contrary, one might say that social systems/human societies often seem to be eternally close/next to points of critical *no return*. No one—I believe—can deny that this is certainly one of the most critical aspects/factors, even within the non-linear and complex sequence of systemic emergencies that have been following one another for decades, both locally and globally. According to *Complexity Science*, when a system draws close to a critical turning point, the degree of connectivity (interconnection/interdependence) between its sub-units always becomes extremely high. Everything depends/interdepends on everything else.[45] In other words, *everything is interdependent* and—I repeat—everything *is relationship*. The "disjunctive thinking", the logic of separation and the "false dichotomies" that have always profoundly marked our educational and training institutions and processes not permit us to grasp a constitutive dimension of this kind. Moreover, these statements and theories are sometimes/often trivialized, not only in media or social terms, by attributing meanings to the epistemology of complexity, to the culture of complexity, to the very concept of "complexity", meanings that associate complexity with confusion, with a lack of definition and clarity, even with irresponsibility.

[44] See, in particular Diamond J. (2005), *Collapse. How Societies Choose to Fail or Succeed*, Italian translation, *Collasso. Come le società scelgono di morire o vivere*, Einaudi, Turin 2005; Diamond J. (1997), *Guns, Germs, and Steel. The Fates of Human Societies*, Italian translation, *Armi, acciaio e malattie. Breve storia del mondo negli ultimi tredicimila anni*, Einaudi, Turin 1998 (See the 2006 edition); Graeber D.—Wengrow D. (2021), *The Dawn of Everything*, Italian translation, *L'alba di tutto. Una njuova storia dell'Umanità*, Rizzoli, Milan 2022.

[45] "Gin 'waadluwaan gud ahl kwaagiidang": the original concept of interconnectedness as expressed by the people of the Haida Nation, Canada—https://www.dfo-mpo.gc.ca/oceans/publications/sk-b-managementplan-plangestion/page04eng.html

The currently adopted/accepted perspective recognizes that complex systems are/are suspended *more or less* constantly between order and chaos. One is always striving to measure the *distance* between the *sub-units* so as to determine the length of "maximum correlation" beyond which they cannot/will not be able to influence/condition one another anymore.[46]

However, in my opinion, although *distance* within social systems and human relationships is important, it is not/has never been the only determining factor in the *non-linear evolutionary dynamics* of social systems, regarding either the quantity of the relationships or the quality of the interactions, which self-generate exponentially, even more so in this present-day hyper-connected civilization of ours. They are linked within time and not only within physical space. It is *the intensity of the interaction*, which is capable of crossing both distances (space) and time, that ensures that the subunits always remain interlinked.

Thus, what we might define the "correlation coefficient" can never reach a maximum point: connectivity is hugely amplified and, as I said before, one small change or *disturbance* in a system—even if located in one sub-unit only—can lead to massive/profound/substantial systemic change. This has profound implications also for the way(s) that systems are modelled, visualized and represented. And also perceived.

First of all, it is often hard to configure the problem itself or determine its characteristics. Furthermore, the individuals involved are unique, each with ideas, values, fears, and desires that may be at loggerheads with one other. The consequences of every action or event are irreversible, the *ramifications* within the entire system intricate and confusing, and many of the problems are simply unsolvable.

One of the most significant errors/fallacies which challenge current *thinking*, and contemporary thought most, is the (useless, illusory) quest—at times even proposed in dichotomous terms—for *simple/linear solutions* to *complex problems.*[47]

A problem can/may be simple, complicated or complex, but complex problems cannot be solved by adopting *linear* solutions, be they complicated or simple. Furthermore, anyone striving to find a solution to any kind of (complex) problem is inevitably a constituent element/part of the problem itself. It is perfectly useless to try to distinguish/separate the thorniest situations, even by giving them a drole name like "*wicked problems*". With regard to social systems, in particular, what is defined as *absurdity* is a set of properties intrinsic to the complexity of human society.

Precisely because it is difficult to configure the problem itself or determine its characteristics and dimensions and due to the fact the individuals involved are "unique" (unrepeatable, non-replicable), each with his/her own, even conflicting, psychological makeup, ideas, values, expectations, lifestyles, objectives, fears and the consequences of single actions or events are, as always irreversible, non-reproducible and/or non-replicable. In other words, since the *ramifications* of the

[46] Rickles, D., Hawe P., Shiell A., *A Simple Guide to Chaos and Complexity*, Journal of Epidemiology & Community Health, (BMJ) 2007 Nov; 61(11): 993–937. https://www.ncbi.nlm.nih.gov/pmc/articles/PMC2465602/

[47] See Dominici P. (2017, 2020, 2021), *op.cit.*

whole system are extremely intricate/tangled, often confused, it happens that many problems are simply unsolvable.[48] I do not think there is anything absurd or negative/evil about all of this: we have simply highlighted some of the characteristics of a complex human/social system. And among the unavoidable dimensions of the problem, what stands out is the one linked, as always, to the naive desire to find/build a theory—a Theory of Everything?—that simplifies all phenomena by reducing them to a series of *relationships* between elements that lie at the base of a hierarchical scale that may then be applied to all the rest.[49]

It is obvious that all of this cannot/will not work, not only because *complexity* is unmanageable, ineliminable, uncontrollable, irreversible and unpredictable, but also because, as I said before, we continue to neglect the innumerable *qualitative aspects* and *non-standard dimensions*,[50] which, to say the least, play a strategic role. On the contrary, varieties of *reductionism* and determinism continue to dominate along with *hegemonic narratives* that try to convince us that reality consists exclusively of *measurable*, empirical dimensions and of irrefutable, self-evident quantitative, statistical, "objective" data capable of supporting and fueling what I have called the "*tyranny of concreteness*" (1998).

A "tyranny of concreteness" that sometimes translates all too 'concretely' into an obsession with facts, figures, statistics, algorithms and *quantities*, *a culture of data* and *standardization* that induces us to believe, with virtually blind faith, that the data can tell us everything, that our assessments are, always and without fail, objective and impartial, with no need for interaction or for recognition/actuation of any relationship among the levels of connection and interdependence. What is absolutely necessary, however, precisely in this perspective, is a systemic view and an epistemology of interconnection/interdependence[51] and of *inter-independence*

[48] McCall R.; Burge J., *Untangling wicked problems in. Artif. Intell. Eng. Des. Anal. Manuf.* 2016, *30*, 200–210.

[49] Turner J.R., Baker R.M., *Complexity Theory: An Overview with Potential Applications for the Social Sciences*. Systems 2019 7(1) (MDPI Journals), https://www.mdpi.com/2079-8954/7/1 / 4 / htm.

[50] See, in particular:, Dominici P., "*Educating for the Future in the Age of Obsolescence*", in CADMUS, Volume 4—Issue 3, November 2020, pp. 93–109; Dominici P., *A.A.A. cercansi manager della complessità* [Complexity Manager], in "Business People", 2019a; Dominici P., *La complessità della complessità e l'ultimate error*, in Enciclopedia Italiana "Treccani", Treccani, Anno 2019c. http://www.treccani.it/magazine/lingua_italiana/speciali/digitale/5_Dominici.html

[51] Regarding these concepts and operational definitions, see: Dominici P., *op.cit.*, 1995/1996–2021; on these questions, I also refer to the fruitful collaboration between the scientist of quantum mechanics Wolfgang Pauli and the psychologist Carl Gustav Jung. W. Pauli, however, dedicated some of his scientific theories (among others, "The principle of exclusion") to Jung and worked on the epistemology of "synchronicity" to redefine the relationship between mind and environment. In this sense, the proposals relating to the *epistemology* of synchronicity and the *Scientific Participatory Action Method* are also significant. Very interesting, not only for our perspective and approach, is the concept itself of "synchronicity". See also: Lakatos I.—Musgrave A. (1970), *Critica e crescita della conoscenza*, Feltrinelli, Milan 1976.

(as of Panikkar's concept of *inter-independence*[52]). In other words, the challenge now is no longer that of becoming increasingly adept/sophisticated/accurate in observing, describing and measuring units, components and individual processes, but that of questioning ourselves about what we intend to *know* about the relationship/rapport between these very elements/aspects, variables, factors. Above all, the most challenging challenge is to try to understand the dimensions, variables and qualitative factors[53] of the reality of the phenomena that we investigate, reflecting at length on what Edgar Morin, friend and mentor, has called *the knowledge of knowledge*.

Dimensions that take on even more strategic importance within today's hypertechnological and hyperconnected civilization—based on illusory certainties (including that of being able to make everything "transparent", "knowable", "controllable" and "accessible")—where, moreover, human beings have not as yet acquired awareness (on the contrary!) that the attainment of the power to control and/or pre-determine their lives does not mean/represent/constitute/is not, in any way, the achievement of a presumed, and eternally coveted form of omnipotence. On the contrary, it is the achievement of an objective that calls into question certain fundamental rights, an objective that is very distant—and that distances us — from our *being human*.

3.5 Cognitive and Epistemological *Cages*

Meanwhile, therefore, we humans go on- as stated before—attempting to *restrain*, cage and control human and social complexity in its entirety, along with the vitality of the spirit and the "*non-observable*", using algorithms, models, mathematical formulae, infinite sequences of data and numbers, molecules, synapses, hormones and chemical reactions. It would seem possible to reduce (and simplify) human beings and social systems to physical and chemical *mechanisms* and functions, even to apparently infallible sets of cogs and gears, almost perfect mechanisms which, at least in theory, do not permit error, anomaly and/or deviation to occur. Doing this we also try to observe/see/visualize what cannot be seen/visualized, not only to re-affirm, in some way, the principle of "observability", but also because we fear, as we always have, the idea that something may escape our continuous "scrutiny" or *control*. Behind this attitude lies our not well-concealed (potentially lethal) illusion of being able to *pre-determine everything, even the colour... of swans*. It is precisely this attitude that condemns us to remain in a perpetual state of apprehension and expectation of *black swans* (an ancient metaphor and classical form of retrospective rationalization, as we said before), which we find frightening yet at the same time, reassuring. Even after having gone through almost three years of yet another

[52] See Panikkar, R., *The Rhythm of Being: The Gifford Lectures,* Orbis Book Published, New York, 1989.

[53] See Dominici P. (2019, 2020), *op. cit.*

systemic, global, and to some extent, unprecedented emergency, we still find it very hard to realize, blocked/trapped/imprisoned as we are in our *"cognitive, epistemological cages"*,[54] that *emergency* is a connotative element, neither temporary nor provisional, typical of complex systems, a (complex) dimension that cannot be expelled/eliminated/re-dimensioned. Nothing that is *vital* or *human*—complex systems—can be broken down, reduced or simplified, nor can it be measured, predicted, managed accurately and/or completely. Except by resorting to extremely conventional and arbitrary models, accepted uncritically, perhaps, within groups, organizations, communities and social systems. Therefore, rather than isolating, separating and breaking them down into parts/units/elements, we should strive to recognize their complexity, unite/recompose what *appears* to be separate (education, training and research), recognise and highlight their connections and systemic relationships (knowledge).

On the contrary, we continue to extend/broaden, fail to heal certain *fractures that* never lead to knowledge, to any field of knowledge, which serve, above all, only to reassure us as we face the uncertainties of life and the indeterminacy of reality. It is most likely that in the past these selfsame *fractures* were, for several reasons (context, environment, ecosystems), far less evident and, as a result, (apparently) less harmful at least in terms of individual and collective perception. The time has come, as we go through this phase of anthropological transformation—a much more radical one than any of the changes that have occurred in the past—to begin to *heal* these fractures, fully aware of our incompleteness and of the urgent need to learn (the strategic relevance of the error) to *inhabit hypercomplexity* (1996) and its dilemmas, paradoxes and contradictions. Learning *to inhabit chaos* and emergent emergency.

Here we are again, therefore, caught between *Order* and *Chaos*: it is no longer sufficient to try to distinguish between the two when attempting to restore lost control and balance. Because in complex systems, described as situated, almost *perching*, on the "edge of chaos",[55] order and chaos are co-existent and integrated within a *systemic living structure of life*. Even less can we afford not to acknowledge how "the social" turns out to be constantly dynamic and unstable, never totally observable, measurable, predictable or understandable, except by recourse to the usual reductionist and deterministic explanations, which, in some cases, are purely conventional and/or arbitrary. *Simple solutions to complex problems* yet again!

It is increasingly clear, as mentioned before, that the hyper-technological and hyper-connected civilization has been co-constructed—erroneously—on the premises of a progressive relegation/marginalization of the Human—and of human responsibility—and on an exponential, inexorable growth of the dimensions of technological control. A civilization founded on programming, automation and (hyper) simulation of processes and dynamics, which repropose/reaffirm a series of

---

[54] See Dominici P. (1998–2017), *op. cit.*

[55] See Rickles, D., Hawe P., Shiell A., *A Simple Guide to Chaos and Complexity, Journal of Epidemiology & Community Health*, (BMJ) 2007 Nov; 61 (11): 993–937. Https://www.ncbi.nlm.nih.gov/pmc/articles/PMC2465602/

dangerous illusions, leaving (above all) our educational and training institutions unfit to face the growing hypercomplexity of the ongoing changes, which are becoming more and more evident and recognizable.

Faced with this exponential growth of interdependence/interconnection/ interaction and conditioning—mechanisms that have constituted and intensified the *neural networks* of local and global phenomena and processes—what we are now (and have been, paradoxically, for some time) witnessing is the paradoxical return of the hegemony/dominance of reductionist and deterministic analyses and explanations and a neo-positivist view/conception of reality and of what is "real". These are dynamics and processes which, as stated before, materialize, on the one hand, in an obsessive quest for *simplification at all costs* (one should bear in mind that the reverse/opposite of complexity is not simplification, but *reductionism*), even when it might even prove dangerous to "simplify" (for example, in education, communication, emergency and risk management, strategy and policy planning, and democracy).[56] On the other hand, among what I have called the *grand illusions of the hypertechnological civilization*, we find the illusions of rationality, control and predictability, and measurability, as well as the most dangerous and misleading illusion of them all, that of being able to *eliminate error* and the unforeseen from organisations and our lives.[57] This, translated into choices and strategies, essentially means/involves trying to eliminate what makes us "human beings", or, in better words, "free human beings".

A series of illusions—I repeat—further strengthened by the "carte blanche" offered/granted systematically to technology/technologies: a series of illusions (I insist) that is the result of that "great mistake"[58] made by the hypertechnological and hyperconnected civilization (as I have so often defined it) which, even within the contexts of labor and professions systematically recurs and depends on reductionist and deterministic approaches, firmly based on the exclusive involvement of technical knowledge and skills: the same elements which appear best able to sustain and guarantee those very illusions.

We are still insufficiently aware of *the ambivalent nature of complexity*, of the "complexity of complexity" and, to quote Edgar Morin, once again, of how relevant—and, at the same time, underestimated—knowledge of knowledge is: a cognitive, subjective, social, ecological and ethical complexity that is also linguistic and communicative and which, evidently, concerns our "communication culture"[59] as well. A (hyper)complexity made even more problematic due to the lack of a *thought system* and a theoretical-interpretative model capable of observing, recognising and (attempting) to understand the (hyper)complexity and chaos that have definitively broken into every dimension of the Vital and the Social. New technologies and the

---

[56] See Dominici P. (1996–2020), *op.cit.*

[57] *Ibidem.*

[58] *Ibidem.*

[59] See P.Dominici (1996), *Per un'etica dei new media. Elementi per una discussione critica*, Firenze Libri Ed. Florence 1998.

digital world have made these interconnections/interdependencies more and more visible.

It is a hyper-complexity related, ultimately, to the encounter/confrontation with the *Other than us*—"within" a systemic area of communication and relationships—and, at the same time, to the progress made by mankind in obtaining control of the levers and mechanisms of our own cultural and biological evolution. Dynamics and processes that radically question traditional codes, languages, aesthetics, models and categories, forcing us to reform/reformulate the concept itself and the definition of "human being", person, knowledge, Science, Life, etc. Hypercomplexity, as I have often repeated, is not an option, it is a "fact". The real problem is that we have not been educated and trained to observe it, recognize it, much less, as mentioned before, *inhabit* it.[60] It is important to keep in mind that complexity is a structural feature of human groups, relationships and social systems and, with some nuances, of the biological world.

The problem, once again, is that we insist on trying to understand reality by following logics and models that reduce its variety, heterogeneity, unpredictability and capacity for self-organization. Technology and digital culture can be of great help when trying to heal the *fractures* we have discussed widely. However, it is not from here that we must start. What is still wanting are *thinking*, thought and a thought system.

In this sense, it is essential to make a new start, beginning with the nourishment and cultivation of thought, with the urgent need to "rethink how we think" and define a thought system fit for the (hyper)complexity of the current moment of transformation. In addition to all this, it is time to rethink the very idea and definition of "Science" by addressing the epistemological and methodological challenge of observation (while making a genuine effort to apprehend even what is "non-observable"), and to trace the pathways and perspectives of Science and research, with the (far from tacit) objective of understanding the meaning of what cannot be measured quantitatively. In other words, it is necessary to retake the first steps on a (complex) pathway, starting from a renewed awareness that the idea/vision/perspective of transforming/translating/simplifying or reducing the "qualitative" elements (the social and human complexity of life) into quantitative data, is, on any level of praxis, deceptive, misleading and, above all, presumptuous (just one of the many forms of reductionism currently adhered to).

Tracing and recognizing the bonds, connections and interdependencies between the parts, processes, phenomena, and people: that is the *ultimate challenge (mainly an educational challenge) and the ultimate dare*: a dare to err and a quest for Error (source of knowledge and vitality), for the Other, for the Human and for Life. Education and training. Authenticity and communication. A culture of error and unpredictability, because *freedom is responsibility*. Being human, human beings.

---

[60] *Ibidem*.

3.6 A "New *Nature*"...

As I often find myself repeating, (hyper)complexity is quite a complex issue! We are still little aware of its *nature*, which is complex (of course), ambivalent, dynamic, unstable, irreversible, interdependent and inter-independent. A "non-observable" form of nature (→"emerging systems/ecosystems"), at times throughout all its (dissipative and *emerging*) dimensions and properties; a kind of complexity that is cognitive, subjective, social and ethical, but also linguistic, *descriptive*[61]and communicative. Moreover, we are still unaware of how to address and attempt to know, interact with, *breathe* and inhabit—all at the same time—what is close to being some kind of "*new nature*": that of artificial technologies of (hyper)connection, interconnected/hyperconnected environments and, specifically, of an even more invasive and pervasive nature, capable of demolishing every theoretical, methodological, epistemological *boundary* of artificial intelligence systems.

A kind of complexity that constantly and continuously self-regenerates and self-reproduces, even under the most adverse and unfavourable conditions. It is able to create and *co-create* without pause, even if characterised by variables, interactions, levels of interconnection and interdependence, whose functions do not seem to further order, organization and/or balance, but the disaggregation and dispersion of the systems/subsystems that constitute and innervate it. This complexity (typical of living, social and *(hyper)complex systems*) is capable of making dichotomies, dilemmas, contradictions, conflicts, paradoxes, and dialectics coexist on all levels of individual and social theory and practice. In this type of complexity "chance" plays—or can play—a decisive role, and fulfills—or can fulfill—a decisive function.[62] It is a (hyper)complexity characterized by many multiple levels of

[61] See, in particular Rescher N., *Nature and understanding. The metaphysics and method of science*, Clarendon Press, Oxford 2000. Rescher, referring to the reality of phenomena, speaks of "infinite descriptive depth" which, for many reasons, we always try to simplify, with risks and implications in terms of *methodological simplification* and *presumption of simplicity*. On these topics, see also the interesting work by McFadden J. (2021), *Life is Simple. How Occam's Razor Set Science Free and Unlocked the Universe*, Italian translation, *La vita è semplice. Come il Rasoio di Occam ha liberato la Scienza e modellato l'universo*, Bollati Boringhieri, Turin, 2021.

[62] On this particular issue, see an authentic classic, published at a time of great scientific and cultural ferment, when humans were beginning to understand fundamental biological and evolutionary mechanisms: Monod J. (1970), *Le hazard et la nécessité*, Italian translation, *Il caso e la necessità. Saggio sulla filosofia naturale della biologia contemporanea*, Arnoldo Mondadori Ed., Milan 1997. In extreme synthesis, for Monod, the nature of Life, of living systems, is based on a number of fundamental principles such as invariance, intended as the ability to conserve, during the transition between generations, one's own structural norm and teleonomy, considered the ability to transmit *its content of invariance to future generations*; something that might lead one to think that there is little room for the evolution of the organisms themselves, though this is not the case. Organisms/living beings evolve and "mutations intended as initial events, occur "fortuitously"." In other words, "chance" plays a determining role, influencing the subsequent *processes of selection (necessity)*. Again, on these issues, I also suggest consulting the interesting work carried out by Friedrich Cramer, who also directed the Max Planck Institute for Experimental Medicine in Göttingen: Cramer F. (1988), *Chaos und Ordnung. Die Complexe Struktur des Lebendigen*, Italian translation, *Caos e Ordine. La complessa struttura del vivente*, Bollati Boringhieri, Turin 1994. See also: Cramer F., *Fundamental Complexity. A Concept in Biological*

interconnection and interdependence that need to be observed and investigated as a whole, paying attention, among the many variables, to the quality and intensity of the relationships/interactions involved. At the same time, it is a complexity marked also by great "*fragility*" (Morin) which, defined in conceptual and operational terms, continues very often to be confused with "complication", with a higher/more severe degree of "difficulty" and "complication". Having made these considerations, our analysis cannot fail to develop, however briefly, a series of premises which we will only be able to examine in a non-exhaustive manner. It is by no way superfluous to point out that, in order to further pursue the theses and topics of this work, we should first of all agree on the meanings of the terms themselves (at the fundamental level of *operational definitions).* We also need to agree on the disciplinary fields/sectors of reference involved, deeply aware of the need to have them engage in dialogue and communicate, possibly in methodological terms as well. In any case, I repeat, we remain unaware that complexity is a *structural/natural characteristic of human groups, relationships, the social system, the biological world and organisms*. At the same time, we are still unaware of the profound epistemological, methodological and ethical implications this way of thinking, this thought system, approach and "view of reality"—how we "view" what we define and recognize as "reality"—involve, imply, almost determine even if not in a linear and/or causal way.

As far as the world of *objects and things* is concerned, however, we need to/should speak of "complicated", not "complex systems", since we are well able to distinguish, break down and analyse their *parts/units/sections* individually, in order to understand their behaviour and functioning. These are systems characterized by phenomena and processes that are substantially observable, *linear*, and to some extent, therefore, predictable (not always), controllable and replicable, describable by mathematical formulas and models/data sequences. Once more, with reference to complicated systems (mechanical, artificial systems, etc.), we should also introduce the (crucial) issue of the *reference system*, of net force, of weight-force, of the "principle of action and reaction", or else we should speak of mass and so on, hence once again speaking of "substantially" linear phenomena and processes described effectively by means of formulae, equations and functions, capable, in many cases, of predicting behaviour/modes of behaviour with relative accuracy. On the contrary, the complexity that concerns, in a specific and peculiar manner, society, organizations and human groups (and with some nuances, biological systems[63] as well), is a

---

Sciences and Beyond, "Interdisciplinary Sciences Review", 4, 1979, pp. 132–139. Cramer deals with *increasing levels of complexity*, distinguishing between subcritical, critical and fundamental complexity. On this point, see, in particular, chapter 9, p. 278 and following.

[63] See, in particular, among the "classics", Schrödinger E. (1944), *What is Life? The Physical Aspect of the Living Cell*, Italian translation, *Che cos'è la vita?,* Adelphi, Milan 1995. I refer also to another great "classic" already cited, which aroused widespread debate in its time: Monod J. (1970), *op.cit.* Among more recent works we find Al-Khalili J., McFadden J. (2014), *Life on the Edge. The Coming of Age of Quantum Biology*, Italian translation, *La fisica della vita. La nuova scienza della biologia quantistica*, Bollati Boringhieri, Turin 2015; Catania K., (2020), *Great Adaptations*, Italian translation, *Adattamentei meravigliosi. Sette irresistibili misteri dell'evoluzione*, Bollati Boringhieri, Turin 2021; Zimmer C., *Life's Edge: The Search for What it Means to Be Alive*, Palgrave Macmillan, 2021.

very particular kind of complexity because, in addition to being perpetually unstable and dynamic, it is marked by irreversible dynamics and "inter-independent units" that cannot be plotted or interpreted on the basis of linear models (cause-effect, stimulus-response)—even less so by way of mathematics and statistics (however necessary); above all, it is not *visible*, let alone *observable* in all its dimensions.

These are systems, phenomena and processes that are not characterized "exclusively" by certain variables/contributing causes; (not that I am implying that complicated systems are "simple" or easy to study, on the contrary) and/or by the response(s)/feedback/interaction to/with forces/stimuli or systems of forces/stimuli (about which much could be said). What we are dealing with here are systems, phenomena and processes, which come about/ evolve / are transformed by means of their *ability to self-organize*, to be open and dissipative, to exhibit "emergent properties", at the most unexpected times and in the most unforeseen ways, within the framework of (constant) systemic relationships comprising multiple connections and levels of interconnection/interdependence.

3.7 Anthropological and Epistemological Transformations. The Transition to *Hypercomplexity*

The paradigm shift and the ongoing profound anthropological transformation (1996)—only partially "technological"—not only (re)define, but create new dimensions, openings and *epistemological implications* that require new thought-frames, new ways of thinking, and (the integration of) different approaches and methods, as well as new tools for detection and analysis. In all fields of research and action and, with various nuances, within the various disciplinary sectors, one should no longer seek—as in the past—universal, inexorable laws of nature, under the banner of reductionist and deterministic approaches. It is now acknowledged, on the basis of studies, theories and fundamental scientific discoveries (not only among the so-called "hard sciences"), that the analysis and explanation/interpretation of a phenomenon, a process or of the complexity of a system, are no longer referable/reducible to empirical observation and knowledge of the *simple* properties (evident and *measurable* in quantitative terms) which characterise the constituent units of the system observed. We have become aware of the increasingly strategic and relevant conditioning and *conditioned* role of the *observer* and of an empirical, "scientific" type of observation, which can no longer be considered "neutral", "impartial" "external", totally "non-evaluative", with respect to the reference system and the object-system observed. This kind of observation can only be *participatory* and, at the same time, *participated*. Along with awareness has come the acknowledgement that isolated "objects" (isolated from contexts) can no longer be observed, even in *ideal conditions*. What can be observed are "systems", "relationships" and "dynamics", characterised by levels of interconnection and interdependence that urgently call for a systemic, multi/inter/trans-disciplinary approach to complexity. The objective, therefore, was /is/will no longer be that of identifying and recognizing *analogies* and common, structural elements within more or less complex phenomena, but rather that of highlighting how phenomena and dynamics that are apparently the same/identical /similar or, in any case, very similar, actually present non-linear, irregular and unpredictable styles of behaviour and reactions. Order, balance, simplicity, linearity, causality, regularity, dichotomies, dialectical relationships—to provide only a few examples—are, if not actually replaced, flanked by new key concepts—such as complexity, chaos, non-linearity, disorder, entropy, irregularity, dynamism, variety etc.—which open up, disclosing extraordinary and fascinating scenarios and trajectories, which are at the same time, however, difficult to read. Above all, they open to reveal new paths of study and research, which cannot fail to correspond to new epistemologies and, fittingly, epistemological transformations.

This way, the "real" and "reality" are shown to be increasingly more complex and unpredictable than the scientific paradigms that have/had been used to observe and investigate them up to that moment. The words of the great scientist, mathematician and physicist, Poincaré, are emblematic:

> If we knew exactly the laws of nature and the situation of the universe at the initial moment, we could predict exactly the situation of that same universe at a succeeding moment. But even if it were the case that the natural laws had no longer any secret for us, we could still only know the initial situation approximately. If that enabled us to predict the succeeding situation with the same approximation, that is all we require, and we should say that the phenomenon had been predicted, that it is governed by laws. But it is not always so; it may happen that small differences in the initial conditions produce very great ones in the final phenomena. A small error in the former will produce an enormous error in the latter. Prediction becomes impossible, and we have the fortuitous phenomenon.[64]

Order and disorder, order and chaos, linearity and non-linearity,[65] predictability and unpredictability, balance and dynamism/entropy, not only blend and mix, but coexist. In short, complexity and chaotic nature have been acknowledged, albeit with a certain tardiness, as constitutive dimension of the Vital, the Human and the Social, despite the fact that human beings have never lost the *illusory aspirations of begin able to bestow order upon chaos*. Complexity never abandons the dynamics of its transformation and of its complex, non-linear evolution. The dialectical relationships between norm/regularity and case/exception are complete toppled, with profound implications, not only for all disciplinary fields but also, and above all, for common-sense knowledge, scientific knowledge, the life of People, social systems and democracies.

In the light of what has been said, and with due awareness of the (fundamental) limits of knowledge, of the quota of ignorance implicit even in knowledge itself

---

[64] See Poincaré J.H. (1908), *Science et méthode*, Italian translation, *Scienza e metodo*, Einaudi, Turin 1995 (ed. C. Bartocci), p. 56.

[65] See, in particular, the interesting work by Bertuglia C.S., Vaio F., *Non linearità, caos, complessità*, Bollati Boringhieri, Turin 2003. See also Strogatz H.S., *Sync: How Order Emerges from Chaos in the Universe, Nature, and Daily Life*, Hachette Book Group, 2004.

(Morin) and of the scope of the epistemological transformations in progress, it is useful to recall some concepts and definitions, first of all clarifying what we mean by "simple", "complicated", "complex" and "linear". As we know, the sciences and the numerous disciplinary fields/sectors have often provided/provide non-overlapping, and in any case, alternative and complementary definitions. In this sense, much could be said/written regarding the importance of the "principle of (weak and strong) causality", determinism and predictability, the concept and theory of *deterministic chaos*, questions concerning *probability*,[66] etc.

In any case, there are formulae, equations and functions capable of describing, even of predicting how certain (complicated) objects/systems behave. Although with regard to certain "natural" phenomena, for example, we must always be aware of the profound implications of "deterministic chaos", the theory of which has highlighted the existence of deterministic and (relatively) predictable phenomena for which, despite identification of a law that describes their temporal evolution, the models and "linear" explanations are neither adequate nor functional. These are systems whose behaviours turn out to be extremely irregular and unpredictable, though in theory only they tend constantly towards order and equilibrium (staticity). These can be exemplified by meteorological phenomena and the difficulties, despite extremely sophisticated technological tools, of making meteorological forecasts with absolute accuracy and precision. Other examples are phenomena such as earthquakes, hurricanes and/or a falling meteorite. In the current state of the art, their behaviour can be observed and predicted in probabilistic terms only, even though they can be traced back to known scientific laws and mathematical formulae. In these cases, we find ourselves in conditions of limited *predictability* and a restricted *horizon of predictability*. "Determinism"[67] and "predictability" have been separated, so to speak, by a veritable *epistemological fracture:* the theory of chaos.[68]

---

[66] On the concept of and on topics and issues related to "probability", as well as the "nature of probabilistic thinking", see: Rota G.-C., Kung J.P.S. (1980), *Probabilità*, in Enciclopedia del Novecento Treccani, Istituto della Enciclopedia Italiana, Chivasso 2019.

[67] On determinism and the epistemological questions associated with it, I refer to three great "classics" relating to the areas of both science and philosophy: Bernard C. (1878), *Le science expérimentale*, Italian translation, *Un determinismo armoniosamente subordinato. Epistemologia, fisiologia e definizione della vita*, Mimesis, Milan-Udine 2014; Cassirer E. (1937), *Determinismus und Indeterminismus,* Italian translation, *Determinismo e indeterminismo nella fisica moderna*, Mimesis Edizioni, Milan 2020; Kojève A. (1931–1932), *L'idée du determinisme dans la physique classique et dans la physique modern*, Italian translation, *L'idea di determinismo nella fisica classica e nella fisica moderna,* Adelphi, Milan, 2018.

[68] Within the relevant scientific literature, see in particular Lorenz E.N., *The Essence of Chaos*, Univ. of Wash Press, Seattle, 1963; also by Lorenz, *Three approaches to atmospheric predictability*, "Bulletin of the American Meteorological Society", 50, 5, 1969; Gleick J. (1987), *Chaos*, Italian translation, *Caos*, Rizzoli, Milan 1989; Waldrop W. M. (1992), *Complexity. The Emerging Science at the Edge of Order and Chaos*, Italian translation, *Complessità. Uomini e idee al confine tra ordine e caos*, Instar Libri, Turin 1995; Prigogine I. (1992), *Le leggi del caos*, Laterza, Bari-Rome 1993 (see also the 2003 edition). See the excellent work by Cramer F. (1988), *Chaos und Ordnung. Die Complexe Struktur des Lebendigen*, Italian translation, *Caos e Ordine. La complessa struttura del vivente*, Turin: Bollati Boringhieri 1994. See also Ruelle D. (1991), *Hasard et Chaos*,

It is probable that, in the future, the immense availability of data and information, together with progressively more sophisticated tools, will prove increasingly strategic and vital in the attempt—not to be taken for granted—to succeed in defining/recognizing as predictable and linear, phenomena which, though explicable according to certain laws of physics, behave differently. In actual fact, theories on chaos and "deterministic chaos" have contributed, in a decisive way, to define that new paradigm, whose epistemological implications, cognitive horizons and territories of open research we are as yet incapable of evaluating fully. It is becoming more and more difficult, at least for the moment, to make exact predictions regarding so-called *non-linear dynamic systems*.

Furthermore, in the same way, complexity itself is definable in so many ways in relation to diverse, multiple fields of knowledge and disciplines (and not only). Presenting itself in multiple forms, complexity obliges us to *rethink* everything, not only how we think, but also the conceptual categories and their related operational definitions themselves, even more so within the hypertechnological (and hyperconnected) civilization in which science and technology, besides having done away with the boundaries between "natural" and "artificial completely",[69] seem to be able to bestow life and intelligence (?) upon "matter".

We are, therefore, confronted with unpredictable (hyper)complexity—the question of predictability, not only of human, social, cultural behaviour, is crucial and strategic (cultural models also serve this purpose) and is related, in the first place, to the "non-observability" and (consequent) "non-measurability" of all the dimensions involved/shown—and *non-replicable* (*replicability*, as we know, is an important requisite of science and permits us to even speak of "scientificity"). We have to/should, above all, observe and understand the multiple levels of the connections existing between the processes and the parts/objects themselves. To do so, we need a systemic view of processes, phenomena and dynamics, a systemic view involving a completely different way of observing "objects", not just as a set and a *whole*, keeping in mind, in any case, that the product is never the sum and/or totality of the parts. However, there exists a further element of complexity: the fact that we are dealing with *adaptive complex systems,*[70] capable of changing to meet new conditions and/or requisites. These are systems whose constituent parts are not "inani-

Italian translation, *Caso e Caos. La scoperta della complessità ai confine tra matematica e fisica*, Bollati Boringhieri, Turin 1992; Rickles, D., Hawe P., Shiell A., A Simple Guide to Chaos and Complexity, *Journal of Epidemiology & Community Health,* (BMJ) 2007 Nov.; 61(11): 993–937. https://www.ncbi.nlm.nih.gov/pmc/articles/PMC2465602/; Sellow T.L., Seeger M. W., Ulmer R.R., *Chaos Theory, Informational Needs and National Disasters*, https://www.orau.gov/cdcynergy/erc/content/activeinformation/resources/chaostheory.pdf; Strogatz H.S., *Sync: How Order Emerges from Chaos in the Universe, Nature, and Daily Life*, Hachette Book Group, 2004; Weinberger D. (2019), *Everyday Chaos. Technology, Complexity and How We're Thriving in a New World of Possibility*, Italian translation, *Caos quotidiano. Un nuovo mondo di possibilità*, Codice Edizioni, Turin 2020.

[69] See Dominici P. (1995, 1998, 2003, 2011 and following.), *op.cit.*

[70] See Miller J. H.—Page S.E., *Complex adaptive systems: an introduction to computational models of social life*, Princeton University Press, Princeton 2007.

mate", passive, or neutral, reacting solely to certain stimuli in a predictable fashion; they are individual items, entities, relationships that make a constant contribution by changing and co-creating the conditions of interaction, of the reference environment, of the ecosystem to which they belong. If we observe a social organization, or simply a group of people, not only does the totality of the people involved not constitute a whole, not only will I be unable to understand the dynamics of that group by isolating single people or limiting my *field of observation*. I will also have to come to realize that those very same people/individuals/entities constantly contribute to modifying—or co-creating, co-constructing—the social environment in which they are immersed. Besides, my very presence and the fact that I am engaged in observation will also modify the conditions and the levels of interaction, exchange and sharing involved. If I really want to observe and understand constantly evolving *relationships* and dynamics, I shall need to observe the whole, the totality, the connections and the systemic relationships that constitute them. An interdisciplinary, multidisciplinary, transdisciplinary approach (Piaget) is required—in addition to the aforementioned systemic view.

As the transition from *simplicity*[71] to *complexity* and from this to hypercomplexity can be explained in terms of the variables and the contributory causes and parameters involved, we need to use these and consider them when seeking to observe reality. Therefore, we must distinguish the "simple" from the "complex" on the basis of the variables, contributing causes and parameters involved, according to their number and quality. In addition to the fundamental distinction to be made with reference to the connections, to the different levels of connection and interdependence/inter-independence existing between the parts/units involved (single or, in turn, complex as well), and in relation to the "systemic", non-linear, "nature" of their interactions/relationships. I might add that, in the present-day phase of global change, the transition from complexity to hypercomplexity is determined, in par-

[71] On "complexity" and "simplicity" see in particular James W., *The Principles of Psychology*, 2 vols., Macmillan and Co., London 1901; James believed that the process of reducing complexity to simplicity was the fundamental goal of Science; Bachelard G., (1934), *Le Nouvel Esprit scientifique*, Italian translation, *Il nuovo spirito scientifico*, Laterza, Bari, 1978 (new revised ed.), with its interesting preface by Ludovico Geymonat and Pietro Redondi. Gaston Bachelard, a prominent figure and famous epistemologist, speaking of complexity and simplicity, dealt with the issue claiming that in reality there were no simple phenomena, that phenomena were a web of relationships, that there was no such thing as a simple nature or a simple substance because substance was made up of a whole fabric of attributes, that there were no simple ideas [...]. The chapters of the book are very interesting, dedicated to "Determinism and indeterminism. The notion of the object" (chapter 5) and "non-Cartesian epistemology" (chapter 6). By the same author, see also: Bachelard G., *The formation of the scientific spirit*, Italian translation *La formazione dello spirito scientifico*, Italian translation by E. Castelli, Cortina, Milan 1995; Bachelard G., *La filosofia del non*, trans. by G. Quarta, Armando, Rome 1998. At times, within the scientific and multidisciplinary literature relating to these issues and approaches (as well as methods/methodologies), certain boundaries/margins, like that between "complexity" and "simplicity", despite the variety/diversity/peculiarity of the approaches and "objects of study", are really very subtle. See Waldrop W. M. (1992), *Complexity. The Emerging Science at the Edge of Order and Chaos*, Italian translation, *Complessità. Uomini e idee al confine tra ordine e caos*, Instar Libri, Turin 1995.

ticular, by two complex "variables": the first is technological innovation, in particular, the so-called digital revolution which, unlike other phases of the industrial revolution, introduces "new velocity" to the social, economic, cultural processes that characterize the current change. This "new velocity" produces new criticalities and problems of control, as in all phases of acceleration and industrial transformation; the second concerns the increasingly strategic role played by communication, not only with reference to processes of education and socialization but also within the dynamics and processes of social representation and perception, so vitally important for social systems, in particular, for the survival of contemporary democracies. A new *virality of communication*, favoured by "connective technologies" (2014) and by new hyper-connected environments, which should not be identified/confused with the virality of new communication and digital tools/environments/processes. This virality of communication has brought out of the traditional "ivory towers" and the equally age-old "centres" used for building social consensus, topics and issues that had always been the exclusive domain of élites and small groups of experts. Between processes (and the rhetoric) of *disintermediation* and *re-intermediation.*[72]

3.8 Beyond "False Dichotomies"

Nature versus culture; natural versus artificial; Human versus technological, culture versus technology, theory versus research/practice; epistemology versus ethics; scientific versus humanistic training; thought and reason versus emotions[73]; thought versus action; reason versus creativity and imagination; body versus mind; complexity versus specialisation; interdisciplinarity versus specialisation; knowledge versus competence; form/s versus content; hard skills versus soft skills; Real vs Virtual. Many years ago, I called these "false dichotomies".

In attempting to observe, describe, recognize and understand complexity, the human, life, the vitality of the spirit, that very essence that is (always) "invisible to the eye" (cit.), we invariably resort to division, separation, distinctions and fractures that often do not lead to knowledge, but, rather, to a sense of orientation and reassurance that characterizes all cultures (bearers of identity, effectively), with respect to the uncertainty and variability of life and reality. The isolation, separation and restriction of knowledge, experiences and life, is a complex operation that has always marked the evolution of social systems, organisations and social action. In any case, these are precisely the strategic functions that cultural models perform.

We go on believing we can see, observe and try to understand reality on the basis of the logics, models, schemes and infinite sequences of data which, in addition to

---

[72] See Dominici P., *op.cit.*, 1995, 1998, 2003, 2011, 2019 and following.

[73] On the substantial continuity between cognitive aspects and emotional elements, see Damasio A.R. (1994), *L'errore di Cartesio. Emozione, ragione e cervello umano*, Adelphi, Milan 2009.

failing to identify/recognise relationships and connections reduce (apparently), their variety, unpredictability and wealth, convinced as we are that we can cage all vitality of spirit and all human complexity[74] in mathematical formulae and infinite sequences of data and numbers. Convinced as we are that we can even measure "quality" in impartial, objective, scientific terms—in my opinion, this is a contradiction in terms—by resorting exclusively to quantitative data and tools, and in all areas of practice and material and intellectual production, resorting to (eternally seductive) simplifications presented, once again, as "facts". *We continue to seek the kind of knowledge that confirms our beliefs, our initial hypotheses, our cultural and educational models, our prejudices and stereotypes.*

3.9 Inhabiting Hypercomplexity

Hence, we find ourselves *hurled* into *hyper-complexity*, in an attempt to inhabit the hypertechnological, hyperconnected civilization, a civilization of intelligent machines (?). A civilization founded on programming, automation and the (hyper) simulation of processes and dynamics, marked by a progressive, exponential growth of the dimension of the technologically controlled which, besides diminishing/marginalizing the spaces of the Human and of responsibility (at least apparently), continues to foster an age-old, counterproductive illusion: that of being able to oust/eliminate error (a fundamental prerequisite of knowledge, life and freedom itself) and unpredictability, not only from education and training, but also from complex systems themselves, as well as from the social and organizational processes that characterize them (Dominici 1995 and following). From this perspective, the challenges of change basically stem from the urgency to rethink/redefine the centrality of the Person and the Human within environments and ecosystems where there is no longer any boundary/limit between the natural and the artificial, as well as between nature and culture (an ancient "false dichotomy"). Today, perhaps as never before, it appears evident that hypercomplexity is no longer an option (it never was), but a "fact of life", which obliges us to recover what I had defined in the past as *the*

[74] In addition to the works already mentioned on these topics, see: Gazzaniga M.S. (2008), *Human*, Italian translation, *Human. Quel che ci rende unici*, Raffaello Cortina Ed., Milan 2009; Bertrando P. & Bianciardi M., *La natura sistemica dell'uomo. Attualità del pensiero di Gregory Bateson*, Raffaello Cortina Ed., Milan 2009; Fry H. (2018), *Hello World. How to Be Human in the Age of the Machine*, Italian translation, *Hello World. Essere umani nell'era delle macchine*, Bollati Boringhieri, Turin 2019; M. Tegmark (2017), *Life 3.0. Being Human in the Age of Artificial Intelligence*, Italian translation, *Vita 3.0. Essere umani nell'era dell'intelligenza artificiale*, Raffaello Cortina Ed, Milan 2018; Dominici P., *Dentro la "civiltà senza errore". Sintesi complesse: l'Umano, il legame sociale e la civiltà ipertecnologica*, in A. De Cesaris (Ed.) *Vite digitali. Essere umani nella società del XXI Secolo*, FrancoAngeli, Milan 2020; again on the subject of the "complexity of the human", for further insights, also with reference to the questions of the "mind" and "consciousness": Damasio A.R. (2021), *Feeling & Knowing. Making Minds Conscious*, Italian translation, *Sentire e conoscere*, Adelphi, Milan 2022.

complex dimensions of educational complexity. It is time to imagine, design, and create complex systems, seeing them for what they are, that is, as organisms and not as *mechanisms*. *Complex*, not complicated *systems*, in fact. The recovery of these dimensions is of vital importance also, and above all, in consideration of the fact that the extraordinary scientific discoveries and technological innovations, the speed and the intrinsic dynamism of the ongoing transformations, are not "leading" us towards simplification, quite the opposite! In this sense, it is necessary to recover the awareness that, precisely in the era of disintermediation, social and professional figures, institutions, processes and mechanisms of mediation must recommence fulfilling their function, a strategic function, to say the least. The role of figures engaged in mediation, in particular, becomes even more strategic, but they need to be educated, prepared, trained, and constantly updated to recognize and deal with this hypercomplexity, with the wealth of systemic relationships and levels of connection that characterize, not only the hyper-technological civilization, but life itself. In this regard, I spoke, long before it became a buzz-word, of the urgent need to educate and train "hybrid figures",[75] managers of complexity (a definition that I have used—in the past in m works and publications—for the sake of simplicity, perfectly aware that *"managing complexity" is practically an oxymoron*; even more so if we are referring to social, relational, human complexity). It is not for building a "culture of control", (it is not within a culture of control) that these *hybrid figures* need to be educated and trained, but rather to interact with that unpredictability which is an essential connotative element of social, human and vital systems. *Hybrid figures* with a strong logical, epistemological and methodological basis and evolution. Moreover—at the risk of repeating myself yet again– unless we put our hands to radically transforming education and training, we will never be able to address or interact with this hypercomplexity. It will not be the technological and the digital to create the conditions that permit this to happen. Likewise, it will not be technology to reconstitute social bonds, to re-activate the social mechanisms of trust and cooperation or to determine the conditions of a truly inclusive kind of innovation. It is necessary to commit ourselves to re-constituting and re-mediating social bonds. Managing information and communication, in this type of context, actually means (trying to) manage/govern complexity (an oxymoron that I use only for the sake of synthesis), taking great care not to fall into the rhetoric of disintermediation and simplification intended as an absolute value (Fig. 3.1).

We can no longer afford to continue to perpetuate the "ultimate error": treating complex systems as if they were complicated ones (1995). Because, as I have tried to argue along a series of study and research pathways I have pursued, everything *is interdependent and interconnected,* "everything is in a relationship". No "objects" are isolated and/or isolable from situations and contexts, except by recourse to theories and paradigms that have amply demonstrated all their weaknesses, inadequacies and incompleteness.

[75] See Dominici P., *op.cit.*, 1995–1996 and following.

Complicated, Complex and HyperComplex Systems

Complicated systems (closed)	**Complex systems (open)**	**Factors & variables of transformation** →	**HyperComplex systems*** *Paradigm shift*
Typical of artificial and mechanical systems Produced by **hierarchical** know-how / knowledge / structures; **categorical approach** Structured; stable **Closed pathways** **Linear** **Reducible/ divisible / dismountable** into separate parts and/or stages; The function of each part or process is definable **Predictable and reproducible: steady and regular** **Sum of parts = whole** **Cause and effect, stimulus and reaction are linear** and chronological Reductionist Deterministic Can be observed by an external observer The dynamics of the system can be analyzed by **isolating the single elements and/or by narrowing the fields of observation** **Systems perceived as objects;** can be considered "machines" **Output** depends on **input** (predictable and predicted) The parts of the system are **in equilibrium** Parts can be replaced / repaired to **recover the initial state** Small/large changes produce small/large effects respectively	Typical of natural, biological, human and social systems, **capable of self-organization** **Shared knowledge, networked structures** and **processes**; perception of surroundings and sensations; **inter/multi/transdisciplinary** approach Chaotic; flowing **Open, intersecting pathways and trajectories, at times discontinuous** Non-linear and **dynamic** **Multiple levels of connection between processes and parts capable of self-modification**; The function of the parts are interdependent and inseparable; their union forms a new organism/system **Unpredictable and irreproducible:** many degrees of freedom among numerous sub-units- no control center **Sum of parts < whole** Impossibility of reducing or interpreting through linear models (cause-effect, stimulus-reaction) – behavior emerges from interaction & feedback among the parts and with the environment: **Emergent Properties** Exponential Non-deterministic; interconnected contexts and processes The observer **affects and is affected** by what is observed. The observer is always an **observer/participant** **Adaptive systems**: elements within the system modify, co-create and co-construct their own interactions and the system itself and adapt to new requirements or conditions **Objects perceived as systems;** should be considered organisms/living networks Importance of social & cultural factors Capable of **hierarchical self-organization** and **emergence** **Irreversible** in time; **dissipative** system, **far from equilibrium** The system has a **history** and **memory;** but does not return to initial state or allow predictability of future states **Small variations** can trigger enormous changes throughout the **entire system (butterfly effect)**	**More intervening variables and concauses: systemic relationship: simplicity → complexity quantity → quality** → **Technological (and digital) innovation** **"New Speed":** ***Acceleration of dynamics and processes*** → **Communication** ***Playing an ever more strategic role*** - in processes of education and socialization - in representation and perception of dynamics and processes of systemic evolution - *New virality* of **communication** → **Increase of the parameters to be considered** → **Organizational cultures:** The culture of complexity is **NOT** a culture of "problem solving" Grand illusions of the hyper-technological civilization: Total Control, Rationality, Measurability, Predictability, Elimination of Error →	**A hypercomplexity** ***that includes all properties of complex systems*** and is also ***cognitive, linguistic, social, subjective, ethical, and communicative:*** *Cultural evolution* is determining *biological evolution* **(the Overturn*)** The quantum leap in acceleration and technologies of connection → irreversible **anthropological transformation*** ***Doubt and uncertainty, chaos and disorder*** **are intrinsic elements of the** ***life-worlds*** Increasing importance: - *of* ***epistemological & ethical implications*** - *of* ***social & cultural factors*** ***Mathematical formulas*** **and** ***scientific methods*** **applied to systems for predicting actions/ outcomes or describing scenarios can no longer produce satisfactory results** *- In (complex) biological systems this already holds true, but its effects have been amplified and accelerated by the latest evolutions in human communication and technology* *** all concepts © P. Dominici 1996-2020***

Fig. 3.1 Complicated, Complex and HyperComplex Systems (Dominici P., 1996–2020)

Chapter 4
Dilemmas and Paradoxes of the Hypercomplex Society

Abstract Faced with a cultural evolution that conditions biological evolution, the need to overcome the false nature/culture dichotomy and to enhance the crucial issues of inter/multi-disciplinarity emerges with considerable force, as humanistic and scientific cultures can no longer be kept separate. Furthermore, the dominant individualism of present-day "asymmetric" social systems, which is the outcome of the ambivalent process of emancipation carried out during Modernity, has, on the one hand, increased freedom and led to the recognition of some fundamental rights (at least in theory); on the other, it has paradoxically weakened the constraints and bonds of community belonging, as well as the mechanisms of trust and cooperation, at a time of maximum expansion of the new global communications ecosystem— at least in terms of connectivity— while the actual *sharing* of knowledge, which should define communication, has been too limited to further inclusion or to shake up the age-old logics of control. The economistic paradigm must be questioned, keeping in mind, however, that the crisis is far from being solely economic.

Keywords Cultural Evolution · Biological evolution · Interdisciplinarity · Multi-disciplinarity · Trans-disciplinarity · Communications ecosystem · Economistic paradigm

Within a world system characterized by increasingly interconnected and interdependent processes and dynamics, we are going through an extremely complex and sensitive phase of global change. This *hyper-technological civilization of automation and simulation* of ours has created the conditions for a radical and unprecedented paradigm shift. This cultural evolution—which also includes technique and technologies—is capable of influencing, as never before,

P. Dominici, *Beyond Black Swans: Inhabiting Indeterminacy*, Philosophical Studies Series 160, https://doi.org/10.1007/978-3-032-09029-4_4

biological evolution. Let us think, for example, of the extraordinary discoveries in the field of genetic manipulation, robotics and artificial intelligence. Human beings are (seem?) now able to (pre)determine the unfurling of events, even the beginning, reproduction and end of a single life, modifying every one of its aspects profoundly. Consequently, in the rediscovered and, at the same time, renewed illusion of total control over the environment and the ecosystems that surround it, the crucial issue of the paradigm shift, in addition to not being a simple slogan used for effect, becomes a priority, a keenly-felt urgency, given that we are forced to rethink the concepts themselves, the operational definitions and categories that frame them. On the other hand, from the point of view of theoretical models, for some time now we have failed to define/produce *a thought system* fit for a (hyper)complex and continually evolving reality, which, among the many implications involved, favours the *rapid obsolescence of knowledge*, skills, theories and working hypotheses, even of operational tools. We therefore find ourselves forced to *rethink thinking and how we think*, as well as re-defining everything inside the so-called "toolbox" a little, endeavouring to go beyond the "false dichotomies" that we have referred to several times and will be taking up again later. Beyond the traditional distinctions made between *natural and artificial*, between natural and cultural, we are obliged to ask ourselves, for example, *what life is*, and in what intelligence, thinking, thought, *consciousness,* (hyper) complex, non-simplifiable and/or replicable "dimensions" may consist, as well as what it means to be *Persons* and/or Citizens today. The crisis of the current era, from the point of view of paradigms, knowledge and intellectual production, represents an extremely problematic moment of transition, which finds us in great difficulty and assumes the form, above all, of a crisis of culture and civilization which is only partly economic.

When tackling this crisis, we find ourselves in difficulty also because we have concentrated solely on the methods and tools of detection, without taking into the least consideration the importance and strategic value of theory (often confused with the history of thought) and of theoretical-interpretative models which, inevitably, are needed to support us both in daily practice and in "scientific" observation of phenomena. Faced with a cultural development that conditions biological evolution, the need to overcome the nature/culture dichotomy appears more urgent than ever before while, at the same time, the demand for a systemic approach to complexity,[1] such as to enhance the crucial issues of interdisciplinarity and multidisciplinarity, emerges with considerable force. This necessity is dictated by the impossibility of continuing to separate, and to keep separate, humanistic and scientific culture,[2] and the two evolutionary "fields".

[1] See Emery, 2001, Bateson G., 1972 et al.

[2] On these issues, see also: Consilience. The Unity of Knowledge Klein É. (2018), *Matière à contredire. Essai de philo-physique*, Italian translation, *Filosofisica,* Rome, Carocci, 2020. Of interest is also the concept of "consilience", proposed by one of the pioneers of sociobiology and biodiversity: Wilson E.O., *Consilience. The Unity of Knowledge, Alfred A. Knopf, New York 1998.*

The thesis that we are attempting to defend here is also the following: the dominant individualism we find in present-day social systems is the outcome, to some extent inevitable, of the complex process/project of emancipation carried out during Modernity. This process/project is characterized by a profound degree of ambivalence. On the one hand, it has actually increased spaces of freedom and led to recognition of some fundamental rights (at least in theory), on the other, it has contributed, almost paradoxically, to weakening the constraints and bonds of community membership, at a time when the bonds of interdependence have increased and the potential of communications has increased enormously.

Advanced societies appear to be increasingly characterized by instances of profound *social asymmetry*, as well as processes of individualization and fragmentation, which trigger dynamics of conflict capable of testing the resilience of social and organizational systems, based on a *limited form of rationality,* arising, among other things, from the weakening of mechanisms of trust and cooperation. Radical and complex modernity, from this point of view, seems to have further radicalized the processes of *detachment* of the individual from the group, generating opposition between freedom and responsibility. We can see the effects and consequences of the current long phase of global, systemic emergency. In a context of this kind, communication, intended as a social process of sharing knowledge (power)·[3] seems to have assumed an ultimate and strategically central role in every dimension of organizational and social practice.

In the era of "biopower" and "biopolitics" (Foucault)[4] the hypertrophy of bureaucratic apparatuses, the progressive dissolution of public spaces and the evolution of democracies, founded on transparency, access, the concept of *popular sovereignty* and, from a cultural point of view, on economic individualism (hegemonic compared to democratic) have caused a *radical politicization of the public sphere*, whose operational space has been reduced so significantly that its sole role is that of representation as the "handmaiden of the power system".[5] The process of evolution of the new-born democratic regimes, often based culturally on the concept of

[3] An operational definition proposed in Dominici P. (1996), *Per un'etica dei new media. Elementi per una discussione critica*, Firenze Libri Ed., Florence 1998.

[4] On these issues see, in particular Focault M. (1976), *Histoire de la sexualité 1. La Volonté de savoir*, trad. it. *La volontà di sapere*, Feltrinelli, Milan 1991; Foucault M., *Nascita della biopolitica. Corso al Collège de France 1978–1979*, Feltrinelli, Milan2005; Foucault M. (1975), *Surveiller et punir. Naissance de la prison*, Italian translation *Sorvegliare e punire. Nascita della prigione*, Turin: Einaudi, 1976; Foucault M. (1981), *Omnes et singulatim: Towards a Criticicism of 'Political Reason'*, The Tanner Lectures on Human Values edited by S. McMurrin, Salt Lake City, University of Utah Press, vol. II, 225–254; also by the same author see *Sécurité, territoire, population. Cours au Collège de France 1977–1978* (2004), trad. it. *Sicurezza, territorio, popolazione*, ed. M. Senellart, Feltrinelli, Milan 2005. See also Nancy J.-L. (2002), *Note sur le terme "biopolitique"*, in *La création du monde ou la mondialisation*, Paris, Galilée, 134–143.

[5] See Dominici P., *op.cit.*, 1995–1996 and following.

popular sovereignty—intended as hegemony or dominance of majorities—and on the failure to define the relationships between the founding values of freedom and equality,[6] has driven the public sphere, which has subsequently branched out into political institutions and new social environments in search of public recognition and operational conversion into legal norms, to increasingly assume the form of an autopoietic system. At this point, to quote Habermas (1981), the level of mediation between the system and the world of life, based on a type of communicative action capable of critically thematizing social instances and opinions generated within the world of life and civil society, has bestowed full legitimacy as well as public relevance on them.

The current social systems, so chaotic and disorderly, are going through a further (critical) phase of change ushered in by the advent of the interconnected economy, which also foregrounds new issues regarding citizenship and inclusion. Under examination within the "asymmetrical society"—marked by new forms of inequality and new *rules of engagement for the citizenry* (a global issue today)—are new (potential) opportunities for emancipation, available through *the spread of knowledge* and "knowledge sharing", which fuel the networks of social protection and promotion, intensifying the bonds of interdependence and interconnection, even as some observers continue to hypothesize the possible end of social bonds. An interesting process that is well worth underlining is the growth of social movements and pressure groups which, feeling unrepresented by an increasingly distant political system, assume the responsibility of rendering social demands that spring from below visible to both the Sovereign and public opinion. The old industrial model, comprising consolidated structures, hierarchies, and logics of control and closure against change, seems to find itself on the verge of being undermined by the dynamics sparked by the new knowledge ecosystem. Knowledge (too) is (finally) starting to be recognized as a common good and a resource capable of (re)establishing less asymmetrical social relationships and favouring greater inclusion. However, the new forms of social production of knowledge can be decisive only if the actors performing within the public arena know what to do with knowledge, social networks and the media and, more in general, with technology. This brings us to the urgent need for a total reform of (complex) thinking and knowledge.

We live, however, in an era marked, more and more, by the demolition of systems of belonging and belief—the true producers of individual and collective identities—and by the consequent affirmation of individualistic and utilitarian values. It is by no fluke that the *tyranny of the individual* which has been a matter of debate in many spheres, is deemed a veritable centrifugal force capable of corroding the bonds of social systems. This process of progressive weakening and disintegration finds further confirmation in the widespread deficit of social and political participation,

[6] See, in particular, Bobbio N. (1984), *Il futuro della democrazia*, Einaudi, Turin 1995; Bobbio N. (1995). *Eguaglianza e Libertà*, Einaudi, Turin

fuelled, in turn, by a climate of general mistrust towards all institutions (formal and informal), solely responsible in the past for the transmission and orientation of systems of value and cognition.

In the wake of the loss of credibility and authority of politics, this extremely complex scenario, so difficult to read, has left the field open to the far from novel hypothesis of *democracy beyond political parties*, as these are no longer able to intercept consensus and *mediate* new forms of conflict as they had in the past. In the background lurks a crisis of the party form itself[7] which adds to the much more profound one regarding *representation*.

Politics, being more and more exposed from this perspective, increasingly risks being reduced to "technocracy"—one of the many processes of reduction and simplification functional to the creation/building of the Mechanism-Society—should and must take action so that the social media and networks become *technologies of cooperation*, not of control/surveillance (as is happening, at least for the moment), open to the experimentation of new forms of democratic participation and the power of the mobile and intelligent *multitudes.*[8]

The dominant individualism of our social systems is the outcome, inevitable to some extent, of the process/project of emancipation effected during modernity, first, a process of emancipation of the masses, and later of the Subject.[9] On the one hand, this process has increased the spaces of freedom available and led to the recognition of some fundamental rights (at least in theory); on the other hand, it has contributed to weakening the ties and bonds of community membership. The triumph of the Subject, not only *free to* but also *free from,* has resulted, paradoxically, in an epoch that seems to be marked not only by greater opportunities for emancipation and extraordinary communicative potential—even though, in my opinion, a great deal of confusion is made between communication and connection—but also by a breakdown in the social fabric, comprising people who are increasingly left to deal alone with this complexity. The contemporary era is one in which the social mechanisms of trust and cooperation—the supporting structure, along with economic and power relations—have been put to the test by *processes of precarisation* which have made *instability an existential condition*. The underlying hypothesis is that beyond the

[7] See Revelli M. (2003), *La politica perduta*, Turin, Einaudi; Revelli M. (2013), *Finale di partito*, Turin, Einaudi; by the same author, enlarging on the topic: Revelli M., *Oltre il Novecento. La politica, le ideologie e le insidie del lavoro*, Einaudi, Turin 2001.

[8] See Rheingold H. (1993), *The Virtual Community*, Italian translation, *Comunità virtuali*, Sperling & Kupfer, Milan 1994; Rheingold H. (2002), *Smart Mobs*, Italian translation, *Smart mobs. Tecnologie senza fili, la rivoluzione sociale prossima ventura*, Raffaello Cortina Ed., Milan 2003.

[9] On the topic of the "Subject", see Crozier M., Friedberg E. (1977), *L'Acteur et le Système. Les contraintes del'action collective*, Italian translation, *Attore sociale e sistema. Sociologia dell'azione organizzata*, ETAS Libri, Milan 1978; Ardigò A. (1989), *La sociologia dell'attore sociale di fronte alla sfida dell'ipercomplessità societaria*, in, Studi di Sociologia—Anno 27, Fasc. 3 (July–September 1989), pp. 267–280; Cipolla C. (1992), *Oltre il soggetto per il soggetto. Due saggi sul metodo fenomenologico e sull'approccio biografico*, Milan, FrancoAngeli; Cesareo V., Vaccarini I., (2006), *La libertà responsabile. Soggettività e mutamento sociale*, Vita e Pensiero, Milan.

profound economic crisis (whose roots are not only economic, quite the contrary!), the phase we are experiencing is particularly critical and, in some ways, dramatic as well, above all because people clearly feel the fragility of their condition and the concrete risk of the end of social bonds. We are well aware of how much people's (individual and collective) perception matters, their feeling of being *alone* in the face of the insecurity and precariousness of life. I hasten to add that this drama of *loneliness*, this void of meaning, is a matter of the difficulty encountered when striving to *bestow meaning on what is real* which we also find in the obsessive need to communicate (or to put it more accurately, to be connected) and be visible at all costs, so as not to be forgotten (an issue of identity and recognition). The civilization of automation and hyperconnection appears significantly more efficient than previous ones, yet it is unable to generate the conditions required to become even more open and inclusive, independently. Within this highly complex, uncertain and multiform scenario, communication, the means of communication (the mass and so-called new media), the information system, the Internet and, in particular, Web 2.0 seem to have definitively occupied, not to say hegemonized, not only the *public space* of discussion and the formation of public opinion, but also the decisive area of social practice once controlled by traditional *agencies of socialization*. In this regard, it is of fundamental importance to keep the two levels of discourse and analysis clearly distinct: on the one hand, technologies and/or means of communication, on the other, communication itself, which is a *social process* characterized by the presence of (new) *ethical subjectivities* (1996) and actors/individuals who confront each other on the terrain of more or less symmetrical power relations. In this sense, an essential presupposition of this analysis is also the firm conviction that only the affirmation of a "culture of communication" (intended as the sharing of knowledge) in general, and in social systems in particular, within and outside of complex organizations (concept of organization as an open system) is able to effectively create the conditions for the realization and concrete application of those fundamental rights/duties of citizenship without which social actors (in their multiple roles as citizen-user -consumer) cannot, evidently, find any type of legitimation/recognition of their requests. This means that people find themselves, in actual fact, in a condition of subjugation, within an already shrinking and completely inconsistent public sphere. Knowledge and skills, in this sense, can determine, to an increasingly significant extent, the balance of power in every sphere of social, organizational and systemic life, with clear implications for citizenship and democratic regimes.

4.1 Radical Modernity and Its Evolution

Modernity profoundly disrupts processes, structures, hierarchies and experiences. It is the epoch of the eternal present, with its increasingly fragmentary sense of experience which undermines the certainties of the Subject, determining a process of disembedding that produces anomie. The bourgeois world, despite being aware of the crisis of modern humanity, has turned out to be totally incapable of dealing with a *new complexity* full of contradictions and incurable dialectics. The often mentioned "death of God" and the crisis of the great "metanarratives"[10]—an element, to be precise, generally associated with postmodernity but which, in our opinion, was already present within modernity—that is, the radical criticism of all dogmas and acquired certainties is what determined the crisis, on the one hand, but on the other, the radical acceleration of all the dynamics involved. The project of modernity—as we know—was developed during the Enlightenment. It was a project with an ambivalent character which, according to the analyses of Adorno and Horkheimer (1947), produced, not emancipation, but oppression and the domination of "man" over "man", though at the basis of this thinking lay the profound conviction that knowledge based on reason, once freed from the irrational constraints of myth and religion, would finally break the chains of the irrational that held the Subject hostage to his/her fears.

The *unitary, harmonious* view of reality—a reality based on order—entered into crisis simultaneously with the crisis of the idealistic and positivistic conceptions of history, time and reality. The Subject, having witnessed the devastation of all the constraints of tradition and the disintegration of the apparatuses of consensus, finds him/herself "alone", devoid of the coordinates capable of indicating a valid route, which, until then, had been defined and guaranteed by the conviction of the *perfect coincidence between the real and the rational.* Modernity,[11] even during its most

---

[10] See Lyotard J.F. (1979), *La condition postmoderne. Rapport sur le savoir*, Italian translation *La condizione postmoderna*, Milan, Feltrinelli, 1982.

[11] See, in particular, Aron R. (1966), *Trois essais sur l'âge industrielle*, Italian translation, *La società industriale*, Comunità, Milan 1983; Aron R. (1969), *Les désillusions du progrés*, Italian translation *Le delusioni del progresso. Saggio sulla dialettica della modernità*, Armando, Rome 1991; Berman M. (1982), *All that is Solid Melts into Air. The Experience of Modernity*, Italian translation, *L'esperienza della modernità*, Bologna, Il Mulino, 1985; Harvey D. (1990), *The Condition of Postmodernity*, Italian translation *La crisi della modernità*, Il Saggiatore, Milan 1993; Frisby D. (1985), *Fragments of Modernity. Theories of Modernity in the Work of Simmel, Kracauer and Benjamin*, Italian translation, *Frammenti di modernità. Simmel, Kracauer, Benjamin*, Bologna, Il Mulino, 1992; Beck U., Giddens A., Lash S. (1994), *Reflexive Modernization*, Italian translation, *Modernizzazione riflessiva. Politica, tradizione ed estetica nell'ordine sociale della modernità*, Asterios, Trieste 1999; Giddens A. (1990), *The Consequences of Modernity*, Italian translation *Le conseguenze della modernità*, Bologna, Il Mulino, 1994; Giddens A. (1971), *Capitalism and Modern Social Theory*, Italian translation, *Capitalismo e teoria sociale. Marx, Durkheim, Weber*, Il Saggiatore, Milan 1975; Giddens A. (1999), *Runaway World. How Globalization is Reshaping our Lives*, London, Profile Books, Italian translation *Il mondo che cambia. Come la globalizzazione ridisegna la nostra vita*, Il Mulino, Bologna 2000; Giddens A. (2000), *The Third Way and its Critics*, Italian translation, *Cogliere l'occasione. Le sfide di un mondo che cambia*, Carocci, Rome 2000; Giddens A. (2007), *Europe in the Global Age*, Italian translation, *L'Europa nell'età globale*, Laterza, Rome -Bari 2007; Giddens A., "Vivere in una società post-tradizionale", in Beck U., Giddens A., Lash S. (1994), pp. 101–159; Giddens A., *Modernity and Self-Identity: Self and Society in the Late Modern Age*, Polity Press, London 1991; Appadurai A. (1996), *Modernity at Large: Cultural Dimensions of Globalization, Italian translation, Modernità in polvere,* Meltemi, Rome 2001; Bauman Z. (2000a), *Liquid Modernity*, Italian translation *Modernità liquida*, Laterza, Rome -Bari 2002; Bauman Z. (2001), *The Individualized Society*, Italian translation, *La società individualizzata. Come cambia la nostra esperienza*, Bologna, Il Mulino, 2002.

extreme and radical phase (globalization), has been confirmed as the age of ambivalence and paradox, of open conflict, of dialectics without synthesis, of the crisis of ideologies and scientific paradigms, an age of collective disillusionment that has pushed, almost flung, the individual outside of the orbit of society. This is a secular age, marked by a continuous process of redefinition of social imaginaries and moral horizons, a sort of *Great Disembedding*[12] which has questioned the very concept of identity. It is a phase of change that seems to have taken shape more as an era of the triumph of the plurality of linguistic games—having social action as their fulcrum—than as one of the global and the homogeneous (Lyotard vs Habermas).

Modern and contemporary thinking seems, therefore, to stem precisely from awareness of this crisis, from the fact that absolute and indisputable knowledge, predominant cultures, absolute values and incontrovertible truths no longer exist, but rather probabilistically and statistically reliable knowledge, relative values and complex explanations instead. In other words, it is acknowledged that knowledge, in addition to being the result of a complex *process of intersubjective acquisition*, constitutes the far from obvious outcome of a pathway that develops, not so much through logical deduction or simple linear accumulation of information, as through processes of trial and error (random or systematic) capable of advancing thinking and research. We are speaking, fundamentally, of a crisis of Western rationality and the forms of life produced by it. This crisis coincides with an autopoietic moment of self-reproduction and self-renewal. Modern social systems, based on the concept of trust, the value of knowledge, scientific knowledge and risk control, find themselves—to use a metaphor cherished by Karl Marx—in conditions similar to those of a magician who is no longer able to control the powers of the *underworld he has himself evoked*.

The progressive and, at this point in time, almost capillary diffusion of new (hyperconnected) communication environments, with their high degree of technological innovation, is upsetting social and political architectures, favouring the affirmation of a new mode of economic production based entirely on ownership and the ability to process and spread knowledge. The so-called knowledge society/economy, which is progressively replacing material with immaterial resources, is defining new forms of social exchange and social asymmetries, which Politics, increasingly scaled down by economic and financial practice, seems unable to manage any longer.

The intrinsic dynamism that has consequently arisen is the profoundly unequal process of development (globalization) which has taken the shape of new forms of interdependence that have a global impact, made (self) evident by reflective knowledge. Contrary to a concept common to postmodernity,[13] globalization itself,

---

[12] See Taylor C. (2007), *A Secular Age*, Italian translation, *L'età secolare*, Milan, Feltrinelli, 2009; see also, by the same author: Taylor C., *The Malaise of Modernity*, Italian translation *Il disagio della modernità*, Laterza, Rome -Bari 1994.

[13] See, in particular Lash S. (1990), *The sociology of Postmodernism*, Italian translation, *Modernismo e postmodernismo. I mutamenti culturali delle società complesse*, Rome, Armando, 2000; Kumar K. (1995), *From Post-Industrial to Post-Modern Society. New Theories of the*

in our opinion, has never been shown to be a fracture from so-called *early modernity*. On the contrary, it has constantly encompassed, maintained—and allowed to coexist—all the contradictions typical of modernity, extending them on a global scale and radicalizing their effects.[14]

The global knowledge economy continues to maintain two forces within itself, that were already present in the modern world and which confront each other openly in a dialectical manner. On the one hand, we find economic and technological interdependence (and interconnection), on the other, social, political and cultural fragmentation. In any case, at the basis of these dynamics lies well-known awareness of the crisis of thought, which is no longer able to provide models of problems and *acceptable solutions* (Kuhn).

The process of reflexive modernization[15] continues to be marked profoundly by this *assumption of the inadequacy of the old paradigms* and cognitive schemes, as well as by awareness of the fact that strategies and possible ways of solving the vulnerability of systems need to be sought within modernity itself, in any case, and at all times. Within modern complex social systems, the dimensions of communication and the social production of knowledge have assumed extraordinary importance, although it remains the task of politics, despite the profound crisis it finds itself in, to identify and develop the strategies best able to ensure that all subjects be truly included, counteracting the present widespread perception of chaotic isolation, of the vulnerability and precariousness of existences, belongingness and social experiences. This is the case also because the *homo faber* has striven to exercise his will to power in an inestimable manner. However, not only has this failed to lead to new opportunities but it has also created new, dramatic forms of conflict and inequality, marked further by limited access or lack of shared resources, knowledge and information. It is in this direction that Politics can/needs to work to recover its space, currently invaded and hegemonised by the economy and by finance.

Communication, in the present as in the past, has always been the decisive variable for the complementary development of social systems in tandem with the development of productive forces. The improvement of flows of communication, from the summit to the base of human societies, has always represented progress, or at least, a moment of transition towards new forms of sociality and new ways by which to mediate interests and resolve conflict. The birth of democratic systems, diplomacy in international relations and bureaucracy in the rapport between the

Contemporary World, Italian translation, *Le nuove teorie del mondo contemporaneo. Dalla società post-industriale alla società post-moderna*, Turin, Einaudi, 2000; Inglehart R. (1996), *Modernization and postmodernization. Cultural, Economic and Political change in 43 societies*, Italian translation, *La società postmoderna. Mutamento, valori e ideologie in 43 paesi*, Rome, Editori Riuniti, 1998; Bauman Z. (2000), *Ponowoczesność. Jako źródlo cierpień*, Italian translation, *Il disagio della postmodernità*, Milan, Bruno Mondadori, 2002; Bauman Z. (2000a), *Liquid Modernity*, Italian translation *Modernità liquida*, Laterza, Rome-Bari 2002.

[14] Think also of the concept of *hypermodernity*, an alternative to that of postmodernity.

[15] See Beck U., Giddens A., Lash S. (1994), *Reflexive Modernization*, Italian translation, *Modernizzazione riflessiva. Politica, tradizione ed estetica nell'ordine sociale della modernità*, Asterios, Trieste 1999.

citizenry and the state are paradigmatic examples. In the current phase of change, also marked by (an evidently) profound (not only economic) crisis, communication and social knowledge could also make a concrete contribution to a process of rapprochement between the system of power and civil society, defining a new symmetry of social relations, with inevitable reconfiguration and repositioning of the public sphere. In practical terms, this would mean a strengthening of public (local and global) opinion that would become increasingly critical and informed and, therefore, involved more and more in the definition of political choices as well as being their active recipient. This could well be the true added value of *radical modernity*, after the grand illusion of postmodernism. From this point of view, the ecosystem of new knowledge finds in the interconnected economy potential opportunities for the democratization of knowledge and cultural processes capable of dismantling, definitively, the old industrial model comprising consolidated structures, hierarchies, the logic of control and closure to change.

The logic of the *self-regulated free market*—along with the hegemony of the economistic paradigm (Fig. 4.1)—is never seriously questioned (and—let us be clear –I am not referring here to the world of academia and research alone) and continues to play a decisive role, but the socio-cultural dimension still remains absolutely strategic when interpreting economic phenomena and processes. In this sense, we cannot fail to note how the hyper-complex global society has been shaped by the values of an individualism that is, at times, exasperated—even by postmodern rhetoric itself—and by the myth of productivity without workers, which, despite all, continues to fascinate and strongly appeal to political *decision-makers*. In our opinion, in terms of educational processes and certain pedagogical approaches as well—what has been created is almost *mythology:* the myth of the

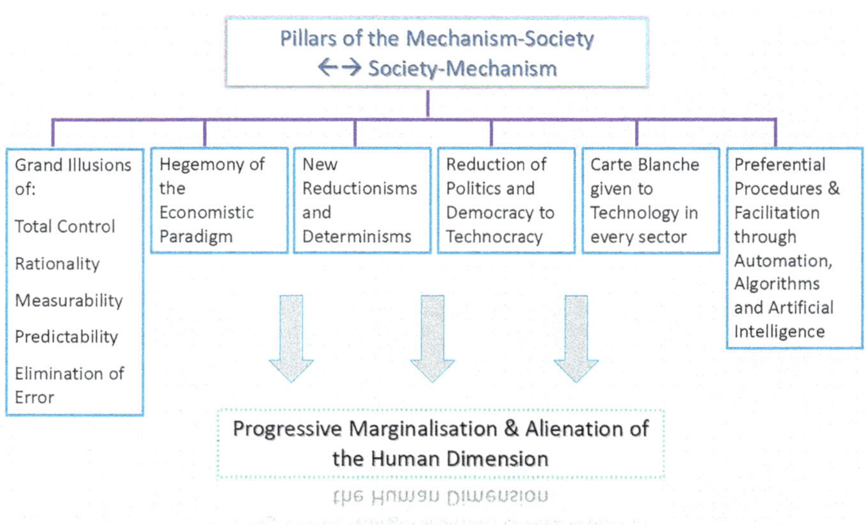

Fig. 4.1 Pillars of the Mechanism-Society ←→ Society-Mechanism

autonomous Individual, free from all bonds, an individual who apparently does not have to account for anything or to anyone for his or her actions. This is a far cry from the well-known distinction between the ethics of intention and the ethics of responsibility. We have gone well beyond all legal and/or cultural constraints. In any case, what always counts is usefulness, the goals to be achieved at all costs (even human) are money and consumption, to the point that the only (micro)power of citizens lies now and to all intents and purposes, in their being "consumers".

These dimensions, along with the "void of meaning" inherited from the crisis of ideologies, have produced, among several other highly significant consequences, a kind of general decline in/*disarming of* morals and—I repeat—an overall weakening of social bonds which, for some time now, has been fueling the *society of irresponsibility*[16]: a type of society that is incapable, even within the agencies of socialization and the educational and training institutions, of producing and/or co-constructing social and cultural antibodies, especially those antibodies that are decisive and particularly important during moments of epochal transition and/or global, systemic crisis (as the past *pandemic* had amply shown). The Weberian *ideal type* of Society has lost its profound meaning of being "around the hearth" (social bonds) and, beyond all regulatory and ethical frameworks and almost total delegation to the Law/laws when it comes to solving problems and resolving conflict, now sees that the dimensions of ethics and responsibility—*which cannot be imposed by law*—though constantly "talked about" and debated, are scarcely put into practice. Once again, we see the "Organism-Society" being managed (?) and governed (believing this can be done) as if it were a "Mechanism-Society".

The myth of the sovereign individual,[17] fully independent and responsible (?) when it comes to making choices, the bearer of rights but not of duties, has caused damage that is difficult to calculate/evaluate, especially with regards to respect/the value of the common good and "public affairs", but also to ways of perceiving and observing norms, values, rules, models of behaviour, etc. This is a myth or, to put it better, a narrative that has produced, among other effects, a delegitimization of regulatory systems and a de-responsibilization of social actors at all levels. From this perspective, too, it would be of fundamental importance to overcome this phase of *playing things by ear,* where the links between the individual and the institutions, between the individual and the traditional agencies of socialization (family, school, religion etc.), between politics and citizens, have weakened considerably. The distance thus created has certainly favoured the increasingly massive and decisive involvement of the media—specifically the Internet and social media—in the formation of individual and collective identities and, even in the recognition and establishment of working definitions for addressing social claims and petitions to Politics. The proliferation of educational and training centres and, more generally, of *arenas* for the purpose of actualizing thought and designing praxes, proceeds hand in hand with the crisis of communication that has affected the institutions and traditional

[16] See Dominici P. (2010), *La società dell'irresponsabilità*, FrancoAngeli, Milan 2010.

[17] See Davidson J.D., W. Rees-Mogg, *The Sovereign Individual*, Simon and Shuster, New York 1997.

actors of the educational process, suspended between excessive information and fear of disconnection.

The exponential growth of financial power has produced extremely negative consequences for the world economy, above all, for people's lives, by producing a virtual space (so that, today, we speak again of a "metaverse"[18]), where economic and information flows proceed at enormous speed, depriving Politics and power systems of control over their *bodies,* separating them further from civil society and individual social actors.

The transformation of the mode of economic production and the labour market, the radicalization of the *social division of labour*, the birth of new disparities, also in terms of *starting opportunities*, new forms of exploitation and, therefore, new conflicts; the weakening of the functions of traditional forms of political participation, illusorily replaced (at least for the moment) by the narratives of/about disintermediation and the *utopias of direct online democracy,*[19] have done the rest, making the existence of social actors (individuals) profoundly uncertain because they are required to be flexible in every aspect of their lives without, however, providing any guarantees in exchange for their availability.[20] The *society of individuals* emancipated from the constraints of tradition and, in a certain sense, at the mercy of the expansion of the potential of a purpose-oriented type of rationality, is obliged to face the exponential growth of productive forces which makes the process of modernization reflexive, that is, a topic and a problem in itself.

The advantage is linked, without any doubt, to the fact that these risks can no longer—as in the past—be ignored by the public sphere and public opinions. It is precisely this perspective within which, the ever-valid analysis carried out by John Tomlinson (1999) regarding globalization is framed. It should be understood, first of all, as a cultural phenomenon comprising a network of experiences which, by means of mechanisms of the disaggregation of space-time, profoundly modifies the perception of the physical places in which we interact with the Other, extending the effects on a global scale of choices adopted locally. Culture is configured increasingly as a transnational, above all, potentially inexhaustible and uncontrollable resource.

---

[18] See, in particular, the novel by Neal Stephenson (1992), *Snow Crash*, Penguin Book, London 1993, where the concept of "metaverse" was used to indicate a virtual three-dimensional world (Internet), inhabited by *digital humanoids* along with the possibility of connecting minds.

[19] See Dominici P. (1995–96) *op.cit.,* and following.

[20] See Sennett R. (1998), *The Corrosion of Character: The Personal Consequences of Work in the New Capitalism*, Italian translation, *L'uomo flessibile: le conseguenze del nuovo capitalismo sulla vita personale*, Feltrinelli, Milan 2000; Gallino L. (2000), *Globalizzazione e disuguaglianze*, Laterza, Rome-Bari; Gallino L. (2001), *Il costo umano della flessibilità*, Laterza, Rome-Bari.

4.2 The New Global Communications Ecosystem

The society of knowledge triggers a revolutionary change in all socio-political dynamics and processes of production and culture, which makes intellectual capital and the production and distribution of knowledge its strong points. This is yet another extraordinary technological revolution—the origins of which are to be found in the last decades of the nineteenth century—generated by the rapid progress of industrialization which, in turn, had already led to a crisis of control. The global communications ecosystem and the very means/environments that constitute and innervate it have become the absolute protagonists of the processes of production and diffusion of knowledge. Moreover, over time, they have fertilised the terrain in which to plant the new real-time, interconnected world system.

In addition to the oft-mentioned marginality of politics, the liberalization of the markets highlights even more the absence of truly functioning, operational global institutions. The global economy, therefore, is facing a process of radical restructuration which implies the downsizing of physical capital and the triumph of the supply of services over the sale of goods and exchanges of property, making access the new measure of social standings. Furthermore, following the recent worldwide economic crisis, global and techno-nihilistic capitalism,[21] characterized by the progressive acquisition of the social experiences of each individual citizen/consumer, also seems to be on the verge of legitimizing new models of social exchange. This means that individuals and institutions are involved in the process of commercializing all praxes, something which outlines a somewhat disturbing scenario, where the traditional structures of social bonds are called into question.

The global information and communications system is the decisive variable which, in addition to having deeply affected the structure of social systems,[22] has made the increasingly massive involvement of entire areas of civil society possible. Beyond the various critical issues, the media have increased the level of knowledge and awareness of social problems available even among the weakest social classes,

---

[21] See Magatti M., (2009), *op.cit.*

[22] On the concept of "civil society" see the work of the Scottish moralist A. Ferguson (1767), *An Essay on the History of Civil Society*, Italian translation, *Saggio sulla storia della società civile*, Vallecchi, Florence 1973. Ferguson analyzed the process of the evolution of human societies and identified the factors that determine the advent of civil society in the birth of private property and private interest. Social systems had, he argued, evolved also thanks to the development of the division of labour, which had also caused alienation and a weakening of the sense of community. Well in advance of other thinkers, Ferguson believed that human societies would be able to avoid probable decline only by returning to the community model, based on the active and conscious participation of the individuals belonging to a commonly-shared political kind of life. Man, according to Ferguson, was the sole creator of the reality in which he lived and all social situations arose from his activity. Society itself was the work of man. Ferguson's ideas, as one can imagine, also greatly influenced Karl Marx.

thus producing a common-sense discourse which, availing itself of communication networks, is progressively transnationalising, in some cases, producing autonomous mobilizations and political action detached from politics.

The age of *globalized radical modernity*, of reflective modernization and of systemic and total communication, presents itself, therefore, as a time of indeterminacy and the abandonment of hierarchy and order; an Age of the Present: a continuous, almost dilated, ever-present present, which places new political, social and economic questions on the agenda at a much higher magnitude than in previous eras, thus effectively creating a primary need for relocating or even reinventing Politics. This is a sphere of discourse that needs to regain its decision-making spaces, having been reduced too far, at times humiliated, in its role and functions, by the domination of the economy, by *technocracy,*[23] by the Weberian petrification of mechanization and, more in general, by the progressive confinement of life and social experiences to institutions and procedures.

The complexity inherent in the process of globalization obliges us to reformulate all categories of political action, broadening the horizons of our thinking and actions, to develop a policy that is not limited to the observation of rules only, but rather to attempts to change them, as well, so that the vast majority of these same rules may be defined within the context of a strong nation-state. This too, because the world market cannot, as has happened to date, be left to drift without an authoritative and credible project of global development, abandoned, as Weber has said, to its *self-normativity,* in which the only dignity acknowledged is that of objects rather than people and the only imperatives recognized are economic, rather than the moral imperatives of brotherhood and solidarity, thus emptying communities of the original human relationships of which they are composed..[24]

An overall rethinking of theory and practice is required, one framed in the perspective of a radical modernity where the dimension of reflexivity[25]—intended as self-analysis, as the acknowledgment of both an increased complexity and the existence of other cultures, as well as the possibility that certain elements belong to certain cultures[26]—places the institutions of control and protection into a state of further crisis. While the inadequacy of old conceptual schemes and the increased complexity of social systems, make us aware of how modern society has managed

[23] See Easterly W. (2013), *The Tyranny of Experts. Economists, Dictators, and The Forgetten Rights of the Poor*, Italian translation, *La tirannia degli esperti. Economisti, dittatori e diritti negate dei poveri*, Laterza, Rome-Bari 2015.

[24] See Weber M. (1922b), *Wirthschaft und Gesellschaft. Grundriß der verstehenden Soziologie*, Italian translation, *Economia e società*, Vol. II, Comunità, Milan 1961, p. 314. Also from this great German author, see: Weber M. (1922a), *Gesammelte Aufsätze zur Wissenschaftslehre*, Italian translation, *Il metodo delle scienze storico-sociali*, Einaudi, Turin 1958.

[25] See, in particular Donati P. (2011), *Sociologia della riflessività. Come si entra nel dopo-moderno*, Bologna, Il Mulino; Bovone L., *Tra riflessività e ascolto. L'attualità della sociologia*, Armando Ed., Rome 2010.

[26] See Luhmann N., op.cit., 1992.

to equip itself with *means of self-description*, superior from a qualitative point of view, compared to those of the past.

However, despite their present-day strategic function, communication and *the social production of knowledge* can only accompany—making it visible and the subject of public discussion—this eruption of a world of artificiality and technology no longer subordinated to nature and which appears, to all effects and purposes, as "nature" difficult to manage. In other words, in this society of hypercomplexity (2003), socially and culturally marked by *fragile social bonds* and the hegemony of individualistic values, the dimension of what is technically/technologically controlled has become hypertrophic compared to that of the non-technically-controlled; but—and this is the hub of the matter –these processes and dynamics have not contributed to the reactivation of the fundamental social mechanisms of trust and cooperation.

Furthermore, in this age of total communication, (individual and collective) theory and practice are no longer affected only by the historical-social conditioning specific to each individual context or reference group but are also strongly conditioned by the weight of the representation (images of reality) provided by the media (active metaphors) which, in a certain sense, force social actors to incorporate the whole of humanity within themselves, creating moreover, instantly and constantly, a total field of interdependent events,[27] which has an influence on their perception and action.

Radical modernity is an era folded and compressed upon itself, like the knowledge and social knowledge that tries to interpret, metabolize, and elaborate on the new modalities of conflict[28]and the thousands of contradictions of the world system, which are revealed when compressed into this continuous present. In a certain sense, the society of individuals,[29] the result of the radicalization of the process of modernization, turns out to be a "asymptomatic carrier" of extremely contradictory phenomena which may, at any moment, lead to its implosion. Which is why it has to take itself into account. As to Politics, I repeat, having become more and more the "handmaid" of economic power, characterized increasingly by evident drifts towards technocracy and the power of so-called *experts*, it is called to face this umpteenth challenge which, contrary to what is actually the case, presumes the reconquest of the territory and praxis.

The categories used to interpret industrial societies have, for some time, been encountering certain difficulties in the attempt to understand radical modernity and the process of globalization, because we have moved from a traditional linear

[27] Dayan D., Katz E. (1992), *Media Events. The live broadcasting of history*, Italian translation, *Le grandi cerimonie dei media. La Storia in diretta*, Bologna, Baskerville, 1993.

[28] See Honneth A. (1992), *Kampf um Anerkennung. Grammatik sozialer Konflikte*, Italian translation, *Lotta per il riconoscimento. Proposta per un'etica del conflitto*, Milan, Il Saggiatore, 2002.

[29] See, in particular, Elias N. (1987), *Die Gesellschaft der Individuen*, Frankfurt, Suhrkamp, Italian translation, *La società degli individui*, Il Mulino, Bologna 1990; Bauman Z. (2001), *The Individualized Society*, Italian translation, *La società individualizzata. Come cambia la nostra esperienza*, Bologna, Il Mulino, 2002.

modernization to one that is reflexive, industrial and post-industrial social, characterized, precisely, by the escalation of the political dimension.

In the era of the global market and the knowledge society, the social production of capital goes hand in hand with the *social production of risks*. As a matter of fact, the power relations between these two kinds of logic have been reversed, precisely because of *reflexivity*, which denies the productive forces the opportunity to hide their more latent side effects. The complex and multi-branched mechanisms linked to the social production of knowledge, supported by an ecosystem containing other ecosystems, by platforms and social media, reveal this dimension by relocating it within the public sphere which has, despite a thousand difficulties, the opportunity to thematise issues and requests, including those originating from below, initially excluded from the public debate.

The *society of individuals*, among its many characteristics, presents the homo faber (*sapiens* and *demens* at the same time) who, with his typical attitudes, constitutes one of the icons of modernity and has managed to exploit the whole of reality. Driven by trust in the all-encompassing scope of the category of means and ends, he has convinced himself that he can find a solution to any problem, uncritically equating production/manufacturing with action. Today, the social actors of this *eternal present* interpret any kind of behaviour according to the principle of utility and are almost obsessed with the need to produce and, above all, consume material objects, often confusing intelligence and ingenuity.[30] So it happens that risk,[31] emergency, complexity, which are also read/interpreted as *deviations* from the norm, have now become dimensions intrinsic to modern nation-states and politics, as well as to social systems. Thus, Politics is obliged, more and more, to make decisions (centrality of the public sphere) that have repercussions, not only at a local level, because "[r]isk communication has become reflexive and thus universal. Refusing to assume risks or demanding their rejection has become dangerous behaviour".[32]

---

[30] See Arendt H. (1958), *The Human Condition*, Italian translation, *Vita activa. La condizione umana*, Bompiani, Milan 1964, p. 227 and following..

[31] Luhmann N., (1991) *Soziologie des Risikos*, Italian translation, *Sociologia del rischio*, Bruno Mondadori, Milan 2009; Lupton D. (1999), *Risk*, Italian translation, *Il rischio. Percezione, simboli, culture*, Bologna, Il Mulino, 2003; Lupton D. (ed.), *Risk and Sociocultural Theory: New Directions and Perspectives*. Cambridge: Cambridge University Press, 1999b; Lupton D. (2001), *Surveillance Society. Monitoring Everyday Life*, Italian translation, *La società sorvegliata. Tecnologie di controllo della vita quotidiana*, Milan, Feltrinelli, 2002; Lupton D., Tulloch J. (2003), *Risk and Everyday Life*, Sage, London.

[32] See Luhmann N., (1991), *op.cit.*, p. 5.

4.3 Amid Age-Old Logics of Control and Opportunities for Inclusion

The interconnected/hyperconnected society and connective technologies, in addition to progressively doing away with (dis-intermediating) every kind of mechanism of political and/or social mediation, have the power, perhaps even unlimited, to extend the possibilities and opportunities for human communication, facilitating the production, processing and exchange of information and knowledge between individuals.

In the hypercomplex society, what the individual as a *homo sapiens*, not merely a *homo faber*, appears to desire is to once again place his trust in technical powers, those same powers, what is more, that have seduced and abandoned him several times in the past.

This renewed trust in *rationality* finds a fundamental foothold in the Internet, in AI systems and in new interactive technologies, capable of making an enormous mass of accumulated knowledge[33] available to all online. Yet, all these extraordinary opportunities continue to reveal themselves as such only for small groups and *élites* who, besides the question of whether they may be more or less enlightened, often do not pursue the common good and/or the general interest.

Furthermore, our analysis requires us to consider another fundamental aspect: the *vulnerability* to which the Subject is exposed ("new subjectivities" are exposed) during these processes. This kind of vulnerability goes hand-in-hand with the well-known fragility, as well as the ontological incompleteness, of the Subject. This aspect cannot but be evaluated, above all because it increases, once again, the *complexity* of our analysis, not only in terms of "variables" to be considered. In actual fact, the extraordinary—but also widespread and invasive—enhancement of communication methods and the radical differentiation of the educational channels offered have led, at least in theory, to a *growing capacity for self-determination* in terms of the Subject's choices, values, models of behaviour and cognitive schemes. What has been shaken up, on the whole, is the entire shared symbolic system along with the (medium) language, which attempts to describe and represent this umpteenth change.

The civilization of risk and continuous emergency, on the other hand, despite presenting itself as an extraordinary opportunity for economic, political and social development, has significantly augmented, in actual fact, the sense of insecurity and vulnerability[34] that lies within social systems, feeding a climate of fear (and/or alarm) at both a local and global level, which radically casts doubt upon the

[33] See Hess C., Ostrom E. (2007), *Understanding Knowledge As a Commons*, Italian translation, *La conoscenza come bene comune. Dalla teoria alla pratica*, Bruno Mondadori, Milan 2009.

[34] See Ceri P., *La società vulnerabile. Quale sicurezza, quale libertà*, Laterza, Rome-Bari 2003.

principle of precaution[35] itself, increasingly disconnected from the social demand for protection. The world system and the new informational, global and interconnected economy require a new sensitivity towards issues concerning the Subject, social relationships and, above all, the spaces of knowledge, with a view to reinforcing the transnational political public sphere.

The knowledge society and the technologies of (hyper)connection have seemed for some time now to be on the verge of creating *unlimited communities of communication* open to all communicating subjects, where there is no room for the usual forms of conditioning, where everyone potentially has the possibility of producing and exchanging knowledge within the framework of a system in which communication has become totally *horizontal,* without any type of filter. Expectations, not only unfulfilled, but which have proven, in many ways, to be utopian.

From a (hyper)complex reality like the current one, which offers the possibility of choosing between a multitude of possible *systems of cognitive* and ethical *orientations*, a *new, more autonomous* (but my no means more responsible) *Subject,* disconnected from others, seems capable of emerging, free from the constraints of his/her context and/or reference groups. A Subject who has become aware of the disintegration of social bonds and religious morality as well as of the *dissolution* of traditional ethical bonds. And it could not be otherwise, since historical phases of social change have always been accompanied by great uncertainty and a general sense of disorientation caused by the absence of a strong cultural model.

The problem continues to consist also in the attempt to understand whether "behind" the hyper-connected global society and the interconnected economy, which, at least initially, still seemed capable of guaranteeing greater opportunities for equal starting conditions for all social actors, in reality there stands the extremely concrete risk of a further weakening of the connective tissue of social systems and of a *generalized passivity* on the part of individuals (people/social actors), convinced that the virtual is real, beyond any operational translation capable of producing change and political decisions.

However, there also exists the danger of cultural homologation, a truly fertile terrain for *a civilization of total social control*, capable of reducing the citizen/consumer's margins of freedom. Therefore, although there is no doubt that the Network Society represents, concretely speaking, an extraordinary possibility for the emancipation and liberation of the forces and energies of the social and global fabric, the nation-states must take care to ensure that the Great Network, in addition to increasing communication and cognitive possibilities, also contribute to creating a type of humanity that some consider *the end of the social.*[36]

---

[35] See Sunstein C. (2001), *Republic.com, Italian translation, Republic.com. Cittadini informati o consumatori di informazioni?*, Bologna, Il Mulino, 2003; Sunstein C. (2005), *Laws of Fear. Beyond the Precautionary Principle*, Italian translation, *Il diritto della paura. Oltre il principio di precauzione*, Bologna, Il Mulino, 2010.

[36] See Touraine A. (2004), *Un nouveau paradigme. Pour comprendre le monde aujourd'hui*, Italian translation, *La globalizzazione e la fine del sociale. Per comprendere il mondo contemporaneo*, Il Saggiatore, Milan 2008.

Communication, as we know, since the dawn of pre-complex societies, has incessantly fuelled the system of social relations, representing the connective tissue of social systems. But this assumption does not prevent us from noticing how the Internet and social media are determining a quantum leap with reference precisely to social action and communicative practice, unprecedented when compared to past eras. Likewise, changes in the social mechanisms related to trust and cooperation[37] are taking place much more rapidly, increased, in turn, by the networks of social protection and promotion—the concept of *social capital*[38]; the bonds of interdependence and interconnection that innervate the world system are intensified, even though some scholars hypothesize the end of the social bond,[39] alerting us about new forms of socialization and the virtuality of reality and social life, the many implications of which have still to be explored and assessed. But the fundamental issue lies in the fact that the knowledge society appears in every aspect to be a global-risk society which has extended, beyond all borders and limits, the dynamics of conflict, risks, emergencies (real and potential) and systemic anomalies. This dimension intercepts the equally crucial dimension of trust which continues to be a fundamental social mechanism capable of reducing complexity[40] and making the acceptance of risk sustainable. Trust maintains a close link with the problem of knowledge and information. Modern social systems, frequently the orphans of a strong cultural model, are now characterized by instability and a high coefficient of unpredictability of actions and processes, a fact that makes the knowledge-sharing option even more strategic and urgent.

The *virgin soil of current communication practice* seems to be marked also, and above all, by extraordinary opportunities for the democratization of knowledge and cultural processes, capable of dismantling definitively not only the old industrial model but—as mentioned above- also consolidated hierarchies, logics of control and resistance to change. What is at stake, therefore, is a redefinition of the structure of the social relations of power, with all that this implies. Certainly, beyond our most personal opinions/impressions, we cannot help but note, as described by Hess and Ostrom, that the Internet cannot simply be considered the successor of historical devices designed to convey information at a distance (from the first telegraph systems to television and computing). Its interconnected, interactive, interceptive and interdependent features in today's digital world consecrate it as a universal knowledge/communications system which acts as an extensor of social relationships, endowed with accelerating, calculating, processing and recognitive powers unprecedented in human history. It is at once a radial and convergent figure that

[37] See, in particular: Coleman, J.S. (1990), *Fondamenti di teoria sociale*, Il Mulino, Bologna 2005.

[38] Putnam R.D. (2000), *Capitale sociale e individualismo. Crisi e rinascita della cultura civica in America,* Il Mulino, Bologna 2004.

[39] Breton P. (2001), *Il culto di Internet. L'interconnessione globale e la fine del legame sociale*, Testo&Immagine Ed., Turin 2001.

[40] See Luhmann N. (1968), *Vertrauen. Ein Mechanismus der Reduktion sozialer Komplexität,* Italian translation, *La fiducia*, Il Mulino, Bologna 2002.

transfigures and transgresses all concepts of boundaries, an unlimited territory of collective cognition.[41]

The hyper-complex architectures that support the hyper-technological and hyper-connected civilisation, further confirm, strengthen—by sweeping away all perplexities—its character as a common good associated with the resource called knowledge, the only truly strategic resource, derived from intersubjective processes of acquisition, capable also of fuelling, bottom-up, ganglia and tissues that innervate the interconnected economy, thus creating the basic conditions for the creation and evolution of that cognitive ecosystem based upon more open, dynamic and democratic processes than in the past. What comes to mind is the idea, or, rather, the ambitious project (no longer utopian, today) of creating an open-source global network, aimed not only at the production/processing of knowledge, but also at the control of information. An ecosystem of knowledge that, structured as a web, is destined to significantly facilitate the processes of reducing complexity, during a phase of transition to hypermodernity characterized, more and more, by an increase in indeterminacy and disorder within systems, whose resilience is constantly being put to the test. On the other hand, indeterminacy and disorder can trigger the jamming (or, on the contrary, the entropy) of all the organizational mechanisms and related subsystems charged with the task of reacting/responding to the unforeseeable. Emergence from this state of affairs is always linked to concrete possibilities of accessing and processing information/knowledge (priority needs). In the so-called *Knowledge Society*, the hierarchic criteria of access to information are tested daily as well as being culturally weakened by the new spirit of the information age[42] and by the multiple forms of mediated interaction that seem to have eliminated, definitively, the boundaries between the public and private spheres. However, the decisive thrust against the ivory tower of knowledge (power) in the traditional sense, – already besieged for quite some time—came from Web 2.0 (and subsequent versions), the engine of collective intelligence, the amplifier of social cooperation and authentic global relational network capable, thanks to its participatory architecture, of rendering network users truly proactive subjects. Always provided that the critical issues linked to the digital and cultural divides are resolved, at least in part, along with the relative crucial issues of skills and social asymmetries.

Within this highly complex scenario—marked by a change that is not only structural—social networks and connection technologies, integrated and supported by AI systems, are configured as processes to be developed and not as simple tools or hyperconnected environments to be surfed. Benkler argues that two hundred years of reciprocal evolution in markets and liberal democracies have been radically transformed by digital technologies and AI, echoed by mutations in economic and social organizational and productive praxes, which have been affected by the new

[41] See Hess C., Ostrom E. (2007) *Understanding Knowledge As a Commons*, Italian translation, *La conoscenza come bene comune. Dalla teoria alla pratica*, Bruno Mondadori, Milan 2009, XIII.

[42] Himanen P. (2001), *The Hacker Ethic and the Spirit of the Information Age*, Italian translation, *L'etica hacker e lo spirito dell'età dell'informazione*, Feltrinelli, Milan 2001 (with an epilogue by Manuel Castells).

opportunities in the way information, knowledge and culture are being invented, designed and exchanged, opening pathways for enterprises devoid of owners and for-profit production, and increasing the possibilities of varying degrees of collaboration between individuals and groups.[43] Even within a context of *weakened social ties*, the society of networks and automation seems capable of defining new logics and production practices within a new global information environment/ecosystem, which presents considerably greater margins of freedom of action—at the moment and who knows for how long?—but not for all social actors (on the contrary!). In this regard, we ought not to forget that the social asymmetries of an economic, cognitive and cultural nature are still markedly evident, as has been documented, moreover, by numerous studies and research reports.

Despite these dynamics and the further critical issues linked to the advent of the so-called *platform society*,[44] in the long term—providing we focus concretely on education, training and research—the new *public sphere* might also regain a certain degree of autonomy from politics. For the moment, the *Great Network of networks*, for manifold reasons, among which, incorrect strategic choices and particularistic views, beyond the great narratives relating to the processes of *disintermediation* and (simulated) horizontalisation, has failed to be/to take shape, thoroughly, as a "cooperative technology" and, far less, as the infrastructure/platform functional to a more participatory and inclusive democracy, even if the assertion of a non-commercial type of production of information and culture is, in any case, destined to generate situations of conflict, especially with companies structured according to the industrial information-economy model.

Our new communicative *habitat*/ecosystem is based, as we know, on interconnected microprocessors –which nowadays are affordable and endowed with high computing power, rendering them capable, at least potentially, of triggering processes of cognitive democratization and emancipation, capable of driving—almost of forcing—complex organizations and systems towards greater openness and transparency. Dynamics and *drifts* that might also have further implications. Against the background of these dynamics, which see the emergence—at least at an *elitist* level—of a more autonomous (though not because of this, more responsible) social actor, free to take part in the process of the creation of knowledge and the production of information (strategic, once again, is the role of education and training systems), as well as the advent of a new type of (networked) public sphere. This hypothesis, widespread in the literature of the field, can be reformulated in brief as follows: beyond the new asymmetries and inequalities, access to information and knowledge—together with the possibility of their autonomous creation—will also continue to manifest itself, in the near future, as a powerful factor of emancipation, justice and human development. In a certain sense, the project of modernity itself,

---

[43] See Benkler Y. (2006), *The Wealth of Networks. How Social Production Transforms Markets and Freedom*, Italian translation, *La ricchezza della Rete. La produzione sociale trasforma il mercato e aumenta le libertà*, Università Bocconi Ed., Milan 2007, pp. 1–2.

[44] See, in particular Van Dijck, Poell T., De Waal M., *Platform Society. Valori pubblici e società connessa*, Guerini Scientifica, Milan 2019.

updated and rethought, might return in the background.[45] Once again, it is Benkler who provides us with confirmation of great importance to our analysis, when he states that individual actors can now, through the Internet, find their places in a new public sphere, in which a relatively tiny group known as the "media" are no longer privileged with the exclusive power to issue top-down declarations to unempowered recipients. Today, the individuals previously limited to receiving these communications are capable of open participation in public discussions, where formerly unidirectional assertions are now seen as invitations to participate in both creating and engaging in public debate. Private opinions can now be formed, stimulated and reinforced by participation in a public dialogue that continues to become more and more extensive and encompassing.[46]

The current (global) ecosystem of de-structured knowledge dismantles, from its foundations, the traditional industrial model that has accompanied the evolution of the mass-media system, guaranteeing, to some extent, its control by power groups and, more in general, by the managerial classes. This means that the *media platform* that shaped and bestowed meaning on the old, though still extant, public sphere intended, as Habermas explains, as that structure of intermediation between the political system, on the one hand, and the private spheres of the world-life and functionally specialized systems, on the other.

The old centralized, hierarchical structures, typical of industrial society, are on the verge of being redefined by the new *distributive architecture of the Web* which will mark a transition from linearity to circularity, from unidirectional to bidirectional, and later, multidirectional connections, creating new space for opinions and actions alternative to those more visible and hegemonic within the media arena, short-circuiting the old media system. Consequently, networked *public spheres*, also envisage social actors who are not involved in the market as protagonists capable of forming a *critical mass* with regard to poorly recognizable social issues, bringing pressure to bear on the system of power and politics. The capitalist system which characterizes the hyper-connected society, reveals itself, therefore, as a *Knowledge-based economy* destined to determine an irreversible transformation of knowledge into *social knowledge*, capable of providing constantly reusable *knowledge* that goes beyond the constraints of exclusive proprietary knowledge, introducing numerous discontinuities and asymmetries into social complexity. The absolute protagonist becomes, after a long and far from concluded non-linear evolutionary process, *horizontal production* and a model that tends increasingly towards

---

[45] See Ferrara A. (1998), *Reflective Authenticity, Rethinking the Project of Modernity*, Italian translation, *Autenticità riflessiva. Il progetto della modernità dopo la svolta linguistica*, Feltrinelli, Milan 1999; by the same author, see also: Ferrara A. (ed.) *Comunitarismo e liberalismo*, Editori Riuniti, Rome 2000; Ferrara A., *Etica e teoria sociale: tre punti di intersezione*, in F. Crespi (Ed.) *Etica e scienze sociali*, Rosenberg & Sellier, Turin 1991; Ferrara A., *Pensare la società. L'idea di una filosofia sociale*, Laterza, Rome-Bari 2001.

[46] See Benkler Y. (2006), *The Wealth of Networks. How Social Production Transforms Markets and Freedom*, Italian translation, *La ricchezza della Rete. La produzione sociale trasforma il mercato e aumenta le libertà*, Università Bocconi Ed., Milan 2007, pp. 229.

cooperation which, however, can never be fully realized under the banner of "technological solutionism" and the "carte blanche" delegated to technology alone.

The new public sphere, *glocal* and hyperconnected, seemed capable, at least initially, of greater potential for the inclusion of new subjects and social actors freed from traditional hierarchies and/or the logics of power, broadening the base of those who were/are inside the *agora* and favouring the multiplication of every kind of opinion—unfortunately, of *disinformation* as well—all opinions, not only the most common or conformist ones. Yet it is necessary to take into consideration that it is only an infrastructure, upon which flesh-and-blood social actors (individual and collective) can define, plan and develop social change. The hope is that a change of this type may be based on *collective knowledge* and on the—unrealized—transition from a (still existing) mass culture that is, simultaneously, "closed" and exclusive when it comes to a (concretely) participatory form of *culture*. It appears evident, therefore, that communication—increasingly strategic during this phase of *systemic and existential pandemics*—and the social production of knowledge have their roots, precisely, in the project of Modernity. It is the task of Politics, even in conditions of marginality, to reformulate them, redefining their priorities in the interest of the common good. Besides radically questioning the developmental model centred entirely on consumption—and the economistic paradigm that justifies and supports it, starting from the incorrect assumption that Economics (as a discipline) is an "exact" and not a "social science" and, at the same time, that Society is (only) a subsystem of the Economy—it is of crucial importance to work to re-build the social and cultural conditions capable of reforging social bonds, in the direction of a *new Humanism*—a condition necessary for truly democratic regimes. In addition to taking care to avoid re-proposing the idea/view of the centrality of the human being which characterized the original humanistic movement, which has had so many negative consequences for ecosystems, this New Humanism should be capable, above all, of including the weak: that majority of the world population penalized by the mechanisms of the world economy and its *hypercompetitive growth* devoid of rules. In this sense, from a certain point of view, we also need to take note of some positive signs of the awakening of a global consciousness attributable to the myriad problems of hypermodernity (identity, gender, precariousness and work, rights of citizenship, human rights, the environment etc.), also demonstrated by an extremely interesting phenomenon that has been under way for some years now: new social movements (local and global) that find/have found/seem to have found in the Internet and in the opportunities of new global ecosystem, the infrastructure and the ideal driving force to coordinate and fuel their projects. The worldwide communications ecosystem, despite presenting numerous criticisable and, in some cases, disturbing aspects,[47] seems capable of providing an extraordinary opportunity for the development of views of the world (scientific and otherwise) alternative to those of a hegemonic and/or hetero-produced nature, even if the Internet and the

[47] See Morozov E. (2011), *The Net Delusion. The Dark Side of Internet Freedom*, Italian translation, *L'ingenuità della rete. Il lato oscuro della libertà di internet*, Codice ed., Turin.

social networks often reproduce dynamics and effects similar to those of the mass society, not only in terms of the *spiral of silence* but, above all, of *agenda-setting*.

The expansion of the praxes we are witnessing, and the progressive dissolution of public space, oblige the scientific community, therefore, to rethink, if not actually reformulate, those very conceptual categories that, for a long time, permitted the de-codification of socio-cultural changes including that of the *public sphere.*[48]

Contrary to what has occurred in recent decades, the choice of focusing on strategies aimed at *sharing knowledge* can only play a decisive role in the reduction of social inequality and asymmetry, in *the governance of social selfishness* and, consequently, in the productive mediation/re-settlement of conflict. This now seems to be a given, even if it is far from materialising. More generally, we can argue that only knowledge and culture have proven to be truly capable of nourishing and fueling systemic change, well beyond the innovative, technological and economic processes (which, moreover, stem from these very essences).

Our arguments are strengthened by a piece of evidence quite difficult to refute: the global market—and this, in a certain sense, is a dramatic paradox—needs to be unfair and fuel inequality in order to grow, even more so, in an era of profound welfare crisis like this, radicalized by the present-day systemic and global emergency. This is an era that has transformed uncertainty and precariousness into "existential conditions", where communications and the *social production of knowledge* can constitute the true antidotes to unequal global development and to technological progress which, despite being fundamental for the development and progress of social systems, always determines the assertion of new inequalities, as shown by recent research.[49] In this sense, communications and the social production of knowledge are destined to increase the options and spaces of freedom available to all social actors (provided that they are increasingly informed and competent), and guarantee them greater opportunities for emancipation than in the past. Authentic emancipation, based on access to shared and intersubjectively developed knowledge, which, if achieved, is destined to constitute the supportive structure of a "new social contract".[50]

---

[48] See Privitera W., op.cit., 2001; see also Dominici, P. (2008), *Sfera pubblica e società della conoscenza*, in Morcellini M., Mazza B., (Eds.) *Oltre l'individualismo. Comunicazione, nuovi diritti e capitale sociale*, Franco Angeli, Milan, Dominici P. *The Hypercomplex Society and the Development of a New Global Public Sphere: Elements for a Critical Analysis*, in, RAZÓN Y PALABRA, Vol. 21, No. 2_97, Abril-junio 2017—ISSN: 1605-4806, pp. 380–405.

[49] See, in particular, the OECD Report: Education at a Glance 2011; see also the following year's reports up until the most recent one: *Education at a Glance 2021*. See in particular: OECD (2021), *Education at a Glance 2021: OECD Indicators*, OECD Publishing, Paris, https://doi.org/10.1787/b35a14e5-en

[50] A concept and operational definition proposed already in the 1990s (Dominici P., 1995–1996) and also taken up by important transnational organizations such as the UNESCO, in the context of global initiatives such as the world conference "The Future of Education. A new social contract". Link: https://en.unesco.org/futuresofeducation/. Initiatives involving my participation as a social scientist and researcher, as well as official Delegate of the World Academy of Art and Science. Confirming how, even at an international level, we are now taking note of how "Society" can be

The challenge, not only cognitive, to this type of (hyper)complexity needs to be met—in our opinion -, first of all, by emerging from the old *ivory towers* and forsaking the equally old paradigms of *monocausal determinism* to embrace definitively and bravely, a *systemic perspective of and regarding complexity.*[51]

4.4 Hypercomplexity Is Not Solely an Economic Crisis. The Urgency of Questioning the "Economistic Paradigm"

Technological and economic progress is necessary, but it is not sufficient to deal with a crisis which, in addition to being systemic, is also, and above all, cultural and civilizational, rooted, amid the multiple dimensions of the problem, in our inability to concretely question that *economistic paradigm* which is still hegemonic and recognized as absolute and continues—at different levels of organizational and social practice—to make us see/identify Economics as an "exact" and not as a "social science". At the same time, this paradigm continues to define and consider "Society" as a sub-system of the Economy, with all the consequences and repercussions that this implies, including reductionist and deterministic analyses of the ongoing global and systemic transition. This is a key point among the issues we are addressing, a point that has been decelerating the complex and urgent process of serious and radical rethinking of a "developmental model" for some time.

Individualism, the end of social bonds and of community belonging. Stories of precariousness and uncertainty have now become existential conditions for many. The current crisis is not only economic, but also one that has been aggravated by the weakening of the social fabric and the social mechanisms of cooperation and trust. In many cases, it is the very perception of forms of loneliness that are invisible and inaudible, along with the absence of a social network of support, that further aggravate this crisis. A crisis—I repeat—that we continue to address from deterministic and reductionist positions, which exclude variables, approaches, knowledge, skills, disciplinary fields, epistemologies and fundamental methodologies grounded in a perspective that can only be multi/inter/transdisciplinary. These are crucial questions that I have been studying for years and which closely concern us one and all. Allow me to simplify—because issues always need to be loosened by highlighting the complexity and the multiple causal links/levels of connection/interdependence/inter-independence: precisely in the so-called hyperconnected society of communications and networks, which are reproductions and extensions of those pre-existing within social systems,[52] we are witnessing—with some timid signs of awakening,

---

re-built—from the roots—only by concretely committing itself to education and the quality (complex concept) of educational and training processes.

[51] See Dominici P. (1995–2018), *op.cit.*

[52] Dominici P., (1995–1996), op.cit.

which promptly occur after crises and/or disasters[53]—the progressive weakening of the social fabric and of those "devices" that make social cohesion possible (including cultures and cultural models). Confirming how the diffusion and pervasiveness of technological infrastructures and hyper-connected ecosystems are insufficient in every way when seeking to re-create and/or *re-mediate* the Social by re-activating certain mechanisms which are—it must be clear—social and cultural.

It is no coincidence that, on the one hand, we had spoken, well before it grew fashionable, of *anthropological transformation*, a new ecosystem (1996) and individuals who are increasingly free and independent (?) when making choices/reaching decisions; on the other hand, we could not but notice how these very same "individuals", Social Actors, are also increasingly "isolated" in their choices and reference to value systems and cognitive orientation, which are, in themselves, indefinite and fragmentary. Conceptual categories such as "anomie" (Durkheim) and the "ethical void" (Jonas) have returned, as relevant as ever and, perhaps, if one tries to understand the complex and ambiguous "nature" of the ongoing change, it is necessary to look back once more at the project of Modernity itself, which produced not only the emancipation of the masses and "new subjectivities", but also disintegrating "centrifugal forces".[54]

It is an interconnected/hyperconnected society (connection vs. communication) based on a political economy of insecurity (policies, welfare etc.) and on opportunities which, for the moment, concern exclusively elites and restricted groups linked to fields of expert knowledge (open architectures, but *closed* networks), which inevitably *reduces* Politics to *technocracy*. Hard as we have been working on this, there is still a long way to go (absence of long-term policies: a radical cultural change at multiple levels is required. Inclusion, digital citizenship, democracy and the knowledge society are still 'far in the distance' and, more and more often, the impression is that we are playing things by ear and what is more, with an alarming reduction in the sphere of individual and collective rights, which goes hand in hand with the increasingly marginal role of Politics, ever less autonomous from the economic-financial sphere. The grave and concrete danger is that of continuing to interpret and address this dramatic crisis relying on mechanical and disjunctive thinking, on reductionist and deterministic explanations and, at the same time, underestimating the "cultural question" and one of the greatest — and closely related — "evils" of our time: *indifference.*

The contemporary crisis closely concerns the systems of value and cognitive orientation, shared beliefs and practices, and the social mechanisms of trust and cooperation, on which not only social bonds but the very idea of civilization, of the Person, of human dignity, citizenship and democracy are based. Furthermore, it is also, and above all, a crisis of communication as we have always intended it: a *(complex) social process of sharing knowledge (= power)*. It is a crisis that affects all levels, from that of social systems to that of complex organizations, up to social relations and forms of symbolic mediation. Identities and subjectivities are also brought into play and, as has been said over and over again in the past, it is paradoxical, to say the least, that all this occurs precisely within the so-called *global society of communication.* This is one of the reasons why I have always preferred to speak

[53] See the scholarly literature on the topic of social capital and trust.

[54] See in particular: Dominici P. (2003, 2005, 2014, 2019), *op.cit.*

of "interconnected/hyperconnected society" (communication vs. connection), in an effort to provide a clear operational definition (otherwise we lapse, as often happens in these areas of study and research, into pure "nominalism" and/or into fashionable key-words/neologisms), which focuses on the strategic value of *relationships*, reduced to *connections*, on the role of information and cognitive asymmetries, on power relations (1995), on the logic of increasingly stringent control and surveillance that characterize what I had defined several years ago as the "hypercomplex society",[55] an operational definition springing from the profound awareness that, in any field of social or human practice, the constantly expanding interdependence/interconnection of systems and the accumulation of information (dis-information) do not determine or guarantee the production and sharing of knowledge, quite the opposite. Nor do they determine/have they determined any reduction (Luhmann) and/or simplification of complexity.

Knowledge, as an indispensable and unique "resource" (it is no coincidence that we use the term "economy of knowledge")—is *faring none too well*, as I put it, due to the radical fragmentation/partitioning of the fields of knowledge and skills and to excessive specialization, both of which have contributed to the isolation and de-responsibilisation of individuals within educational and training institutions and, more generally, within social organizations and systems[56] (complex…organisms). Furthermore, the specialization of disciplines, although useful and functional for various purposes, has, over time, produced a further *compartmentalization* of the fields of knowledge and skills, making it practically impossible to adequately study and address multidimensional, non-linear problems and issues, something which, at the very least, would/require/would have required the integration of approaches, epistemologies and methodologies. I repeat: these are logics of separation and confinement of the fields of knowledge which have always divided and separated experience and experiences.

In this sense, it is of fundamental importance to redefine the territories of knowledge and rethink the "relational space",[57] within which identities are built—which

---

[55] See, in particular: Dominici P. (2003, 2005, 2011), op.cit.

[56] Topics and questions that I have often come back to over the years, which cannot be improvised → school, university, ivory-tower logic, education and training in complexity, knowledge and skills for complexity etc.

[57] See also the following articles, where I addressed these issues in a more popular dissemination format (1) *Rimettere la Persona al centro*—Il Sole 24 Ore http://pierodominici.nova100.ilsole24ore.com/2014/04/27/rimettere-la-persona-al-centro-cultura-e-civilta-della-rete/; (2) *Il diritto alla filosofia per rethink l'educazione, la cittadinanza e l'inclusione*—Il Sole 24 Ore http://pierodominici.nova100.ilsole24ore.com/2017/04/23/il-diritto-alla-filosofia-per-rethink-leducazione-la-cittadinanza-e-linclusione/; (3) *Innovazione e domanda di consapevolezza: la filosofia come "dispositivo" di risposta alla ipercomplessità*—Il Sole 24 Ore—http://pierodominici.nova100.ilsole24ore.com/2016/03/14/innovazione-e-domanda-di-consapevolezza-la-filosofia-come-dispositivo-di-risposta-alla-ipercomplessita/; (4) *Il grande equivoco. Rethink l'educazione (#digitale) per la Società Ipercomplessa*—Il Sole 24 Ore http://pierodominici.nova100.ilsole24ore.com/2016/12/08/il-grande-equivoco-rethink-leducazione-digitale-per-la-societa-ipercomplessa/; (5) *Tra etica ed etichetta: l'urgenza di una "nuova cultura della comunicazione* http://pierodominici.nova100.ilsole24ore.com/2016/11/24/tra-etica-ed-etichetta-lurgenza-di-una-nuova-cultura-della-comunicazione1996/; (6) Un Nuovo Umanesimo per la Società Interconnessa—intervista concessa a Marco Minghetti de *Il Sole24Ore* http://marcominghetti.nova100.ilsole24ore.com/2014/11/17/un-nuovo-umanesimo-per-la-societa-interconnessa/

can never be taken for granted once and for all, as they are always in constant "becoming"—and the subjectivities: a "co-construction" which occurs through dialogue, conversation, reciprocity, empathy, and communication, as said before, = (complex) social process of sharing knowledge (power). We are always a "WE" and not an "I" (identity < > recognition), even if we are not aware of it. *Systemic dimensions* vs. a *linear dimension*. Our existence always takes place, in any case, within a network of systems of conversation and communication. Teaching complexity and critical thinking, immediately: the pathway to *Well-Made Heads* (Montaigne). Because knowing/knowledge means living and vice versa, and these dynamics always arise and evolve solely through OTHERS, in a systemic, as well as relational, key. The "micro" level (relationships and social interaction) and the "macro" level (organizations, systems, nation-states, etc.), are not only inseparable, but, in their interdependence, they constantly influence, connect and relate to each other. Over the years, I have repeatedly spoken in terms of a *dual level of analysis*, which requires an approach to complexity and a *systemic perspective.*[58] The challenge of and to complexity demands that we *rethink the way we think* and that we transform education and training profoundly and radically. It involves rethinking the very concepts of "freedom", "responsibility", "community" and, consequently, "democracy" and, as regards the strictly current situation, rethinking our idea of country, of Europe and of humanity. If that sounds like a classic case of fighting windmills…it is not so and the fight must go on!

What these dimensions, as mentioned, along with the *void of meaning* left by the crisis of ideologies and traditional socialization agencies, as well as by the consequent advent/hegemony of an extremely complex type of *educational polycentrism*, have produced, among various consequences, is a sort of general *moral demobilization* that feeds into the "society of irresponsibility"[59]: a society (US and the social networks of which we are a part) that appears devoid of any sense of belonging to a community or of the ethics of sacrifice. A type of society where, beyond an extremely complex regulatory and ethical framework, ethics and responsibility are the "talk of the town" today, "visible" not only in the media (perhaps more than ever in the past), where they are widely preached but little practised, amid labels, "ethics" and "etiquette".

Broadening our panorama, it is useful to reiterate the risks, in light of the complexity of the issues considered, not only interpretative risks (recourse to the usual reductionist and deterministic explanations)—as ruling-class cultures translate these models and visions into actions and strategies (?)—linked to the widespread belief (common not only to certain economists and technocrats) that technological progress and economic recovery can resolve, sooner or later, all issues. These two factors are necessary conditions, "variables" of fundamental importance, however,

---

[58] To overcome the principle of causality, of any form of mono-causal determinism and reductionism; there are many contributing factors and many variables to consider; systems and organizations evolve and differentiate in a nonlinear manner, etc.

[59] See Dominici P. (2010), *op.cit.*

they are not sufficient to address and counter a crisis which is also, and above all, a cultural and civilisational crisis.

A crisis, even more radical in its effects due to the emergency ecosystem: emergencies like *Covid*, which should be/should have been faced, considering their scope, as an increasingly interdependent and interconnected world system, with strategies "designed" on the basis of a transnational framework, redefining the role of the European Union and the so-called (?) international community, "entities" which are currently inconsistent and devoid of identity, as well as of common strategies. It is, therefore, time to: (1) redesign, from a systemic perspective, *welfare* and cohesion policies in light of the new social risks and new forms of precariousness; (2) define long term/very long term policies for education, training and research, investing *resources* (investments), thus prompting societies to generate "antibodies" capable of stimulating evolution and development, and strengthening social ties (teaching values of citizenship, legality, anti-corruption etc.), whose conditions are increasingly critical, owing to dominant individualistic and egoistic values, and to a lack of co-built cultural models at least partially capable of fostering social cohesion and the "common good". In other words, the lack of shared ethics still appears problematic; (3) it is also necessary to focus seriously and substantially on social and cultural change which, I shall go on repeating ad nauseam (and never tire of doing so), in addition to being "systemic change" (not top-down), can occur exclusively in the long term and, once again, bottom-up, starting from the *centrality and the quality of educational processes*. This is what I have been defining (for years) as the "mandatory pathway", which everybody—not just politics—talks about and with which everyone seems to agree although, to date, I find professed awareness and attention only. We hope this insistence will serve, at least, to increase this "declared" awareness of the importance of "cultural change"—which can only be *systemic*—because what we are experiencing, as I mentioned above, is a cultural crisis, a crisis which the numerous variables and contributing causes involved, have made into a "crisis of civilization".

An eco-systemic, global crisis which, as usual, amid risks and opportunities, obliges us to question ourselves (and to act), on beyond the potential "aftermath", first and foremost, on what it means to be "People", "Citizens", and "Human beings" in this hyper-technological and hyper-connected civilization of ours; a precarious and uncertain transitional phase that radically calls into question the foundational dimension of "trust" and the *security paradigm* itself. It forces us to rethink the way we understand and practice the values/principles of freedom and responsibility (relational concepts[60]); it drives us to redefine our conception of "rationality"—not to mention our idea/conception of "Science"[61]—as well as the codes, models, strategies it produces. A crisis so thorny, on beyond the extraordinary scientific discover-

[60] Dominici P., *op.cit.* 1996, 2995.

[61] See also, on these topics, the interesting work of the science philosopher John Dupré, who, despite defending a rigorously non-secular view of Science, harshly criticizes what he defines "scientific imperialism". See Dupré J. (2001), *Human Nature and the Limits of Science,* Italian translation: Lo scontro della civiltà e il nuovo ordine mondiale, Garzanti, Milan 1997.

ies and technological innovations, as to bring the question of human dignity and rights and (global) citizenship back to the foreground. A complex crisis/transition that urges us to embrace *the indeterminate…*

The *hypercomplex society* (2003) urgently requires education and training in complexity and unpredictability—new epistemologies and a renewed thought system re-founded on other assumptions—necessary for the co-construction in designated "seats", not only People—Citizens (i.e., the conditions of "social bonds") and, subsequently, profiles and skills etc., but above all, a shared ethic of responsibility and understanding. Even more so in a historical context as sensitive as this one, when the danger of the well-known clash between civilizations[62] seems to be raising its head again, through dramatic conflicts and the emergence of new asymmetries and forms of inequality, not attributable to economic indicators and factors alone.

And, again on the subject of "critical thinking" and complexity, education, communication, control and unpredictability, I conclude—recalling Maturana, who has always been an essential reference point (but on these topics the scholarly literature is of high quality and practically endless)—that we need to be aware that education and Politics itself are crucial moments in the inexorable process of *autopoiesis* (self-construction) that marks the lives and practices of human beings: *the Social,* too, has an emotional and not only a rational basis. Maturana defines as "blinkers" any concepts which are unthinkingly accepted as meaningful by everyone. The claim that human beings are characterized by rationality, according to Maturana, is such a blinker, which leaves us unable to perceive that emotion is an essential constituent of human life, or to understand the emotional basis of reason itself. He further supposes that language arose as a way of connecting and coordinating the cohabitation among different generations (adults and their young), a way to make sharing and participating in life together pleasant, simply through conversation.

As Edgar Morin argues, "If teaching is teaching to live, according to Jean-Jacques Rousseau's correct maxim, it is necessary to identify the shortcomings and gaps in our current teaching to address vital problems such as error, illusion, partiality, human understanding, and the uncertainties that every existence implicates" (Edgar Morin)[63].

[62] See Huntington S.P. (1996), *The Clash of Civilizations and the Remaking of World Order*, Italian translation, *Lo scontro delle civiltà e il nuovo ordine mondiale*, Garzanti, Milan 1997.

[63] Morin E. (2014), *Enseigner à vivre*, It. Trans., *Insegnare a vivere. Manifesto per cambiare l'Educazione*, Raffaello Cortina Editore, Milano 2015, p. 7.

Chapter 5
For a "New Social Contract". The Asymmetrical Society and the Strategic Centrality of Education and Training

Once basic needs have been translated by a society into demands for scientifically produced commodities, poverty is defined by standards which the technocrat can modify at will.

Ivan Illich

Abstract A new social contract can only be established if citizens are capable of critical thinking, but a citizen is first and foremost a person, who must be socially and culturally empowered through systemic processes of education, based necessarily on *equality of starting conditions*, without which all discourses on merit are purely rhetorical: an education aimed toward the construction of a society conceived as a living organism, rather than a mechanism-society. If schools (which should be considered constitutional bodies) and universities are to rediscover their original functions as social elevators, if they are to succeed in inspiring a society that intends to eliminate asymmetry, they themselves must be the first to avoid delegating carte blanche to technology, which is "the Great Mistake", as well as the first to avoid fostering a "culture of cunning". Above all, they must activate a culture of responsibility, without which liberty cannot prevail; they must stimulate creative minds toward a kind of innovation that is cultural and *inclusive*—keeping in mind that innovating initially means destabilizing—and they must work actively to create elastic, open-minded, *hybrid figures* capable of building a genuine culture of complexity, and of healing the fractures between fields of knowledge.

Keywords Critical thinking · Inclusive innovation · Starting conditions · Living organisms · Constitutional bodies · Social elevators · Great mistake · Hybrid figures · Liberty-responsibility · Healing fractures

P. Dominici, *Beyond Black Swans: Inhabiting Indeterminacy*, Philosophical Studies Series 160, https://doi.org/10.1007/978-3-032-09029-4_5

5.1 The Importance of "Building" the Person and the Citizen, Socially and Culturally

As can be clearly seen, I consider these issues strategic to say the least but, at the same time, I believe that it is of fundamental importance to frame them in a more complex and general discussion, regarding the rethinking of what I have called the "new social contract" (2003), not to mention the very concept of citizenship. This rethinking/reformulation will need to lead, in turn, to an operational translation functional to the definition, planning and implementation of educational proposals and strategies.

Because, as already said, the issue is educational and cultural and concerns, first and foremost, the teaching of liberty, *which entails responsibility*! In addition to the social, relational, ethical dimension, our young people—from the early years of school—increasingly need to know how to live, practise and apply "logic", to identify and rebuild the connections and the systemic nature of relationships, to develop/verify arguments logically, as well as to already be—at least partially—aware of the epistemological implications of choices, thoughts/actions. This is mainly because—as we know—once they have reached the university years, it is particularly difficult at this level to co-build and/or modify a previously structured *forma mentis*—or as it is put today, *mindset*. Our young people desperately need (allow me to reiterate once again) to know *how to think*, as well as *how to be*. More and more, they require a "method" with which *to develop thought in a "non-linear" way*, to reason, to summarise, to bestow systematicity on the enormous (excessive?) quantity of information received, to question themselves and know how to question not only the "how" of things but also, and above all, the "why" (philosophy...ethics and epistemology). They need an education and preparation in complexity and critical thinking, which should educate and coach—almost "train" them—to identify the connections between phenomena and processes, between the fields of knowledge and life as experience ... so as to enable them, for example, to evaluate the historical-social origins of cultural norms and models critically, to reflect and distinguish between "nature" and "culture" (resulting from convention), a dichotomy that should be overcome once and for all! They need to recognize diversity and pluralism as fundamental "values" and not "dangers". As has been written in the past, to achieve objectives of such complexity, long-term policies, investments and a grand relaunch of humanistic studies and humanistic education, as well as social sciences, are needed at all levels (school, university, research, etc.); the objective must/should also be to try to overcome once and for all—even though far from simple—what I have defined as "false dichotomies" and the logics of separation and confinement of the fields of knowledge and skills that so much damage has caused in recent decades. The rest would arrive, evidently, almost as a consequence.

A humanistic education teaches you *to think with your own head* but, above all, to think with the minds of others, the minds of those whose cultural models differ from yours. From this point of view, the very acquisition and possession of the so-called technical and digital skills (albeit highly important, let it be clear) would have even more significant impacts on "well-made heads" (Montaigne), critically trained and curious about the complexity in which they are immersed.

We urgently need a "New Humanism" that places the Person—who is a relational and systemic subjectivity—along with that person's education and relationship with others, at the centre not only of educational and training processes (a concept I have discussed in other articles and publications that can be referenced as well). What is at stake are identities and subjectivities and, in this perspective, we cannot but notice the absence of theoretical-interpretative and, more in general, cultural models suited to the ongoing change, as well as the substantial inadequacy of the educational pathways of schooling and university training. At this point in time, it seems that everyone now agrees (better late than never!); all voices exalt the value of critical thinking and of an education/training in complexity but, for the moment, the impression is that these are merely formulae and slogans, key words one must adopt in order to ensure success. The problem is that the corresponding actions/strategies—which can only be long-term—are wanting; also because Politics has always thought of the "short term" and pursues very different logics and objectives.

In addressing these issues, we must take care not to fall for the temptations of simple solutions (shortcuts), deterministic explanations and easy reductionisms which, not only as of today, but have always been fueled by a number of great narratives regarding the so-called digital revolution and which mark, ever more profoundly, the hypertechnological civilization of simulation and automation. It is as if nothing had happened in recent decades, as if the ongoing systemic and global emergency had never even begun. We urgently need elucidations and analyses based on data and research, but we are also in dire need of a critical theoretical approach to complexity which enables us to emerge from the quicksand of the *monocausal determinism* mentioned before (2003) but also, for example, at the less demanding level of *acritical neophilia* that, in recent years, has convinced us that every novelty is fantastic and "within reach", only because it is "new" and/or (defined) "digital".

I reiterate and with great emphasis: School and University, (continuous) education and training, scientific research—including that pertaining to the human and social sciences, totally disregarded—must/should be placed, concretely (!), at the core of every innovative project and process (systemic views). When addressing the challenges of (what is now global) citizenship and "inclusive innovation", which are the challenges of (hyper)complexity and also of responsibility, it is of vital importance to be aware, not only in words and in public discourse—that the future, that "true" innovation (which, I insist, is social and cultural innovation), belongs to those who will succeed in healing the fracture between the Human and the Technological, to those who will manage to redefine and rethink the complex relationship between *the natural and the artificial*, to those who will be able to integrate multiple/different approaches, to those who will be able to unite (rather than separate) knowledge and skills, to those who will be able to combine, or even better, *merge* the so-called *two cultures* (humanistic and scientific, according to Snow[1]) both in education and training and in the definition of professional profiles and skills.[2]

In this sense, there is a palpable and urgent need to overcome—in addition to the traditional, and often-mentioned "logics of separation and confinement of the fields of knowledge and skills"—all of what I have defined "false dichotomies": in addition to the above-mentioned (classical) separations between "nature" and "culture" and between the "natural" and the "artificial", I am referring to deeper and more profoundly rooted fractures, also in terms of implications and repercussions: from thought vs. action, theory vs. research/practice; scientific vs. humanistic training; knowledge vs. skills, multi/inter/trans-disciplinarity vs. specialization, hard vs. soft skills to the more recent *real* vs. *virtual*.[3] It is vital to however, when referring to issues concerning School and University more specifically, to carefully avoid the

---

[1] See Snow C.P. (1959), *The Two cultures*, Cambridge University Press, Cambridge 2012 (new ed.).

[2] For further contributions on these issues, please refer to some scholarly and more popular publications: Dominici P. (2014), *Dentro la Società Interconnessa. La cultura della complessità per abitare i confini e le tensioni della civiltà ipertecnologica,* Milan,FrancoAngeli 2019; Dominici P. *Controversies on hypercomplexity and on education in the hypertechnological era*, in, A. Fabris & G. Scarafile, Eds, *Controversies in the Contemporary World*, Amsterdam-Philadelphia: John Benjamins Publishing Company, 2019; Dominici P., "*Educating for the Future in the Age of Obsolescence*", in CADMUS, Volume 4—Issue 3, November 2020, pp. 93–109; See the article published on Il Sole 24 Ore "Competenze e Saperi per la Società Interconnessa. Le due culture e la complessità" http://pierodominici.nova100.ilsole24ore.com/2015/07/28/competenze-e-saperi-per-la-societa-interconnessa-le-due-culture-e-la-complessita/; and the essay, published in the specialist journal "TechEconomy": http://www.techeconomy.it/2016/02/25/uninnovazione-inclusiva-ricomporre-la-frattura-lumano-tecnologico/

[3] Regarding this perspective see in particular, "European Qualifications Framework for Lifelong Learning—EQF" and Dublin Descriptors, important and, at the same time, binding references, but little known and applied, even in the academic field. Among recent publications, see Dominici P., *The Digital Mockingbird: Anthropological Transformation and the "New Nature"*, in "World Futures. The Journal of New Paradigm", Routledge, Taylor & Francis, Feb. 2022; Dominici P., Flores J.M., *La Gran Equivocación: Replantear la educación y la formación virtual para la "sociedad hipercompleja"*, in "Comunicación y Hombre". Número 18. Año 2022; Dominici P., *Beyond the Darkness of our Age. For a Non-Mechanistic View of Complex Organization as Living Organisms*, in "Rivista Trimestrale di Scienze dell'Amministrazione. Studi di Teoria e ricerca sociale", n°2, Anno 2022 http://rtsa.eu/RTSA_2_2022_Dominici.pdf?fs=e&s=cl

continuous temptation of shortcuts, simple solutions, excesses *of reformism* and, even, of the routes followed previously which, for that reason alone, appear reassuring. Such linear routes and pathways often specifically conceal conflicts of interest based on economics and power, along with ideological views rendered visible, as well as acceptable and shareable, by dint of incessant promotion and marketing of events.

In this regard, it is necessary to develop awareness—widespread awareness—that "innovating means destabilizing" and that the process of complex innovation, to be truly systemic, needs to be generated *from below*! not imposed, as has always been the case, by more or less enlightened élites. First of all, however, it is necessary to transform educational and training processes,[4] to educate and train people to think critically with their own heads (to interrogate themselves and ask questions, not to be satisfied with the usual answers/solutions) and, above all, to see "objects" as "systems" (and not vice versa).

From this analytical perspective, the "educational question" has never ceased to be central and strategic and, it is no surprise that we return to speak about a "new social contract" (1996) which is evidently conceivable/possible only as a result of a cultural transformation capable of triggering systemic economic, political and social change; The ability, both theoretical and political, to devise a "long-term" political project, at the centre of which stand educational and training processes, will also be decisive. It is necessary to change, radically, the political choices and strategies that have affected educational and training institutions in recent decades, the choices and strategies, defined and constructed at national and transnational level, all overly centred on the short-sighted and—for many reasons, as already discussed—devastating view, which has always based (and continues to base) the trajectories to follow on the demands/requests/trends of the "market, with the final objective of fully realizing *the neoliberal (hegemonic) paradigm of education*. This paradigm completely lacks both a "long-term" systemic vision and a robust conception of the complex issues to be addressed. Furthermore, it is founded on a "culture of assessment" detached from educational and social practice and centered totally on what I have called *the grand illusions of the hyper-technological civilization*.

[4] Just as I was writing this, intense efforts were at the same moment being put forward among global participants to organize the Transforming Education summit https://www.un.org/en/transforming-education-summit, subsequently held at the head office of the United Nations in New York.

5.2 The Close Correlation Between the Educational Question and the "New Social Contract"

On beyond some of my own profound convictions,[5] which have always characterized my life, not only as a scholar and researcher, it is evident that unequal and *non-quality* schooling has always been the constitutive and fundamental prerequisite of any kind of unequal, asymmetrical, exclusive, closed society, unable to guarantee any type of *vertical social mobility*. If we cannot yet speak of it in terms of causality, we can certainly point out a strong and compelling correlation.

In these precise terms, we are duty-bound to deal with what, in the past, I have called the "Asymmetrical Society", that is, a type of society capable only of guaranteeing and acknowledging citizenship and inclusion from a formal, legalist point of view, insisting. Hence, on the urgency of a "new social contract".[6] In actual fact, what have changed are the conditions and variables that constitute and characterize complex concepts, phenomena and processes, such as those associated with liberty, equality, citizenship, inclusion and democracy. This crucial issue has never ceased to characterize the political agenda (there may still be hope, although optimism appears excessive) and the different *climates of opinion* which, in turn, influence the very practicability of the strategies and policies needed. It is probable that, in our hyper-technological and hyper-connected age, in which Google, the social networks and platforms would seem capable of solving (?) all our problems of information and cognition, the credibility and reputation of scholars and experts (or persons presumed to be such), may well be built in a manner relatively simpler than those who have been researching these topics for a lifetime. But this is a topic which we might decide to explore later on.

For the moment, I shall limit myself to pointing out that, ours is without a doubt an era which, from a cultural point of view, not only does not accept, but actually rejects the complexity and interdisciplinarity of many fields of knowledge, considering theory, critical thinking, reflection—"time" for reflection—completely useless and misleading compared to "know-how" and concreteness, numbers and proofs (?) of life, along with the "new speed" which has also become synonymous with intelligence/capacity, and not just efficiency. Ours is an era which, having rejected theory and critical thinking, as mentioned above, having delegated everything to intelligent machines (?), paradoxically trusts and relies on "gurus", *influencers*, experts in anything and everything, on visionaries of all kinds and, above all, on "vendors of simple, readily-available solutions" to every kind of problem.

---

[5] I am referring, in this case, to the close correlation existing between education and participation, between education and citizenship and, finally, between education and the "quality" of inclusion and democracy. See, in particular Dominici, P. (2017a), *Oltre la libertà ...di "essere sudditi"*, in F. Varanini (Ed.) *Corpi, menti, macchine per pensare*, Casa della Cultura, Anno 2, number 4, Milan; Dominici P., *The weak link of democracy and the challenges of educating towards global citizenship,* in Torres C.A., Gaudelli W. and Bosio E. Eds., *Values, Knowledge and Curriculum in Global Citizenship Education,* Springer Nature, UNESCO, 2022.

[6] Dominici P., (1995–1996), *op.cit.*

Ours is an era in which everything, absolutely everything, is designed and created with an almost obsessive attention to functions and usefulness. On the other hand, as reiterated on other occasions, when it comes to what I have called the Mechanism-Society, what is perceived and emphasized is the exclusive importance of technical fields of knowledge, specialisms and, specifically, of technocrats. Politics itself is being reduced, increasingly, to *technocracy*[7] and to reductionist and deterministic approaches.

5.3 The Educational and Cultural Issue: A Crucial Matter for the Organism-Society

The ongoing *technological transformation*, is above all, an *anthropological transformation* with profound epistemological and ethical implications. It is causing us to progress (?) rapidly towards a redefinition of identities, relational spaces and, above all, social asymmetries. It is characterised by complex dynamics that make *the reformulation of thinking, the supersession of disjunctive thinking* and the traditional architecture of the fields of knowledge more urgent than ever, especially within educational and training institutions, designed and managed, up to now, as clearly distinct and separate "closed systems". The social construction of change and of a more inclusive society cannot but pass through the quality of our educational and training institutions: this is the true strategic "factor" of processes of change and innovation: the cultural "factor", a complex variable capable, in the long term, of triggering and accompanying economic, political and social processes. Despite all, the truly strategic level is once again represented by "educational processes", whose protagonists are (or should be), above all, schools, along with other agencies of socialization, whose spaces for learning and socialization, moreover, have in recent decades been devoured by the media, networks and peer groups. Nevertheless, this strategic level is the crucial node upon which it is possible to educate and train *well-made, not well-filled heads* (Montaigne); it is also the strategic level where it is possible to try (at least) to foster and practise authenticity, empathy, pluralism and recognition of the values of uniqueness and, at the same time, diversity. It is the decisive level upon which to educate and train *recognition and valorisation of fragilities*—of all the fragilities—that characterize us precisely

[7] On "technocracy", a complex and founding dimension of what I have called the "Mechanism -Society", and related issues, see in particular: Berlin I., *Two Concepts of Liberty*, Clarendon, Oxford 1958; Habermas J., "The Scientization of Politics and Public Opinion", In *Toward a Rational Society: Student Protest, Science and Politics*, Beacon, Boston 1970; Fischer F., *Technocracy and the politics of expertise*, Sage, Newbury Park [etc.] 1990; Fisichella D., *L'altro potere: tecnocrazia e gruppi di pressione*, Laterza, Rome -Bari 1997; Foucault M., *Security, Territory, Population: Lectures at the Collège de France, 1977–78*, Palgrave Macmillan, Basingstoke 2007; Foucault M., *The Birth of Biopolitics: Lectures at the Collége de France, 1978–1979*, Palgrave Macmillan, Basingstoke 2008; Friedman J., *Power Without Knowledge: A Critique of Technocracy*, Oxford University Press, Oxford 2019.

as "human beings", and then as People and Citizens[8] as well. To endeavor to co-construct truly *open and inclusive societies*, based on a culture of co-habitation and comprehension capable of enhancing the coexistence of contradictory and/or conflicting elements. At the same time, educational processes are crucial for co-building a culture of legality, prevention, responsibility, respect and non-discrimination. In the end, they constitute the crucial level upon which to seek to determine the socio-cultural conditions required to reduce the hegemony of individualistic and selfish values, which have significantly contributed to the weakening of social bonds and the community. In the long, very long term, also co-building a "culture of sustainability",[9] which can be based solely on a "culture of complexity", on pathways that inevitably intersect, to the point of overlapping, and which simultaneously concern theory and scientific research, schools and universities, citizenship and democracy, equality of starting conditions and inclusion. ***Education and citizenship…Education is citizenship, education is the possibility of participation, education is inclusion.***

5.4 Relational Subjectivities. Systemic Co-building of the Person and the Citizen

If it were still necessary, the correlation between education and citizenship/inclusion would turn out to be, from this point of view, more evident, even logical-consequential. However, what is still missing is the awareness that it is neither technology nor technological and/or digital connections that will define and/or determine the conditions of this new kind of citizenship and inclusion which have

[8] On these themes and issues, so vitally important and full of profound implications, I suggest some multidisciplinary pathways: Demetrio D., *Educatori di professione*, La Nuova Italia, Florence 1990; Levi Montalcini R., *Elogio dell'imperfezione*, Garzanti, Milan 1990; Morin E. *Educare gli educatori. Una reform of thinking per la democrazia cognitiva*, EDUP, Rome 2002; Freire P., *Pedagogia dell'autonomia. Saperi necessari per la pratica educativa*, EGA, Turin 2004; Tramma S., *L'educatore imperfetto. Senso e complessità del lavoro educativo*, Carocci, Rome 2003; Montessori M., *Educare alla libertà*, Mondadori, Milan 2008. In particular, See also the interesting work of the American scholar Gloria Jean Watkins, published under the pseudonym Bell Hooks: Hooks B., *Teaching to Transgress: Education as A Practice of Freedom*, Italian translation, *Insegnare a trasgredire. L'educazione come pratica della libertà*, Meltemi, Milan 2020; the work of Gardner H. (1991), *The Unschooled Mind. How Children think and How Schools should Teach*, Italian translation, *Educare al comprendere. Stereotipi infantile e apprendimento scolastico*, Feltrinelli, Milan 1993; the important work by Andrea Canevaro: Canevaro A., *Scuola inclusiva e mondo più giusto*, Erickson, Trento 2013; by the same author *Nascere fragili. Processi educativi e pratiche di cura*, EDB, Bologna 2015; Canevaro A., Ianes D., *Un'altra didattica è possibile. Esempi e pratiche di ordinaria didattica inclusiva*, Erickson, Trent 2022.

[9] See,iIn the UNESCO document: UNESCO (2017), *Educazione agli Obiettivi di Sviluppo Sostenibile* - Obiettivi di apprendimento: http://unescoblob.blob.core.windows.net/pdf/UploadCKEditor/MANUALE_ITA.pdf.

by now become a transnational, "global question",[10] that it is neither technology nor the so-called digital revolution, that will specifically create Montaigne's "well-made heads", much less reactivate the social mechanisms of trust and cooperation. In this sense, beyond these preliminary considerations, I feel it is important to clarify that, in my opinion, despite the many years that have been spent in pursuit of certain "narratives" regarding digitality, that there is still an extremely concrete risk of believing (and acting accordingly) that the so-called *digital education*—and, with it, that digital culture itself, or rather, *digital cultures*, founded, in the first place, on misleading and fallacious epistemological premises, are merely "technical" issues regarding "technical training", specific "skills" linked (exclusively) to the "nature" of the (new) connective technologies, of the new ecosystems/communications environments (as well as working and professional environments, of course).[11]

On the contrary, within a perspective of complexity and of a critical, systemic view of the processes of education and communication, it is necessary to rethink/re-invent/re-imagine education and, as a result, the dimensions, variables and tools which comprise it; even the (relative) parameters of measurement and the indicators are by no means equivalent, and must be considered as pertaining to a multitude of problematic yet simultaneously interconnected dimensions. This requires *a socio-systemic, relational view* of educational processes, an "Other" culture of evaluation, in particular a recovery of empathy, of critical thinking, and of methodological knowledge and competences which must not be confused in any way with digital skills or the capacity for using various devices/appliances. At the same time, it is vital to work on a renewed awareness of the epistemological implications (for some time now I have been proposing and enouncing the concept of the "epistemology of error") and on a form of education in *communications, that is, in complexity and responsibility*. Communication is, above all, a constituent element, not a distinct and/or superstructural aspect, of complexity and complex systems. They are a determining and decisive assumption and a "fact", given that, according to my theoretical and epistemological proposal, they imply awareness of the fact that the models, schemes, diagrams, and data that concern visualization and processes, the way/ways in which we communicate complexity itself and/or the complexity of complex systems, do not contribute to reducing or simplifying complexity itself. On the contrary, they profoundly affect the (individual and collective) psychological, cognitive, mental, social, cultural and political processes of social representation, *social perception* and, as a result, the processes of "social acceptability". In other words, they deeply impact resilience and our ability as human beings to adapt to and metabolize the indeterminate and unexpected so as to be prepared and ready to *inhabit* unpredictability, uncertainty, and the emergent.

---

[10] See Dominici P., *The weak link of democracy and the challenges of educating towards global citizenship*, in "Prospects", Springer Nature, UNESCO, 2022.

[11] On these specific topics see: Tramma S., *L'educatore imperfetto. Senso e complessità del lavoro educativo*, Carocci, Rome 2003; Pellerey M. (2011), *L'approccio per competenze è un pericolo per l'educazione scolastica*, in "Scuola Democratica", n.s., pp. 37–54; Ribolzi L. (2011), *Tra conoscenza e competenze: tra integrazione e conflitto*, in, "Scuola Democratica", n.s., pp. 26–36.

This is clearly one of the essential points that characterizes/distinguishes the approach I have proposed within the context of the research and studies I have conducted. Currently, at least in terms of public discourse and/or official declarations, many are revising—or have already revised—their former assessment of the "power" of digital systems to generate, in and of themselves, inclusion and opportunities for all (*verba volant, scripta manent*) by *disintermediating* hierarchies, processes and dynamics of all kinds. We are faced with yet another important industrial revolution, more invasive and pervasive than any of the previous ones: a revolution which, unless it puts its hand to education and training, will continue to increase the current *distances* and numerous asymmetries, so marked and evident and made even more evident by global, systemic emergencies such as the 2020–22 pandemic.

It is necessary to start from these premises, mainly because, as I have stated over and over again in recent years, the lengthy crisis we are going through—moments of crisis and of innovation often coincide—is, above all, a cultural and civilizational crisis, a thinking crisis, a thought crisis, a crisis of a *thought system,* which is only partially economic. A phase of transition that with a serious degree of tardiness has brought the educational and cultural question back to the centre of the global debate.

5.5 Guaranteeing *Equality of Starting Conditions*, Beyond All Rhetoric

Furthermore, we have to admit that we find ourselves within a socio-cultural ambit of perspectives—of discourse and action—but, above all, of (short-term) strategies that continue to be based on absolutely partial awareness of the multidimensionality, ambiguity and unpredictability that characterize the processes of innovation and change typical of complex systems. An imperfect awareness, often exclusively proclaimed but not participated, which ends up leading to a reduction, sometimes a trivialization of the very concepts of education, communication, sharing, inclusion, citizenship, democracy, even of complexity itself, with the risk, among many others, of determining, in an irreparable fashion, the structural conditions of a hierarchical and elitist form of innovation. I have gone over this aspect several times, so I will limit myself to reiterating the point that to speak about inclusion, citizenship, democracy, "digital democracy" (sic!) without making an attempt, at least, to counter those phenomena and processes that make them difficult to achieve (hindering open and inclusive innovation), is equivalent to legitimizing the mechanisms of a historical-social context increasingly marked by cognitive and cultural inequalities—at a global level[12]—which clearly define an international kind of social stratification. Until equality of starting conditions is guaranteed—access to quality

[12] In this regard, see the latest reports published by international bodies like UNESCO, the UN and OCSE, for example. The situation, in many cases, especially as regards education and access to (not only) "quality" processes of education and training, is dramatic and further aggravated by war, conflict, environmental and epidemic catastrophes.

education (a complex concept) must/should be considered, to all intents and purposes, a fundamental "human right". Even mere talk of "citizenship" and "inclusion"—let alone "meritocracy" and/or "excellence" (using the usual common rhetorical terms[13])—remains a purely rhetorical exercise, as has been practiced in recent years. Evidently, there can be no "digital citizenship" without guaranteeing and making the rights of plain citizenship effective, practicable and concrete, in addition, evidently, to the rights of each Person as a human being.

5.6 Beyond Idols of Knowledge and the *Confinement* of Fields of Knowledge

Reiterating the assumption that "there are no simple, short-term solutions to complex problems", we can certainly affirm that the system logics and organizational cultures that drive educational and training institutions are still able to hinder, or at least slow down any attempt to carry out innovation and change.

On the other hand, "disjunctive thinking" continues to prevail, despite digital connections and technologies that offer/seem to provide us with infinite opportunities to re-invent, observe interdependencies/interconnections and, as a result, go *beyond*. In the opposite direction, we continue to build, socially and culturally, a series of inadequacies which appear even more evident and paradoxical precisely within the society of interdependence and global interconnection.

Beyond our constitutive fragilities as human beings, (increasingly) *limited rationality* and vulnerability dominate social systems and complex organizations, which prove to be increasingly chaotic, difficult to manage and marked by profound contradictions and the ambivalence of productive, economic and cultural processes. I repeat: there is still little awareness of the fact that these are "organisms", not "mechanisms"; there is still little awareness of being "inside" a *constitutive hyper-complexity* which—it is important to clarify this yet again—are cognitive, social, subjective, ethical, linguistic and communicative, affecting (it is difficult to find the verb capable of describing this condition) every sphere of life and practice, requiring us to rethink/re-imagine/redefine the categories, paradigms, education and, as a result, the "social contract" and the "forms" of citizenship which arise from these dynamics. Paying attention to what I first called—long before it became

[13] Among the numerous contributions, produced over time, relating to these issues, I refer to a recent one I myself have authored on the feudal society of dominion: "La società feudale del dominio e dell'arbitrio iper-normato, published for Il Sole 24 Ore, on the 24th of July 2021. Link: https://pierodominici.nova100.ilsole24ore.com/2021/07/24/la-societa-feudale-del-dominio-e-dellarbitrio-iper-normato-ancora-sulla-vecchia-questione-culturale/ See also, starting from the operational definition, proposed many years earlier: "La società asimmetrica e la centralità della "questione culturale": le resistenze al cambiamento e le "leve" per innescarlo" published on the 23rd of September 2015 in Il Sole 24 Ore. Link: https://pierodominici.nova100.ilsole24ore.com/2015/09/23/la-societa-asimmetrica-e-la-centralita-della-questione-culturale-le-resistenze-al-cambiamento-e-le-leve-per-innescarlo/

fashionable—the "grand illusions" of the hyper-technological and hyper-connected civilization and to the "idols" (Cini) we have built for ourselves—at different levels: that of thinking, of thought/thoughts, and, finally, those of theory and practice—in the attempt to know, understand and inhabit the world.[14]

5.7 Teaching Complexity to Address the Dilemmas of the Hypercomplex Society

The urgent need to provide education/training in complexity, unpredictability and critical thinking (logic) is not only a strategic and determining aspect of the complex process of social and cultural co-building of the Person and the citizen. It also plays a fundamental role in view of the much-evoked "systemic change" and the way local and global contexts of reference are rapidly changing. In actual fact, the evolutionary advance of social ecosystems is proceeding towards a progressive redefinition of relational spaces and asymmetries, which brings with it the need—as already mentioned—for a "new social contract" (2003), built upon a robust foundation implicating the "quality" of educational processes and public, the global recognition of education and knowledge—and access to education and knowledge—as true "*human rights*"[15] capable, if concretely acknowledged and implemented, of defining the structural basis of the *equality of starting conditions* (a transversal, universalizing value), without which no global citizenship or possibility of inclusion can be made available/provided (Fig. 5.1).

As a result, *a reformulation of thinking, of thought*, regarding education, of didactics itself and, more generally, of educational processes and fields of knowledge (in an open and multidisciplinary key) becomes even more urgent. Subsequently, these proposals/projects need to materialize into educational actions and strategies functional to the social construction of change, a change which, let us remember, if imposed exclusively from above is (and will always be) an exclusive, short-term

[14] In the context of a multidisciplinary and, as usual, endless scientific literature, I limit myself to suggesting some non-linear pathways. See, in particular on thinking and "reflective" thinking, capable of freeing us from habit and unconscious action: Dewey J. (1933), *How We Think*, Italian translation, *Come pensiamo,* Raffaello Cortina Ed., Milan 2019; on "idols" of knowledge, common sense, science and speculation of science, on the path towards what we define as "knowledge", see Sini C., *Idoli della conoscenza*, Raffaello Cortina Ed., Milan 2000. To learn more, see also, Sloman S., Fernbach P. (2017), *The Knowledge Illusion. Why We Never Think Alone*, Italian translation, *L'illusione della conoscenza,* Milan: Raffaello Cortina Editore, 2018; Boorstin D. J. (1994), *Cleopatra's Nose. Essays on the Unexpected,* Vintage Books, New York; Boorstin D. J. (1998), *The Seekers. The Story of man's continuing quest to understand the world*, Italian translation, *L'avventura della ricerca*, Raffaello Cortina Ed., Milan 2002; Hustvedt S. (2016), *The Delusions of Certainty*, Italian translation, *Le illusioni della certezza*, Einaudi, Turin 2018.

[15] In this regard, I would like to refer to the action "*The right to education*", carried out at transnational level by UNESCO. Here is the official link: https://www.unesco.org/en/education/right-education

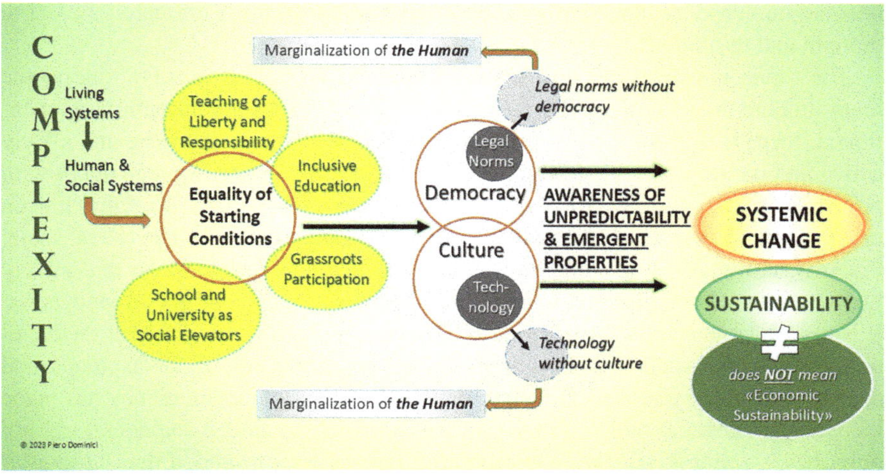

Fig. 5.1 Systemic change in complex social systems

one, favouring only the few. One thing is certain: we are faced with/we find ourselves within a *social complexity* that escapes the traditional devices of control and surveillance, even algorithms themselves (programmed by humans, let us remember this even if there is no need to do so) and which would need, as I have argued several times in the past, a *reformulation of thinking*[16] and a redefinition of the fields of knowledge that might contribute to reducing this complexity (to use an expression cherished by the great Niklas Luhmann), or at least to defining the *minimum conditions of the predictability internal and external to the behaviour* of organizations and systems. In this sense, many years before, Edgar Morin had spoken of "thought reform", telling us that thought reform would initially require a reform of teaching (that would extend from the first years of primary school up through university), which, in turn, would itself require a reform of thinking, that is: thought reform. With this paradoxical cycle, he also warns us that the right to think must of course be fully democratic, which implies the necessity of a revolutionary paradigm shift, giving free reign to complex thinking, allowing it to reorganize and connect fields of knowledge that today are still separated into narrow disciplinary borders. Thought reform is thus an issue that encompasses, for example, both anthropology and history. According to Morin, the Copernican revolution pales with respect to this potential mental overturn. Today, therefore, the responsibilities of thinking have

[16] Just as I was sending the Italian version of the monograph to the press, with reference to these issues, Nuccio Ordine's interesting interview with Edgar Morin appeared in the cultural magazine of the Corriere della Sera "La Lettura" (n.55°, Sunday 21st of August 2022), under the title "Edgar Morin. Questa non è soltanto una crisi economica, è una crisi sociale, è una crisi ecologica. Questa è soprattutto una crisi del pensiero" ("This is not only an economic crisis. It is an ecological crisis and, above all, a thought crisis"). The great French sociologist and philosopher recovered some of the study and research avenues he pursued and concerning which he has published various works over the years.

become immense, as never before. The tragedy of our times is intimately linked to thought and thinking.[17]

The topic, in effect, is extremely sensitive and difficult to resolve, due to the many implications involved. Certainly, we can start from an assumption: as mentioned, there is a close correlation between school/ education/training and a truly active and participatory kind of citizenship (which I have often spoken about with reference to the possibility of a *less asymmetrical relationship with Power*), even more so in social systems characterized by poor (if not non-existent) vertical social mobility and widespread *(im)moral familism* which continue to make this society highly corporate and resistant to (true and profound) change and social innovation. It is no coincidence that for many years I have continued to speak about the hegemony of a feudal social—and cultural—model but, above all, of an "asymmetrical society"[18] which, moreover, the ongoing global and systemic emergency has made increasingly unequal and exclusive. In the so-called advanced societies (and not only these), schools, education and training have always provided the only possibilities/opportunities for social emancipation and improvement of one's starting social condition; even more so, in a rigidly structured society. In short, these are the only "social elevators", now (unfortunately) completely blocked; indeed, for quite some time now, they have no longer been performing this vital function. The crisis of the *welfare* systems completes this extremely problematic picture which, having made precariousness an existential condition of life, has led to an ulterior weakening of the mechanisms of solidarity and cooperation while also calling into question the People's (citizens') right to know.

Remaining within this perspective of discourse and analysis, I shall repeat myself again and say that because "true" innovation, and the cultural change that characterizes it, are complex processes, or rather, they are hypercomplexities requiring governance, instruction, education, training and research must (should) be the supporting axes, not simple "tools" that arrive downstream of the processes of change to correct unexpected and/or unpredictable trajectories and discontinuities. Otherwise, we will always be obliged to chase after the accelerations of technological innovation and of a world system currently "designed" and conceived to enhance competition without rules and, above all, without waiting for those left behind. Only now are we starting to realize how short-sighted and self-defeating this vision and the strategies and policies that accompany it are, both in local and global terms. If we fail to correct our course, we will continue to chase after the changes induced by our hypercomplex and asymmetrical society, with very little chance of being able to manage and, at the same time, metabolize them. On the other hand, I repeat, the strategic issue is still that of education (and training) and educational processes. Because the social and cultural factors, often underestimated, remain decisive.

[17] See Morin E. (1993), *Terre-Patriè*, Italian translation, *Terra-Patria,* Raffaello Cortina Ed., Milan 1996, pp. 170–171.

[18] For clarification, I would like to state that these are concepts and operational definitions that I have proposed in various publications.

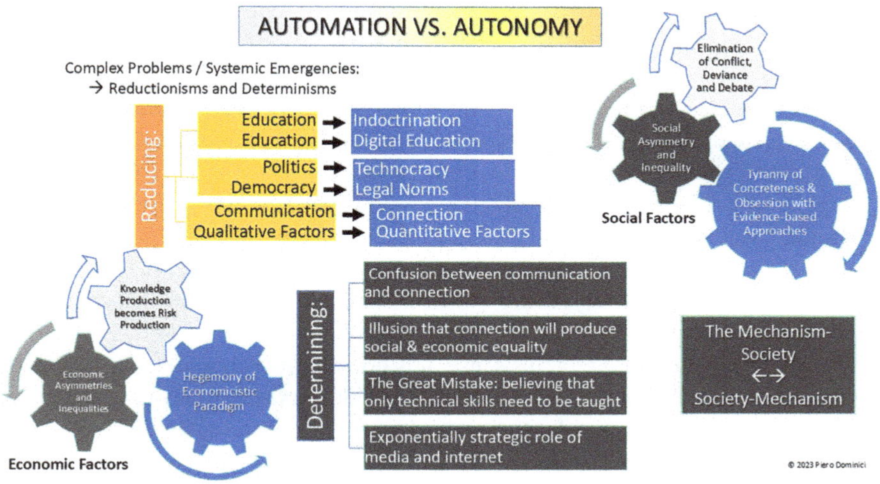

Fig. 5.2 Automation vs. autonomy

Building sustainable ecosystems, rethinking the "social contract" means facing the challenge of and to hypercomplexity, an unavoidable challenge that demands, first and foremost, that we rethink education and instruction, in a profound, radical way. To do this, we also need to rethink the very concepts of freedom and responsibility in a relational key, placing People, educational, relational and communicative spaces at the centre. At the same time, we might ask ourselves whether to choose education or indoctrination, communication or connection, quality or quantity, democracy or legal norms; in short, whether we are headed for autonomy or automation (Fig. 5.2).

Unfortunately, the underlying impression that we are continuing to confuse these issues, in the erroneous belief that processes of automation and simulation will grant us autonomy and freedom.

5.8 Facing the Challenge of the *Paradigm Shift*

When, in the mid-nineteen-nineties, I was studying and exploring the topic of the "paradigm shift"—a theme addressed by the great classic schools of thought and by innumerable multidisciplinary analyses in scientific literature—I declined and defined this question in terms of a process of *progressive reversal of the complex interaction existing between biological evolution and cultural evolution*; a process which, even back then, had demolished all of the boundary lines and/or false dichotomies that rendered our educational and training institutions inadequate and chronically unprepared for change and unpredictability. A paradigm shift which, in addition to bringing with it yet another anthropological transformation, which is much more profound and radical than those of the past, immediately put to the test the stability of the *overall architecture of the fields of knowledge and skills*, also questioning the well-known dichotomy between "facts" and "values" (Max Weber), apparently irredeemable and insurmountable. The objective, even at that time, was—and remains today—that of trying to bring the debate on ethics and anthropological transformation itself back *within* the more purely scientific ambit of epistemology, methodology and the production of knowledge and fields of knowledge.

Themes and issues that closely concern the so-called "cultural question": and, when we say "the question is cultural" (!), merely affirming a principle of this kind obliges us to deal, necessarily, with a systemic view and an approach that cannot but be multidisciplinary.

We ought never to forget, in this sense, it is non-quality schooling and education (a concept in dire need of a solution)—along with university education, which intervenes later and at a higher level—that *create/define the structural conditions favouring an unequal society* incapable of providing the conditions that guarantee equality of starting opportunities. These are the very germs of the "cultural question" and of what I have called, at an operational level as well, the "asymmetrical society".

The "cultural question", as we know, has deep roots, multiple variables and contributing factors that require consideration. In addition to the importance of policies and investments in education, training and research (not slogans), a change of culture and mentality needs to come about—something neither simple nor obvious—which can only be a long-term working process that needs to be undertaken immediately: education and training must be protagonists. And perhaps—but this is a strictly personal opinion—we should stop thinking of education exclusively as a "means" for obtaining a job (the same goes for training, whose aims, in any case, need to be better finalized).

The problem is that we are no longer even able to communicate the importance and value of education and training. Furthermore, there are many reasons why education in the humanities, in particular in the specific fields of the political, social, philosophical and communication sciences, is strategic, especially because it is precisely teaching (that is, the education and training of our young people) and public administration that have always been—and will long continue to be—among its most probable professional career outcomes. It is essential, in this sense, to think seriously about how to regain credibility (but—it must be clear—not just as a question of advertising and/or marketing) for education and training and (in particular), for certain academic degrees, earned in those very disciplinary fields that are erroneously considered less important and above all less "useful", according to the paradigm of the hyper-technological and hyper-connected civilization. It is a question of awareness, above all of political will (?), but also of awareness and responsibility on the part of everyone.

5.9 Elastic, Open Minds Vs. Rigid, Structured Pathways

One fact is certain: we continue to address a strategic issue of this kind in terms that, once again, trivialize the complexity and dimensions involved. We continue to insist on the uselessness of degrees in the humanities and, more generally, of humanistic education, not to mention the continuous revival of the age-old, misleading clash/diatribe between humanistic and scientific culture,[19] which resonates within public—in particular Italian—discourse. Time truly seems to have come to a standstill despite the accelerations and numerous special effects introduced by multimedia didactics, proposed and disseminated without any genuinely robust project "behind it", and, above all, without a methodology and epistemology that would support its introduction into educational processes and training. And it is actually incredible and surreal that there is still a need to discuss/debate this issue and that, when we do, it is perceived as original, almost revolutionary, almost as a "brand new" challenge.

Once again, these trends are strongly indicative of the *cultural delay* that has been underestimated for decades (as it is considered relevant to academics only), whose profound implications impact multiple levels of discourse and practice: education, instruction, training, research on education and thinking, *experts in fields of knowledge* involved in policies, analyses, strategies and the cultures of organisation, safety and prevention, the strategies of resilience, risk/uncertainty management, conflict mediation and management, as well as management (?) and simplification/*reduction* of complexity, etc. To this we need to add a further critical issue: unfortunately, especially in Italy, any talk or debate on any topic tends immediately towards polarization and an ideological and biased interpretation.

Beyond the data often read uncritically or, in some cases, in favour of a given interest, the fundamental problem—I reiterate this point vehemently—is that what we do not need are poorly-educated/trained graduates, what we do not need are graduates who, in some/many cases, are unable to speak and write correctly in their own language or, even worse, who do not know how to "think". It is important to make this clear, that we have no need whatsoever for graduates whose identities/personalities and intellectual maturation have been untouched by the studies they have carried out, who have not been affected in the slightest by the subjects and approaches they have studied/practised. Undoubtedly, some qualifications continue to be more serviceable on the job market, not so much for a question of knowledge and/or specific skills acquired and required—which may be necessary to carry out certain "tasks" (I continue to remain convinced that work is learned only in the workplace). As to the credibility and consideration that—in some cases deservingly so- some have the reputation of being (are "branded" as -and "brands" counts, even in higher education) "difficult degrees", characterized by more demanding and selective study pathways, which better guarantee those who have to select the so-called "human resources". But, in the long run, if only the aforementioned criteria

[19] *Ibidem.*

and quantitative indicators are considered, the level will have to be lowered in these sectors too. If the logic remains that of a purely *quantitative* type aimed at increasing the number of students enrolled and graduating, there will be no way out and, above all, asymmetries and *social immobility* will increase. Therefore, it must be clear, while these objectives are absolutely acceptable, they need, however, to be integrated with strategies that pay greater attention to the quality of graduates (a complex concept, which must be resolved and *rendered operational*). I cannot but repeat myself: increasing the number of graduates, as we have tried to do in recent years, will not suffice. Our young people need to complete a process of intellectual maturation as People first and foremost, because the organizations where they will go to work are "made" of people; and this maturation, evidently, must be based on a serious and rigorous education/training that is able to shape elastic and open minds, capable of grasping uniqueness, of valorising diversity and unpredictability, of recognizing the emergent, *of inhabiting (hyper)complexity*. As mentioned previously, however, the public debate circulated by the media is of no help whatsoever, because it consists of slogans and *emotional outbursts*, simplifications and clichés which, time after time, are further propagated by politicians and so-called experts.

As I have been repeating for years, the complexity of the change underway requires diversified profiles, knowledge and skills (*hybrid figures*), not just "technical", but well-trained and, possibly, passionate young people! Furthermore, a degree needs to return to indicate, always and in any case, a value; beyond the debates on the legal value of a qualification, it needs to certify once more the quality of the training and the intellectual maturation of the Person who obtains it, regardless of the final evaluation. In addressing these issues *without mincing words*, I intend to reaffirm, coherently and forcefully, the strategic relevance of these issues to the so-called "cultural question": education, instruction and training (along with research) have always, always (!) been the only "tools" capable of bestowing dynamism on social systems devoid of *upward social mobility* and of meritocracy, but full of widespread "amoral familism" (Banfield). To this I add that it is also of vital importance to pay attention to the quality of the training of those who choose courses towards diplomas and degrees considered less strategic (humanistic disciplines and political-social sciences), who, upon leaving university, may well end up working in the strategic sectors of public administration and, in particular, in teaching. Not only those in Politics, but each of us is called upon to take on our responsibilities! Because what we are dealing with are not only systems or large organizations.

5.10 The Cultural Issue and the Culture of Cunning

From this point of view, another crucial aspect of the "cultural question" concerns the widespread mentality, especially, though not exclusively, prevalent in Italy, that it is possible to solve any problem whatsoever by resorting to the legislator (once again with the added belief/narration that all problems will be solved digitally), what is more, staking everything on granting *carte blanche* to technology.

Evidently, when dealing with the "cultural question", we are touching a raw nerve not only when it comes to Italy, but also regarding many other "actors" of the so-called international community. In particular, Italy is a country that continues to entertain the delusion that it can solve any and every problem by recourse to the law (criminal law, in particular), to prohibitions, sanctions, and more in general, to repression: necessary but, once again, insufficient, logics and "devices", which are incapable of dealing with the qualitative dimensions of the questions being faced. And thus, "educational emergencies", an ethical void (Jonas), moral torpor, nihilism, a culture of cunning and *legitimate illegality* (Dominici), clientelism, corruption and *amoral familism* (Banfield), are, in one way or another, fast sliding downhill toward us, calling into question social bonds: complex issues that cannot be resolved (merely) by further recourse to increasingly harsh decrees and sanctions (I repeat: necessary but not sufficient). Complex issues that, as I warned many years ago, will not be solved by digitalisation, by IT systems, by artificial intelligence and/or by increasingly sophisticated hyperconnected ecosystems (extremely useful but not decisive). The heart of the problem—and I reiterate this point fervently—concerns education, the educational and training institutions, the cultural models, the thought systems. The heart of the problem lies in thinking itself, in the quest for shared ethics and, on a micro level, in demonstrations of behaviour.

5.11 The Society of Irresponsibility and the Overcoming of the Dichotomy Between Humanistic and Scientific Education

In 2009, precisely within the perspective expounded above, I conducted research following the earthquake of that same year in L'Aquila—a city of which I am particularly fond and where I taught for quite some time—starting from an empirical investigation of how the press and media had reported the earthquake. I chose the title "La società dell'irresponsabilità" [The society of irresponsibility].[20]

This to underline that we are in a context where very serious deeds and facts can take place in full compliance with laws and regulations and perfect formal observance of rules and procedures. A dimension, that of responsibility—which closely concerns liberty; it is necessary to rethink the very concepts of liberty and responsibility in a relational key, so that they presuppose a We, not an I. But a *culture of responsibility* (like ethics) can never be imposed *from above*, nor can it be conveyed by means of sophisticated communication/marketing campaigns. It needs to be built up socially and culturally from the very early years of life and the first years of study. Schools and the other socializing agencies are those truly responsible for this

[20] Here the reference is to a contribution published by LSDI—Libertà di Stampa Diritto all'Informazione: La Società dell'Irresponsabilità http://www.lsdi.it/2016/la-societa-dellirresponsabilita/

very important process which also directly concerns issues relating to legality, corruption, prevention, respect for *the Other than us*, living together, etc.

Responsibility is a core topic, linked to the issues discussed previously. Complexity and systemic thinking presuppose a careful evaluation of the variables involved and their consequences. Being educated in complexity, in the scientific method, in critical and systemic thinking, brings with it an epistemology of uncertainty (Morin) and also attention to the (well-known) principle of precaution. Overcoming the dichotomy between humanistic and scientific education is too important, since we can no longer afford even the luxury of training only technicians; this precisely because we dwell in a hyper-technological civilization and because the implications of hyper-technological civilization are social, political, cultural and concern everyone...identities, subjectivities, life on the whole. At the same time, the question of a *paradigm shift* is no slogan; in a nutshell, we can say that the extraordinary scientific discoveries of recent years—genetic manipulation, artificial intelligence, robotics, the possibility of replacing parts of the body, stem cells -oblige us to revise all our categories and relative operational definitions, starting from those of consciousness and life. What does this highlight and what does it entail? Never before has cultural evolution been able to condition and determine, in a non-linear/non-causal way, biological evolution; the paradigm shift is no longer simply an efficient theory or a successful narrative.

Hence, the need to rethink the fields of knowledge, the pathways and the logics that animate educational institutions and scientific research itself is not linked to somebody's hunch. It is the phenomenal reality that requires us to take a quantum leap. It is the reality that is increasingly interconnected and interdependent, it is the systemic, global reality, that asks us to overcome these logics of separation and confinement of fields of knowledge and skills.

5.12 For a Culture of Complexity in the *Society of Irresponsibility*

Rethinking education by going beyond "false dichotomies" is a matter of crucial importance that can be addressed only from the perspective of complexity, critical thinking and a systemic view. By no means has this perspective been explored or developed by educational-training processes, which are increasingly focused on certain (dramatic and, in some ways, devastating) degrees of separation, most of all between studies in the humanities and sciences, between the human and the technological, between technology and culture. Furthermore, civic sense, an education in citizenship, shared ethics and a strong model of culture and identity are fundamental "devices" for the very survival of social and organizational systems. Also, and above all, *I repeat*: because *ethics and morality cannot be imposed*.

These are processes which—as mentioned above—call into play multiple variables and require profiles and skills built in the field. These issues, however, do not only concern the issues of inclusion and citizenship: for example, I continue to believe that "real" prevention is achieved and built within educational and training processes, particularly at school and that (long-term) policies and substantial, even massive, investments are required.

In keeping with this line of discussion, I propose a number of considerations to be underlined while keeping in mind, once again, the absolute relevance of issues concerning education and instruction. Italy, for example, is a country with a complex and intricate regulatory and legislative framework: there are many laws (too many, perhaps), professional codes, ethical charters, guidelines, systems of formal rules, value and cognitive orientation systems. Yet these "tools" continue to prove necessary though insufficient precisely because of a dimension, crucial and foundational at the same time, called responsibility. This concept, as I mentioned above, urgently needs to be redefined in a relational key. It is a dimension that escapes any type of "cage" and/or control system/device, because it concerns the freedom of the People (another topic to be explored in greater depth, linked to the theme of emancipation in modernity: the concept of "generative freedom" is interesting). From this point of view, how can we disagree with the definition of "society of individuals": a society (us) in which many individuals (precisely) feel that they are not obliged to respond to any of their actions, much less to a "community" whose bonds have been greatly weakened (and, not surprisingly, there are those who speak of the end of social bonds)?

This, in fact, is the principle reason that I entitled the above-mentioned book *The society of irresponsibility,* with the precise intention of connoting this extremely critical situation, which can be only partly linked to the economic crisis (or to economic indicators): the "cultural question" emphasises, once again, not only the crisis of educational and training institutions, but also the weakness of the old apparatuses and logics of control and repression which have never solved underlying problems. The strategies are always "short term" (the culture of emergency vs the culture of prevention, the culture of control/repression vs the culture of responsibility—at all levels and within all sectors of organisational and social practice).

Italy, a country on the brink of paradox, is culturally grounded in the "principle of irresponsibility", as well as in a widespread inconsistency of behaviour (this always calls to mind the formula ethics-vs.-labels). Irresponsibility is widespread in all sectors, including those of communication and information, which are, to say the least, vital for the quality of democracy itself. Widespread irresponsibility is a true feature of the "cultural question", which legitimizes those who circumvent laws, rules and even shared social norms (the culture of cunning). This pervasive irresponsibility finds its ideal ecosystem within a historical context and cultural climate that always exalt those who put particular before general interest and the "common good". In this sense, one need only think of the metastasis of corruption which, as the news of recent decades has highlighted, involves not only the so-called "caste", but large sectors of civil society, which probably continue to believe, despite all, that they can obtain advantages from the "caste" itself. This irresponsibility also expresses itself in ethically incorrect behaviour that takes no heed even of the principle of precaution. Therefore, a widespread irresponsibility which makes the violation of laws and norms socially acceptable, which also has its roots in a certain "culture of cunning", which is developed and spread, at times unconsciously, in the very halls dedicated to the socialisation and education of the Person. Thus, what happens is that the solution to these problems, which are in some ways inevitable (but, evidently, not the only way forward), is always the same: continuous recourse to increasingly rigid and stringent laws and regulations. Although, to be clear, these are often necessary, even at times fundamental, nonetheless. as amply demonstrated by the social, political and economic history not only of Italy, these conditions/factors are insufficient. We are obliged to deal with the intrinsically problematic and complex "nature" of social systems, no longer attributable to the (significant) categories of risk, uncertainty, vulnerability, liquidity, etc. On top of this, almost paradoxically, stands the fact that never before in recent years has there been so much discussion on ethics and responsibility in all fields of social action (from politics to culture, from information to scientific and technological innovation etc.) This paradox might be simplified by using the "formula" of the triumph of labels over ethics. A country of paradoxes and contradictions (not only in cultural terms): on the one hand, for every "new" problem/criticality, new laws, new codes of ethics, new prescriptions, new prohibitions are immediately invoked, on the other, culturally speaking, those same laws, norms, "rules" are often considered as an obstacle to one's self-assertion and social success/prestige. Furthermore, what often appears to be missing is precisely coherence of behaviour which, communicatively speaking, seems to be (is!) much more effective than words and principles, explained by using more or less politically correct language. It seems clear that we are faced with a veritable "educational emergency"—even though I do not like this expression, which represents that still hegemonic mindset that deals with any problem only through exceptional, sporadic measures (from workplace health and safety, from health to violence, from forms of discrimination to bullying, from corruption to illegality)—linked to a multitude of factors and variables, which have led to a profound transformation of socialization processes and a crisis of the traditional agencies/institutions responsible for the internalization of values and the formation of personalities/identities (recognition- respect- altruism- civic sense- citizenship (being a citizen as an experience rather than as a—hetero-directed—subjugation).

5.13 School as a "Constitutional Body": *Vital Infrastructure* of Any Social System

In extremely simple terms, there cannot/will never be any "digital citizenship" unless the minimum conditions of plain "citizenship" are guaranteed first, which naturally take precedence, as they represent, in substance, the most essential guarantees. At the same time there cannot be "true" (social and cultural) innovation without guaranteeing the conditions of a form of inclusion that cannot be exclusive (the inclusivity vs. exclusivity issue returns). Schools have always played a vitally important role in democratic regimes and Piero Calamandrei member of the founding Commission for the Italian Constitution,[21] in a historic speech delivered in 1950 (which I refer to frequently), did not hesitate to refer to school as a "constitutional body". He held that, having read the Constitution, one knows that in the second part, the section entitled "The Order of the State", the bodies expressing the will of the people are described. "Those are the bodies through which politics are converted into law, where the vital, wholesome struggles of politics become legislation." He continues: "if it occurs to you to ask yourselves what the constitutional bodies are, the answer that will come naturally to all is that they are the Parliament, the House of Deputies, the Senate, the President of the Republic and the Judiciary. Yet it would not occur to you to consider the school among these bodies, which is, in actual fact, a vital organ of democracy as we conceive it. If a comparison were to be made between the constitutional and human organisms, one would have to say that schools correspond to those organs which in the human body create blood [...]. School is the central organ of democracy, because it serves to solve what in our opinion is the central problem of democracy: the formation of the governing class. The education and training of the governing class, not only in the sense of that particular political class that sits in Parliament and argues and discusses (even shouts), which is at the apex of specifically political bodies, but also of the governing class in a general cultural and technical sense: those who run factories and companies, those who teach and write: artists, professionals, poets. The problem of democracy is the creation of this class, which must not be a hereditary, closed caste, an oligarchy, a church, a clergy, an order. No. In our idea of democracy, the governing class needs to be open and constantly renewed by the upward influx of the best elements of all classes, of all categories".

These words are so clear and significant that they do not require further comments even with respect to their extraordinary relevance today. Having recognized and accepted (?) the strategic relevance of School and education, we must, however, ask ourselves a question: can we even speak about "digital citizens" or "connected" only, if we do not first educate/train people to be, first and foremost, People, then "citizens"? School—I repeat—is truly strategic, more so than other formal and informal institutions. Certainly, in this sensitive phase of change (paradigm shift, sharing economy, knowledge society, etc.), schools once again take on, or at any rate ought to once again take on a further, even more delicate, role of accompaniment, mediation, preparation and support for the changes brought about by the digital revolution. This aspect must in no way be underestimated, as it also it brings to the fore the "age-old", but constantly current question of *educating the educators*. To be (even) more explicit, with respect to the approach I am proposing and the perspective that marks my studies and research: there is not, nor can there be, any innovation without focusing on education and training and, at the same time, there is not, nor can there be, be any innovation without inclusion. These are questions of

[21] See Calamandrei P., (1950) Speech delivered at the III Congress of the *Associazione a Difesa della Scuola Nazionale* (ADSN), Rome, 11 February 1950.

vital importance which, I repeat, also call into question the "culture of communication" that we shall have to try to build, promote, spread, accept, share etc..

5.14 *Caring and Taking Care*. Education as a Relationship

The "educational and cultural question" that I have referred to several times is linked, as mentioned above, also and above all, to a problem of the interruption/crisis of communication between generations (a concept that needs to be resolved and developed). However, in this analytical perspective, we cannot fail to notice how the media, social networks, the new communicative environments/ecosystems—together with the famous "peer group" phenomenon—have devoured the relational spaces as well as those of communication and knowledge (?) that in the past were managed by the traditional educational and training institutions and agencies. In this thorny and constantly evolving situation, the social actors and professional figures involved in educational and training processes see their responsibilities increased further, also and above all, in terms of the management and development of new relational and communication spaces. I may be radical, but I always prefer to state openly what I think: aware that, in recent decades, schools and universities have been heavily penalized by cuts and counter-reforms, I continue to remain (profoundly) convinced that there are jobs/professions that should be undertaken/chosen when one expressly perceives/recognizes (and values?) a certain "calling". These are trades/jobs/professions which truly place "at the centre" People, their growth and maturation (not only intellectual), the development of their knowledge and skills and their relational and social capacities. These are trades/jobs/professions that cannot be chosen solely as a form of social prestige and/or because perhaps, in one way or another, they permit the exercise of forms of micropower over Others.

The very concept of "taking care" of a Person is a complex concept (systemic, relational dimension) which implicates the necessary ability and culture, as well as the sensitivity and capacity for empathy required to activate social and relational dynamics which can only be/reveal themselves to be systemic and, therefore, complex. Focusing in particular on the (apparently, at least) weaker subjects, who are running a tangible risk of being left behind, which would further fuel the vicious circle that is dramatically increasing new forms of inequality and asymmetries that we have referred to several times. In other words: teaching, educating and training presuppose a relationship, implying and requiring the acts of listening, learning, sharing, cooperating, collaborating, comparing, and engaging in dialogue. It is only to the smallest extent, evidently, that all this means transmitting and/or imparting *notions*, arguably with a less than open and inclusive attitude, something which might one day prove more or less "useful" to those who have introjected and metabolised them. In other words, teaching, educating and training mean, above all, meeting and opening up without prejudices to the OTHER FROM OUSELVES, triggering and fueling "relationships", seeking dialogue and discussion: never ceasing to ask questions and learn, being aware of the many variables and contributing factors that come into play in educational and training processes, making them complex, multidimensional, capable of embracing uncertainty and unpredictability, which constitute the "true" added value and an element of wealth of inestimable value, always provided that we are able, concretely, to enhance them and make them a common and shared value.

Educating is, first of all, "relationship", a relationship that is, once again, complex and demanding. Again, along this line of discussion, moreover, we are well aware of how our children, students and, more generally, young people—let us say –are "on to us", observing how we behave (as People, and not just as educators, teachers, professors, trainers etc.), well beyond what we teach and say to them. In short, what counts are the "facts" and the consistency of behaviour, not "words". Beyond knowledge and skills, what matters—and matters deeply—is our desire to become involved with, meet and come to know them, to "listen" (a process that is anything but simple and obvious), understand their experiences, uncertainties, aspirations, anxieties, even their aspirations and dreams. All this must (should) also happen within the framework of the spheres of education, training, and communication, of and relational spaces, in a more or less intensely codified manner, like that of certain educational and training institutions. That is to say: "taking care of…". Reactivating the relational and systemic dimensions.

For these reasons (and others), I have always spoken of the urgency of rethinking relational and communicative spaces, in particular of educational and training institutions (but the concept can be applied to the entire world of work and to all types of organizations). The credibility and authority of educators, teachers, professors, etc. is based precisely on their behaviours and on the consistency of these behaviours with what they assert/profess (an issue common to Politics as well). If you ask to be "listened to", you need to listen yourself and to know how to listen; if you ask for correctness, you must be correct first; if you demand respect and a sense of responsibility, you must first of all be respectful of the Other and responsible towards the Other, even if the *relationship* is, inevitably, asymmetrical, due to roles and hierarchy. Once again, we come up against the systemic and relational dimension of the Social and of Life. This is why certain "roles" and "activities" require, in my opinion, profound awareness of the responsibilities and implications linked to educational and training processes; but they also require participation, passion, communication (non-technical) and relational skills, and empathy. This means actually putting yourself on the line and placing your stakes on involving and including the OTHER.

5.15 *Fractures* That Need Healing

After having reflected on the systemic and relational dimension of teaching, we cannot shirk from asking ourselves what damage is still being done (and what implications are being produced) by the logics of separation and confinement between the fields of knowledge and skills? And in what sense could/should the prospect of a "new humanism"—reconfigured from an "other" perspective—lead to the overcoming of those very logics, and in particular, of the "false dichotomy" between studies in the humanities and the sciences?

I am fond of repeating this over and over: we can even pretend not to notice, and to delude ourselves that we can keep going forward using "disjunctive thinking" and the consolidated and still hegemonic models and paradigms, but experience and observation, which are never neutral, of phenomenal reality highlight, especially as a consequence of the extraordinary scientific discoveries and the continuous accelerations induced by technological innovation, that the "age-old", atavistic boundaries between nature and culture, and even more so, between "natural" and "artificial", have not so much been blurred as been completely done away with. Consequently, the logics of separation between scientific and humanistic education display even more clearly their substantial and complete inadequacy. "Fractures" that must be reunited/healed and that, for the reasons we have indicated previously, render the urgency of education/training in complexity, critical thinking (logic), and a systemic vision even more unavoidable.

However, in situations and historical-cultural contexts that have aways based their founding points and their reasons for cohesion on habit, conformism and tradition, the resistance to a such a radical change in perspectives like this (models, practices and tools), and approach, is very strong, both from *outside* and from *within* every single ecosystem (internal vs. external: another false dichotomy), and paradoxically, this resistance stems above all from the "halls" where knowledge is produced, processed, and shared (that is, where it *should* be shared!) In other words, this resistance is linked to motivations of various natures: pedagogical and communication cultures, dominant logics, the feudal social model, cultural questions, primacy of politics in every dimension, *amoral familism*, organizational cultures, climates of opinion, etc. Fundamentally because, above all, as I was saying long before it became trendy, in any field of individual and collective practice, innovation means questioning fields of knowledge and consolidated practices, individual and collective imaginaries, unbalancing equilibriums, *breaking the chains of tradition* (cit.), abandoning the certain for the uncertain, even along with the perception of considerably higher risks (opportunities). In other words, it makes the systems and the communicative and relational space that characterizes them more vulnerable, at least temporarily. Accepting unpredictability and opening up to the improbable and the indeterminate.

All the more so, in times of uncertainty and rapid obsolescence of the fields of knowledge and skills, education and training become issues of strategic relevance, more than in the past. Decisive issues, to begin with, for the complex process of social and cultural construction of the Person and the citizen and, secondly, of that public sphere that plays a role of fundamental importance also in consideration of the constant and rapid transformation of the local and global context of reference. In this sense, I continue to be convinced, and it is from this approach that I have developed my research, that technological innovation has always constituted a strategic factor for change in social systems and organizations, but that if not supported by *critical* (as well as systemic) *thinking* and by a culture of complexity but, above all, in practical and operational terms, by long-term policies capable of triggering and supporting cultural change (strategic centrality of schools, education, universities, and research), regularly turns out to be an extraordinary opportunity for the few and/or, better said, for more or less enlightened elites. From this point of view, we shall long continue paying for the substantial inadequacy of our educational and training systems, which are still designed and implemented on the short-sighted, as well as disastrous, separation between the "two cultures", the scientific and the humanistic; a separation which has been opposed, at least up to now, only by "lip service", from schools and universities alike. From this point of view, as regards what I have called "the interconnected and hyperconnected society", the horizontality and democracy of procedures and systems cannot/will not be guaranteed in any way by technology in and of itself, since the human factor and the quality of social relations and bonds of interdependency, inside and outside social systems, will always make the difference; inside and outside complex organizations.

5.16 Cultural Change as Agent/Vector of Social Change

As I have had the opportunity to state several times, the interconnected/hyperconnected (and hypercomplex) society determines a truly revolutionary change in all socio-political dynamics as well as in cultural and productive processes: intellectual capital along with knowledge production and distribution are its strengths. This is yet another extraordinary technological revolution, the origins of which are to be found in the final decades of the nineteenth century, generated by the rapid progress of industrialization which had already determined, even then, a crisis of control, to which the information and knowledge society seems to have offered (possible) solutions (information—knowledge—risk management/complexity/unpredictability-control-trust). A revolution achieved by strengthening the means of information and communication, networks and artificial intelligence systems, which have been capable—first of all—of governing, at least in terms of narratives and representation, the globalised economy; a revolution that is causing an unprecedented upheaval in the history of the evolution of social systems, which even James Beniger appropriately associates with the definition of control revolution (the true "driving force" of the information society). What this amounts to is what he calls:

> …the Control Revolution, a complex of rapid changes in the technological and economic arrangements by which information is collected, stored, processed and communicated, and through which formal or programmed decisions might affect societal control. From its origins in the last decades of the nineteenth century, the Control Revolution has continued unabated, and recently it has been accelerated by the development of microprocessing technologies. In terms of the magnitude and pervasiveness of its impact upon society, intellectual and cultural no less than material, the Control Revolution already appears to be as important to this century as the Industrial Revolution was to the last.
>
> But history alone cannot explain why it is information that increasingly plays the crucial role in economy and society. The answer must be sought in the nature of *all* living systems – ultimately in the relationship between information and control. Life itself implies

> control, after all, in individual cells and organisms no less than in national economies or any other purposive system.
>
> [...] By far the greatest effect of industrialization, from this perspective, was to speed up a society's entire material processing system, thereby precipitating what I call a crisis of control, a period in which innovations in information processing and communication technologies lagged behind those of energy and its application to manufacturing and transportation.[22]

Electronic networks and new social and digital ecosystems favour an almost instantaneous planetary communication (connection) even if transition towards the society of global information/communication (infocosmos) still fails to provide adequate guarantees in terms of a concrete opening up of the systems and social/cultural innovation, with dynamics that oscillate between chaos and order, between inclusion and exclusion. Dialectical and ambivalent processes that have not yet found an opportunity for synthesis, and which make dramatically evident the absence of an adequate theoretical-interpretative model capable of embracing hypercomplexity, defining new problems and not just solutions, new questions and not just answers. There is still very little awareness of the fact that any practice and research activity can only be *conceptually oriented. At the same time*, there is still little awareness that the dimensions of *thinking* and thought are of vital importance.

---

[22] See Beniger J.R. (1986), *The Control Revolution*, Italian translation, *Le origini della società dell'informazione. La rivoluzione del controllo*, Turin: UTET, 1995, pp. XXXVIII–XXXIX.

Chapter 6
The Tyranny of Concreteness and the Urgency of a Radical Breakthrough

> *Probability is expectation founded upon partial knowledge. A perfect acquaintance with all the circumstances affecting the occurrence of an event would change expectation into certainty, and leave neither room nor demand for a theory of probabilities.*
>
> *George Boole*

Abstract Teaching complexity, the scientific method, critical and systemic thinking, nurtures a *multidimensional* kind of thinking and awareness that although data and empirical evidence are indubitably fundamental, they fail to give a complete picture of phenomena, of reality, and of the hypercomplexity of the Social and the Vital. Never before has it been necessary to modify so radically the approach we apply to investigate the reality of the phenomena that surrounds us. In other words, we need to change our *thinking* and the thoughts that permit us to act: thoughts, that are themselves actions, actions that are necessarily based on theory—essential to bring about any form of change—but theory, along with thought, is currently considered useless or of very little use, almost a waste of time, yet another way in which the *tyranny of concreteness* continues to manifest itself, in infinite forms that are conventional and hegemonic: all areas of intellectual and cultural production, even scientific research, have been reduced to one dimension, the apparently neutral and objective dimension of quantity, and rendered functional, not only to profit, but also to the deceptive principle of utility, on which we have based the counter-reform of our educational and training institutions.

Keywords Teaching complexity · Thought system · Systemic thinking · Tyranny-of-concreteness · One dimension · Quantitative · Multidimensional thinking · Rapid obsolescence · Utility principle · Concreteness tyranny

P. Dominici, *Beyond Black Swans: Inhabiting Indeterminacy*, Philosophical Studies Series 160, https://doi.org/10.1007/978-3-032-09029-4_6

Among the profound yet problematic dimensions that must be dealt with, there is certainly a dictatorship (and an obsession), which I have called the "tyranny of concreteness": a dictatorship, an aporetic hegemony which determines the progressive marginalization of thought and of everything that is "theoretical", "abstract", "unobservable", in other words—according to the dominant narratives—"useless".

But before beginning our reflection on this specific point, let us start from the definitions:

The Italian *Dizionario Zingarelli* [Zingarelli Dictionary] defines concreteness in these terms:

Concreteness: characteristic of who and what is concrete.
Concrete: [derivation lat. concrētu(m), past participle of concrēscere 'to condense, harden, coagulate' ☼ av. 1327]

(A) adj.:

1. (lit., rare) Dense, compact, solid, congealed: *I would mix it with concrete hops and chicory sauces* (F. REDI).
2. That which is identifiable through sensory experience: concrete objects; ANT. abstract | That which has a close connection with reality: moving from abstractions to concrete facts | Practical: concrete experience; *he is a concrete man*; ANT. theoretical | Precise, clearly determined: concrete project; concrete idea | in concrete, (ellipt) in a concrete way, from a concrete point of view | concrete noun, in grammar, that which indicates things that are real or imagined as such; ANT. abstract.
3. concrete art, artistic direction born within abstractionism, characterized by a more rationally geometric orientation. SYN. concretism.
4. concrete music, that based on the use of natural noises that have undergone various electroacoustic manipulations.
   || concretely, adv. In a concrete way, from a concrete point of view; practically.

(B) noun:

- That which is concrete: going from the abstract to the concrete; stick to the concrete.

In addition to my academic and research activities, I am often asked to hold tutoring/training/refresher courses for managers, executives, public administrators, school directors, teachers, etc. Along with the relative differences, specificities and nuances, what never fails to capture my attention is that, in general, the unifying element among all these categories, professional and working fields, as well as (and above all) among these experiences, is—apart from the interest and curiosity shown for novelties, trends and possible updates—a sort of *obsession with concreteness*, with "how to do it", solely and exclusively with "solutions". With few exceptions (there are always some, but they remain such), what all the attendees regularly demand are "concreteness" and "facts" only, which they continually remind you of, almost as though they need to feel "reassured". At least until further along, when they begin to appreciate the strategic value (at all levels of analysis and practice) of thought, of theory/theories, of an "other" view, of a different approach to issues and, more generally, to the unpredictability of the social and the human, of the (complex) systemic relationships between People, subsequently realizing that by continuing to proceed in the usual directions, with the usual approach, "tools" and procedures, in the illusion of total control and predictability, they will never be able to bring about any change and will encounter many difficulties when seeking to manage new risks, uncertainties and vulnerabilities. Continuing along these lines, what one ends up with is a far cry from a paradigm shift, a far cry from a new ecosystem, from business networks, *open systems*, sustainability and the "digital revolution".

But let us go back to concreteness. Despite knowing and understanding its logic—currently no one can afford not to be concrete nowadays, (perhaps) not even artists or so-called intellectuals anymore—despite knowing its importance and the whys (as well as the rhetoric), despite having to deal continually with it, the further I go, the more I am convinced that there is no area/field/sector of social action or social, organizational and systemic practice, of production, even of the most creative kind, no "area" of our existence, of the so-called "vital worlds", from work to research, from education to training, that is not managed/controlled/directed by an (often) obsessive, almost compulsive and exclusive quest for concreteness (and utility), of "we only need concrete things/we need to be concrete", of "we must be concrete", of "we only do/are interested in concrete things". It is precisely during this delicate phase of paradigm change and *anthropological transformation*, that this obsession is particularly counterproductive, above all in the sectors of education, training and research.

Paradoxically, no one, at the moment, would dare to say/write the opposite regarding the importance of certain keywords-labels. All are in favour of critical thinking, they are all for imagination and for recovering the importance of the more creative disciplines in educational and training processes, they are all for interdisciplinarity and trans-disciplinarity. In their (thunderstruck) fervour for empathy and unpredictability, they are all talking about the "paradigm shift", having suddenly realized that "know-how" and skills alone do not suffice, even in this hypertechnological and hyperconnected civilization, indeed—as I have maintained for many years—especially in this hypertechnological and hyperconnected civilization. Currently, crucially important questions such as the strategic/vital role of education and training and the importance of imagination are being discussed (?) by everyone, absolutely everyone, but the bottom line is that they are being propagated and fueled mostly by slogans and "buzzwords" or by citing the aphorisms of great/scientists/philosophers/writers: in short, as keywords that have been completely stripped of any effective (that is, truly concrete), operational translation into either pedagogical and cultural models or organizational, social, political and cultural life. In reality, beyond a given type of hegemonic rhetoric and narratives, functional above all to the construction of a positive reputation/image, both as an innovator and as an innovative (and efficient) organization on the whole, it is clear that awareness is still scarce when it comes to the strategic relevance of *thinking* (and the thought system), speculation, the imagination, creativity, of a bent for abstraction, of error and the opportunity to make mistakes in order to learn. In reality, what I have called, and not only in this chapter, the obsession with concreteness, or better, the "tyranny of concreteness", is condemning us never to be able to trigger/determine (?) any real change (only adjustments, at best), to create any "real" innovations, except some of a superficial type. It is also condemning us, who knows for how long, to adapt almost passively to the extraordinary technological transformations taking place, which involve, first of all, an *anthropological transformation* of identities, experiences, epistemologies etc.

This obsession, this tyranny of concreteness is condemning us not to know how to *inhabit hypercomplexity*[1] and the future.[2] It is condemning us to take refuge, once again, in our incompleteness, in our (in)securities, in our inability or, in any case, our lack of courage to abandon the routes and pathways already travelled down a thousand times before. It is condemning us to take refuge in our prejudices and clichés (at all levels), in styles of behaviour and choices that have always worked (?), in the "things" we have always done in a certain way because it was "successful" or produced clear "results" and solutions ("simple solutions to complex problems") (?)...labels vs. ethics. In other words, it is condemning us not to determine, not to attempt to govern in any way the change evoked and desired by all, but to choose to go in the usual direction.

An obsessive and exclusive quest often pursued especially by those who—at the level of public discourse—talk about creativity and "creative thinking" (a term which apparently becomes valid only when expressed in the English language), about innovation, change, transformation, *digital transformation*, multidisciplinary and interdisciplinary approaches, paradigm shifts and complexity (found everywhere). More recently, there has been talk of "digital complexity" (precisely by those who had associated the digital with simplification, with a new "digital citizenship", a new democracy and even more so with the simplicity of certain processes and dynamics). However, against the background of their working and professional lives and even their research, they continue, in many cases, to play things by ear, the same as ever. In short, at least for the moment, we are faced—as mentioned before—with keywords used above all as slogans. Those who *inhabit* educational and training institutions, those who dwell in the world of higher education, are well aware that what are always requested are "facts", "concreteness", "how to do it", "what is useful". They also know all too well that only "solutions" are demanded, (along

[1] See Dominici P. (1995–2019), *op.cit.*

[2] See Kaku M. (2018), *The Future of Humanity*, Italian translation, *Il futuro dell'umanità*, Rizzoli, Milan 2018.

with the proliferation of guidelines, decalogues, manifestos, recipes, new axioms, etc.). They see, too, that (practically) no one is truly interested in "how" the alleged solutions are arrived at, and even less so in "why"; continuing to think and behave as if organizational/social systems were "complicated" and not "complex", with the consequently exclusive involvement of certain experts belonging to specific fields of knowledge. Those who are apparently the best at providing/guaranteeing certainties and rational and "linear" explanations, reinforcing what I have called the "grand illusions of the hypertechnological civilization". In the meantime, thinking, creativity, education, continuous training, learning, imagination, innovation, the centrality of the Person and the human, the importance of error[3] and unpredictability, the sharing of knowledge and fields of knowledge: these are what everyone is speaking and writing about today, without ever having dealt with them, however. The real problem—let it be clear—is not that they talk about these things, but rather that they always present themselves as "experts" in and on everything, as if knowledge, skills, preparation and experience were not required/necessary in these sectors/fields. And yet, even as they are all talking and writing about these matters, in terms of choices and practices, they continue to work/coordinate/direct/operate in their institutions and organizations by seeking, above all, order, stability, balance, conservation, not only for reasons of power or *relative advantage,* but also because, in the hypertechnological civilization, it is precisely through equilibrium and stability that the technical fields of knowledge realize their maximum aspirations and cognitive potential, in terms of control, security, rationality.

Thought, theory/theories, philosophy, even the most creative disciplines, are not "useful", are of no use or, in any case, of little use. Then again, what is surprising about this? Speaking for my own nation, Italy is a country that continues to question the value of "culture", instruction, training and education. A linear and disjunctive kind of "thinking" (?) that has increased to the point of becoming hegemonic even when it comes to the halls of education and training. After all, as I have said above, what matters is "being concrete", knowing "how to do it", finding the "solutions": as usual, the same solutions as before, those that have always worked in the past,

[3] See, in particular Swartz R.M., *Mistakes as an important Part of Learning Process*, in "High School Journal", LIX, n°6, March 1976, pp. 246–257. I shall simply point out, in this case too, some of the "classics" on these topics: Canguilhem G. (1966), *Le normal et le pathologique*, Italian translation, *Il normale e il patologico*, Einaudi, Turin 1998; Oppenheimer J.R, *Scienza e Pensiero comune*, Bollati Boringhieri, Turin 1965: another "classic": Oppenheimer, the great physicist and scientist who speaks of "science" as the "discovery of errors" (pp. 61–63). Among the most recent publications, also introductory to the topic and the numerous implications related: Baldini M., *Epistemologia e pedagogia dell'errore*, La Scuola, Brescia 1986; Benes R. et al., *Per una pedagogia dell'errore*, Asterios, Trieste 2017; Dominici P., *Dentro la "civiltà senza errore". Sintesi complesse: l'Umano, il legame sociale e la civiltà ipertecnologica,* in A. De Cesaris (ed.) *Vite digitali. Essere umani nella società del XXI Secolo*, FrancoAngeli, Milan 2020. Among other "classics", see also: Bruner J.S., *The Process of Education*, Harvard University Press, Cambridge 1961, a work in which, among the numerous issues addressed, error is defined and recognized as a constitutive element of learning.

albeit perhaps in different contexts and situations, or involving different causes, variables, indicators, parameters or people.

And no doubt, once again many will claim that they too have always supported thought, theory/theories, a long-term strategic vision, a different view of reality, and texts/articles will be published underlining their importance. Because everything becomes "fashion", "norm", "mainstream", thus losing any disruptive potential and any possibility of creating a real discontinuity with the hegemonic models and with what has gone before...

I conclude by clarifying that this is a reflection not—evidently – "against concreteness" as such (which—I repeat—we cannot fail to reckon with), but rather, as put forth in the title, against the obsession, against the "tyranny of concreteness". An obsession that materializes as a reductionist and simplistic approach to the problems, uncertainties, unpredictability and variability of social and organizational phenomena; a radical vision of "concreteness", certainly necessary but which, taken to its extreme consequences, appears/reveals itself, at the same time, as an obstacle to true organisational, social, cultural and political change. A *culture of concreteness and of evidence* in many shades of grey; a culture that takes "data" to be "given facts" based on another culture, that of standardization; a "culture of concreteness" which is based on the deceptive and misleading idea that knowledge/fields of knowledge must be "useful". A culture of concreteness and of evidence based on the illusion of being able to eliminate error and unpredictability from social and organizational systems. A culture of concreteness which, evidently, although built starting from the principles (axioms) of rationality and control, presents the characteristics of ambivalence and ambiguity typical of all social and cultural processes.

Overall, there is truly too little awareness that not all (by any means!) is *identifiable or recognizable through sensory experience and/or empirical evidence.* Similarly, there is little awareness of the kind of effort that needs to be made so that not only what is measurable in quantitative terms can stand out and be highlighted.

In perfect continuity with what has been said, and as I have been repeating for many years, we are continuing to educate/train/instruct *mere executors of functions and rules* (1995–1996) who are not even capable of reflecting on the nature of these functions and rules and on the "whys" that govern them. What is even more worrying is that we aim more and more at the construction/formation of a thinking (?) style aimed at/functional, exclusively, to concreteness: a kind of thinking which, in the best hypothesis, can be identified with the calculation and achievement of a result/a solution (*problem solving*). An almost hegemonic "approach", which further concerns and calls directly into question topics and issues relating to learning.

We continue to be unaware: unaware of being "within" complexity, hypercomplexity, variability, the emergent, the constant and the dynamic instability of the relational, the social, the human and the vital, none of which we are yet capable of realizing or recognizing from the inside, because we are within life itself, inhabiting its constitutive dimensions, as well as *the infinite web of non-linear relationships.* Complex dimensions which, moreover, interact with new technological practices and with social systems increasingly innervated by artificial intelligence and systems of automation.

We thus find ourselves in a state of "cultural delay" which hinders education, training, the co-construction of an "other" view and of a systemic vision, of togetherness; a delay that continues to make us see/recognize/design/manage organizations as "machines" (complicated systems) and not as (complex) "living systems". Once again, with scarce awareness that—as I wrote many years ago—"innovating means destabilizing"[4] and that *thought is action* (once more a matter of "false dichotomy"):

> At the same time, we need to be aware that innovation implies a profound change also and above all in the way of seeing, observing, understanding phenomena, processes, objects, "things" (→systemic perspective): innovating (also) means having the courage to destabilize something that – at least in appearance – is profoundly balanced, stable, rooted, orderly and, in some ways, ideal. Innovating is always an (essential) challenge that implicates abandoning certainties, consolidated visions and ritual behaviour, freeing oneself even of the very idea that "things" are to be done in a certain way because they have always worked that way. Historically (there may be exceptions), those who are in power and are responsible for decisions, even in terms of simple organizations (?), will find it hard to be motivated/interested in concrete innovation, precisely because the dynamics of innovative processes cannot but make the situations in which they are involved and the context of reference less stable and controllable. It is – and will always be – the "cultural factor" to determine both the stasis and dynamics of social systems and complex organizations: because, as I always love to repeat, it is "with people's legs" that innovative processes "march on".[5]

Speaking of concreteness and of never abandoned routes and pathways:

> The first years of our lives are characterized by a lengthy critical period of great plasticity of the brain, a period during which this organ is extremely sensitive to experience, changing according to its experiences, as it is schooled by the environment. Subsequently, a slow stabilization of the nervous circuits and a progressive decrease in the number of synaptic contacts takes place. The circuits that have become more stable tend to repeat the same functions, and this means that when faced with certain exogenous or endogenous stimuli, the cerebral machine responds with similar modes of behaviour. This way, mental *routines* are generated and the brain begins to resemble, more and more, a machine whose functions, including behaviour, are at least "partially mechanized" and therefore predictable.[6]

---

[4] See Dominici P., (2008, 2014 e 2019), *op.cit.*

[5] See Dominici P., *op.cit.*, 1995–1996 and following.

[6] Maffei L., *Elogio della ribellione*, 2016, pp. 32–33. On these topics, see also the classic: James W. *The Laws of Habit*, in "The Popular Science Monthly", 30, 1887, p. 433–451. The contribution of William James is particularly significant also with reference to the approach and paradigm of complexity: influenced by Henri Bergson (in particular, by the work *Matter and memory. Essay on the relation of body and spirit,* 1896), he addresses the important question of the integration, in his opinion, total and complete, between the "psychic world" and the "natural world", speaking of radical empiricism. Among his works, see James W. (1890), *The Principles of Psychology,* Henry Holt & Co., New York, 2 vol; James W. (1904), *Essays in Radical Empiricism,* Italian translation, *Saggi di empirismo radicale*, Quodlibet, Macerata 2009. Once more, on the subject of "complexity", see William James, who in *The Principles of Psychology,* argues that "*The aim of science is always to reduce complexity to simplicity"*.

In fact, as I repeat every time, we always stand on the "shoulders of giants":

> If it should turn out to be true that knowledge (in the modern sense of know-how) and thought have parted company for good, then we would indeed become the helpless slaves, not so much of our machines as of our know-how, thoughtless creatures at the mercy of every gadget which is technically possible, no matter how murderous it is.[7]

6.1 The Power and Practice of Theory

> I fully agree with Gell-Mann when he points out the importance of theoretical thinking, not only to discover the "big" answers to all of our queries about the universe, but for countless other pursuits and intentions. A theorist, says Gell-Mann, has a mind-set which well adapts to comprehending the interconnectedness of all phenomena. Naturally, one can separately study each individual phenomena in great detail, but everything changes when we are able to see them as part of a general interconnected whole. It even becomes easier to keep them in mind, for rather than trying to memorize endless sets of single information, looking at their interconnections allows us to posit a more synthetic idea, a theory, a formula which we can use to learn and recall them schematically. That is when the world begins to make sense, and we are able to comprehend so much more. Human beings have an innate skill for recognizing patterns, structures and regularities, and for adapting to them, being ourselves complex adaptive systems. We inherit these tendencies both biologically and culturally. However, at times, our approach to these concepts and patterns is conditioned by selective pressures. This is where science, with its observational approach, can help to clarify these schemes in a more objective manner.[8]

Poof, just like that, it's the power of theory. Yes, that's precisely "it"... Theory: an important concept (and an important "practice" as well) though, to some extent, cumbersome, one for which myriad definitions and nuances might be provided/postulated, linked to different approaches, fields of knowledge and disciplines, as well as, evidently, to a range of historical-social and reference contexts. In defining the term, the *Dizionario Zingarelli* [Zingarelli Dictionary] identifies various aspects and meanings which, once again, highlight the semantic ambiguity and (perhaps) even the origin of certain clichés and stock phrases concerning it, too: (1) Systematic formulation of principles specific to a philosophical doctrine, a scientific section of knowledge, an artistic or cultural movement: the Platonic theory of ideas; the theory of relativity, of sets. CFR. -logy. (2) Complex of precepts that serve as a guide to practice: the theory of rifle handling | (derogatory) Excess of theoretical elaboration: getting lost in theory | in theory, theoretically. ANT. Practice; (3) System, way of thinking: *I don't agree with his theories on friendship.* SYN. idea, opinion. The Treccani Encyclopaedia too provides a definition of "theory" in terms of a "Logically coherent formulation (in terms of more or less abstract concepts and entities) of a set of definitions, principles and general laws that permits one to describe, interpret,

---

[7] See Arendt H. (1958), *The Human Condition*, Italian translation, *Vita activa. La condizione umana*, Bompiani, Milan 1964, p. 3.

[8] Gell-Mann M., *The Quark and the Jaguar*, W.H. Freeman and Co., New York 1994, p. 112.

classify, explain, at various levels of generality, aspects of natural and social reality, and of various forms of human activity".

Likewise, even "scientific" theories do not constitute/are not more or less simple complements, useful at an interpretative level. They represent the backbone of science and scientific research, evidently along with the scientific method. *Theory* not only directs and influences the "scientific" observation of phenomena and processes, but also guides and effects the choices and uses made of investigative tools. This too, because theories, not only "scientific" ones, in addition to the parts comprising hypotheses, are divided into definitions, formulae, arguments, etc. which permit, in some way, the observer/manager/researcher to trace possible pathways towards the verification or falsification (Popper et al.) of the initial hypotheses. Furthermore, theories can be/can prove to be important tools for foretelling[9] and classifying phenomena, and not merely (complex) tools for explaining facts. In this regard, Pierre Duhem stated in his *The aim and structure of physical theory* that a true theory is not one which provides an explanation of physical appearances that conforms to reality; rather, it is a theory that satisfactorily represents a set of experimental laws.

With reference to these issues, we are truly standing "on the shoulders of giants", as we have a vast and detailed scientific literature, also attributable to the so-called hard sciences, well beyond any logic of separation between the fields of knowledge and the disciplines. At the same time, it is necessary to acquire a definite awareness of the fact that all theoretical and experimental knowledge is historically and culturally conditioned, even if in different ways. Hence, once again, I can only strongly reiterate the importance of (1) critical education and training, (2) education in the "scientific method" and the urgency of (3) a systemic approach to complexity, which needs be practised and experimented from the early years of school, facilitating openness and dialogue between fields of knowledge, as well as the famed "collaborative-learning" method.

In short, Theory is more important than ever, also with reference to the possibilities of providing a symbolic and conceptual representation of the evidence and data that have emerged. Yet—it is useful to repeat—theory is that "dimension" (complex and anything but linear) that almost everyone despises, considering it useless, misleading, dispersive (bordering on the paradoxical), especially in an era when the belief/narrative that data tells us everything and that reality itself is made up of "data" which are "actual facts" is proving to be increasingly hegemonic. That this reality is so real (measurable) because it is made up of empirical evidence (quantitative and statistical, big data etc.) which is so self-evident that it does not require any

[9] See, in particular Wiener N., Siegel A., Rankin B., Martin W.T., *Differential Space, Quantum Systems and Prediction*, MIT Press, Cambridge 1976; Batty M., Torrens P.M., *Modeling Complexity. The Limits to prediction*, Working Paper 36, Centre for Advanced Spatial Analysis, University College London: https://discovery.ucl.ac.uk/id/eprint/158379/1/paper36.pdf; see also this interesting publication by Gammaitoni L., Vulpiani A., *Perché è difficile prevedere il futuro. Il sogno più sfuggente dell'uomo sotto la lente della fisica*, Edizioni Dedalo, Bari 2019. On the same topics see also Vespignani A. (2019), *L'algoritmo e l'oracolo. Come la scienza predice il futuro e ci aiuta a cambiarlo*, Il Saggiatore, Milan 2019.

observation and/or interpretation (identification of correlations and/or causal links, observation and recognition of levels of connection etc.).

The very idea of "theory" almost seems to cause unease, because "theory" is held to be—especially in certain areas of the imaginary that generate "commonplaces" and reassuring formulae/slogans, not only for organizational practice- something useless and a waste of time (and money), something meant precisely for those who have "time to waste" (along with money) and/or, even worse, for those who do not "know how to do things" which, currently, is equivalent to saying for those who "have no solutions" (for everything).

How many platitudes and, I repeat, clichés regarding this very important concept-category! The refrains we hear are: "enough theories, facts count/are needed" or "we don't need theorists/scholars/thinkers – who are perhaps able to sustain, in a clear, logical and coherent manner, their own hypotheses/theses (on the other hand, we rely on the so-called" gurus " and/or the" visionaries " who, as they say, are really "in" these days—we need people (hyper-specialised figures or, on the contrary, experts in everything, or highly trained technicians?) who know only how to act/do things". They need to know, perhaps, how to "execute" tasks perfectly, as if action, practice, know-how and doing, even research, were not, in any case, always oriented by concepts and theories. On the contrary, when they are not (that is, not oriented by concepts and theories), the results are plain to be seen and not only in terms of playing things by ear, as it were. Evidently these complex and sensitive issues which concern, also and above all, politics and public life on the whole. Similar to certain analyses and public discourse regarding Culture, which claim that it *does not pay the bills.* This is tantamount to presenting Culture as something profoundly abstract, theoretical, above all "useless" (sic!), incapable of producing wealth and well-being (!). Evidently, behind these conceptions/commonplaces, there is a view, not only of the Economy, of the development model (currently built upon consumption, and not on work and/or People) and of the progress of a country, but also of the whole of Society, civilization, social well-being, innovation (inclusive or exclusive), citizenship and inclusion and the common good.

Education, training and research are/should be the foundations and *architectural supports* of any process of (systemic and long-term) innovation and change but, unfortunately, beyond public, political and media discourse, these assumptions have never been achieved and/or pursued, not only in terms of lack of investments. Although, periodically, the cultural climate regarding these issues seems to be on the verge of changing. But the *route* (I know, an inappropriate metaphor) to pursue remains long and uncertain, full of critical issues and unachieved transitions (over time), also because, as I said earlier, we are obliged to deal with an overall view/idea of "Society" that continues to be considered a "subsystem" of the Economy, which takes shape under the still undisputed and unassailable hegemony of the *economistic paradigm*, based, in turn, on the erroneous and misleading conception of Economics as an "exact science" (completely "rational" and practically infallible, even with regard to its predictions) and not as a "social science" (fallible and capable of providing knowledge and indications of an exclusively probabilistic and statistical character).[10]

And, as I love- and am obliged to—repeat every time, if we do not thoroughly rethink education, training and research, we shall go nowhere: continuing to appear, be and consider ourselves *inadequate* and unprepared for the ongoing anthropological transformation and to the continuous emergence typical of complex systems. We are in the midst of a transformation that is also, and above all, cognitive and epistemological. Despite certain narratives, we will have to keep in mind how incapable we are, not only of managing and/or directing change, but also of endeavouring to *inhabit* the profound transformations underway.

For these (and other) reasons, at the beginning of this section, I chose to paraphrase a few, very few words written by the great scientist and scholar, Murray Gell-Mann (Nobel Prize for Physics in 1969 and founder, in 1984, of the Santa Fe Institute). He is, I confess, one of my favourite authors, whom I began to appreciate as a student because of his ability to sustain and include—which, contrary to one of the most widespread and difficult clichés to eradicate in organizational and social contexts, underlines/reaffirms—in no uncertain terms—the vital importance of theory, on a par with *a systemic approach to complexity,* which can enable us to observe, recognise, describe and understand (hyper)complexity.

Yet, certain clichés are still so strongly rooted, also/especially in our educational and training institutions, that they lay bare all our inadequacies, incompleteness and "limited rationality" even more so today, now that we find ourselves in a hyper-technological and hyper-connected civilization, where everything is "given", where "data" and "facts" are presented (and claimed) to be self-evident. I simply cannot hold myself back from exclaiming how truly incredible it is that so much still needs to be done. Understanding the *complexity of complexit*ies (1996) cannot be considered a mere question of practice and practices (which might not even be conceptually oriented). To synthesize what I have been saying, so much still needs to be done against the obsession with know-how and problem solving, where what counts are only the answers, never the questions, only the solutions, (and, as I always point out, these are inevitably simple solutions to complex problems—Popper even spoke of "dilemmas"), where what matters are only skills and not knowledge—another of the false dichotomies I have been writing about for twenty years. Even allowing for a range of nuances, only quantitative and/or numerically measurable data and "evidence" are considered scientific and therefore "useful".

As I have had the opportunity to write in recent years, the principle of "useful knowledge" represents one of the major errors as well as one of the great deceptions of the modern and contemporary eras: a principle upon which we are counter-reforming (demolishing) the School and University of the future. While –and I never fail to make a point of recalling this—in the courses of research methodology of a few decades ago, which were extremely arduous and demanding, we were taught, among the many topics covered, that "the data never speak for themselves".

[10] See Dominici P. (2005, 2008), *op.cit.*, and following.

As a finishing touch to the misconceptions I have mentioned above, there are those for whom "what counts is exclusively the research itself", who ignore or pretend not to know that there is no such thing as research (or research activity), that is not directed/oriented theoretically and conceptually, as is true for any human activity. To confirm how widespread this mentality/view/narration is (beyond a few of the quotations or phrases used in public), just think of how the word "theory" itself is seen and treated with (great) suspicion. Not to mention the term "theoretical", a term adopted also to discredit people, scholars, experts, incapable (in the opinion of those who use the word in this sense) of providing quick solutions for all and every type of problem. The word "theoretical" is sometimes even used as an insult.

The word "philosophy" and the term "philosopher" often receive similar treatment. In many cases, this derogatory use of words, especially in the Italian historical-cultural context (but it suffices to go abroad a little to realize how different things are elsewhere), is used to indicate figures and activities deemed completely useless, abstract and distant from experience (?) and real life itself and from the solutions (?) they require. The solutions envisaged would need to be simple even in the presence of complex, multidimensional problems.

Almost paradoxically, theory, confused with the "history of thought", is often not considered and/or is despised precisely by those who, beyond the fields in which they operate, have never carried out any true "research", precisely by those who continue to confuse methodology with technology, or, even worse, methodology with digital solutions. Scorn or, in any case, a sort of contempt which, precisely in line with what has been said, also accompanies, more generally speaking, those who study, those who think(?), who reflect on problems, processes, life, cultures, identity, rights, and dilemmas (perhaps using a critical approach); those who ponder the dimensions, which characterize all adaptive complex systems, of the unpredictable and the uncertain, dimensions that are complex and not easily measurable. This same contempt, accompanied by suspicion, has also profoundly characterized the public discourse on "culture" in recent decades. A deviation which neither academic debate nor research (nor distribution of the few available resources) has been able to avoid. A kind of "suspicion" that also affects those whose analyses and studies are aimed at planning "something even grander and more important", who are especially thinking in a long-term perspective (which today, finally, has become the talk of the town, despite the fact that everyone who is talking about it continues to work in the opposite direction).

Again, from this perspective, we can see that, particularly in Italy, the phases and work of conception and design are (almost?) never compensated financially. They are not considered or acknowledged as important, which, for numerous reasons, is truly incredible. Payment is made, in the best-case scenario, for the "finished product"... and that is that! Everyone is talking (and rightly so) about "project culture". All this is anything but random: on the contrary, it is indicative, an indicator of a culture that gives scant importance to creativity, imagination and talent for design and planning. In some cases, it has even discarded them from the halls of education and training.

As I was saying, suspicion and a sort of contempt that tars thinkers, theorists and scholars, not to mention so-called intellectuals (a 'category' that is effectively in decline, for an entire series of reasons): all those who carry out intellectual work, whose "effects", of fundamental importance to communities, public life, the health of democracies, do not appear immediately evident, recognisable, measurable. And often it is precisely those who despise or, in any case, consider certain "figures" useless, that actually seek, exalt and emphasize gurus, visionaries, *entrepreneurs* etc. And I wish to emphasise the fact that this is obviously not a criticism of those who define themselves and/or present themselves this way.

To conclude, all that is left is to do at this point is to add (something that is known and widely so, despite the dominant thinking/narrative…): that it is precisely in the field of scientific research that "theory" and "research" have always nurtured each other. As I have had the opportunity to repeat several times over the years, the challenges of (hyper)complexity are first and foremost educational: they concern learning and educational processes, relational and communicative spaces. These are challenges that require great efforts at all levels. It will not be simple nor will it be a short or medium-term effort. Furthermore, perhaps, until we become aware of our "weaknesses" precisely in theoretical terms, not to mention in our thought systems, we will not go very far, but will continue—as mentioned above- to play things by ear, incapable of facing the challenges of an increasingly uncertain and unpredictable hypercomplexity. My purely personal impression is that—with a few exceptions—we are continuing to educate and train the mere "executors" of functions/tasks mentioned before—highly skilled in "know-how"—within an increasingly (hyper)complex reality whose dominant dimension is that of technological control, whereas what is direly needed are instead elastic minds structured by a mode of thought that needs to be systemic and multidimensional. The implications for crucial issues regarding public life, citizenship, inclusion, democracy, and the common good are profound, even as key concepts, words and fundamental "principles" risk remaining simple slogans.

6.2 In the Age of Rapid Obsolescence

The transition to hypercomplexity—determined, among the many variables/concauses, by today's digital hypervelocity and the virality/pervasiveness of communication—along with the urgency of thinking in a systemic manner about the "long-term" period, are the reasons why it is mandatory to rethink education and, even more so, to rethink *thinking on education*, so as to modify the choices and strategies regarding both teaching and training in this sense.

In light of the previous considerations, never in the past as in the present hypercomplex, hyperconnected society (*Società Ipercomplessa Iperconnessa* 2003), has it been necessary to modify so radically our "view" and the approach/approaches we intend to apply to investigate the reality of the phenomena that surrounds us. In other words, we need to change our *thinking* and the thoughts that permit us to act—thoughts, that are themselves actions. *Teaching complexity* signifies/implicates/requires, to begin with, knowing how to observe, to see "beyond", to imagine, and above all, to be able to co-create connections/correlations/interdependencies, but also to teach the "scientific method" and a systemic vision of problems and phenomena, which are relationships, ecosystems of relationships rather than isolated "objects" (or objects that need to be isolated), so that on an initial plane of action, recognizing this hypercomplexity can enable one to create the conditions for managing it (even though as far as this author is concerned, the idea of managing complexity is in itself a contradiction in terms) and transform it into opportunity (knowledge). Questions of fundamental, decisive, strategic importance for both organizations and democracies, currently marked by a profound crisis. Yet in the *Hypercomplex Society* the risk is that all this may no longer be sufficient. It is of ever more fundamental importance also *to know how to communicate this (hyper) complexity*, bringing the question of knowledge and skills back to the fore (if they had ever really lost this position), besides the urgency of overcoming, once and for all, the "false dichotomies". With the awareness—still lacking—that, often, *communication and information* do not determine, in any way, a "reduction" and/or a "simplification of complexity. They are, rather, constitutive and conditioning elements, both in terms of representation/visualization and of individual and social perception of these same representations or visualizations.[11] In the so-called *knowledge society* which, on the contrary, seems be to more and more of a *no-knowledge Society*[12]- "knowing" and "know-how" no longer suffice: apart from the necessity for education and training to educate and train both to "know" and "know-how", it is now also (and above all) necessary to teach and train "knowing how to communicate knowledge" and "knowing how to communicate know-how". These are the knowledge and skills now required in all professions with high cognitive content, destined to characterize the "knowledge society" and the sharing economy. This is why it is impossible not to return, again and again, to the strategic centrality of schools and universities, to the educational and training pathways they propose and their relative objectives. I do not fear to appear repetitive, because this is the key "question" of all questions. If we do not intervene in a profound and systematic way regarding these dimensions—education and training—we will find ourselves in a problematic condition of perennial *cultural delay* with respect to the complexity, multidimensionality and ambivalence that have always characterized processes of innovation and change. Processes which, moreover, due to the extreme speed and unpredictability that currently characterizes them, to their being in continuous evolution/metamorphosis, in addition to creating the usual problems of control and management—typical of all historical phases of technological change and innovation—determine, and will continue in the future to determine the increasingly rapid

[11] See Dominici P. (1995–1996), (2005), (2008), (2014) and following op. cit.

[12] I coined and proposed this operative definition in Dominici P. (2017), (2019), (2022). Op. cit.

obsolescence of knowledge, skills and curricular profiles,[13] that, in the meantime, have been designed, provided and acknowledged by educational and training institutions. In other words, the problem is how to rethink/transform/re-imagine schools and universities, how to rethink educational-training pathways, how to redefine curricular and professional profiles in a phase marked so deeply by irregular and discontinuous trajectories? And how can we try to do this so as to be able to take into due consideration an increasingly and constantly evolving society and an equally rapidly changing job market? How can we try to change the traditional/consolidated organizational logics and cultures that distinguish our educational and training institutions? This operation is anything but simple, and it also requires a "long-term" view. Scientific research itself is currently based on logics and value cultures (?) that discourage and openly hinder dialogue between fields of knowledge and multi/inter/transdisciplinarity, pre-requisites that are unavoidable and essential if we are to address the dilemmas and challenges of hypercomplexity. Furthermore, it cannot be hidden; this evolution is exposing all our inadequacies, being so rapid and unstoppable as to dramatically shorten the "life cycle" of the knowledge and skills that are needed. It is so rapid and unpredictable (for example, artificial intelligence and robotics permit us, with great difficulty, to catch a glimpse of completely unimaginable scenarios) that it favours the obsolescence of all the decisions taken today regarding the professional profiles and curricula of the near future. Speaking of obsolescence, it is now commonly assumed that AI is slated to become the principal educational tool in the near future, which can only result in teaching students to think like machines, becoming, as said before, mere executors of functions and rules. If we are to avoid *doing things backwards,* we must train and stimulate creative minds through a systemic, transdisciplinary approach, forming hybrid figures who will be capable of inhabiting complexity and of utilizing AI to enhance their own, very human, ideas and imagination (Fig. 6.1).

In this sense, as I have long been pointing out, it can turn out to be extremely risky, as well as misleading, to even think of being able to define educational-training and curricular pathways, or even professional profiles themselves, solely and exclusively on the basis of so-called "market requirements" and/or the increasingly specific requests advanced by businesses. I am well aware that there are many (the majority of professionals and so-called experts) who think in diametrically opposite terms, but I consider their approach extremely fallacious, not only regarding their nature, but also the objectives that educational and training institutions should have.

We are still confined within "short-term" logics, which are those of immediate responses/solutions, of control, of balance at all costs, of emergency. Put more explicitly: does it still make sense to continue chasing after the markets, to still be convinced that it can even be done? An unpredictable, deeply unstable, and constantly evolving market, one that is driving all experts to continually evoke what

[13] See, in particular Dominici P. (2003), *op.cit.* and following. Of the more recent publications see Dominici P., "*Educating for the Future in the Age of Obsolescence*", in CADMUS, Volume 4—Issue 3, November 2020, pp. 93–109.

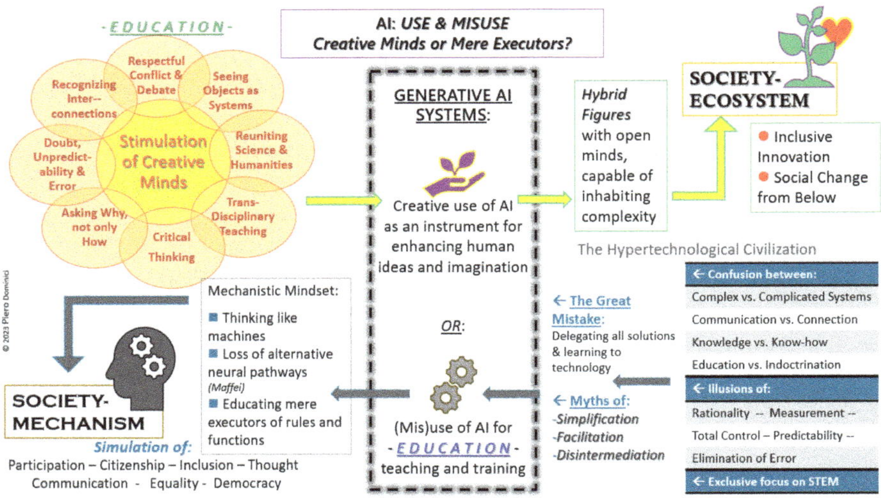

Fig. 6.1 AI—use & misuse—creative minds or mere executors?

is called supply-and-demand *"mismatching"*, as though it were the only variable/theory capable of explaining the multi-dimensional dynamicity of the ongoing change: a mismatching seen and explained as a potentially "controllable" element, if not downright eliminable through appropriately contrasting measures. We have gone completely *off track*, and this did not start happening today. The most substantial risk is still that of continuing to "chase after technological and digital innovation", to find ourselves under its yoke, without even knowing whether there will be sufficient time to adapt to it or try to manage it. From this perspective, beyond the many paradoxes of the change underway, the "great mistake" in/of the hyper-technological and hyper-complex civilization is precisely that of continuing to think of education and educational processes (this regards training too) as "exclusively technical issues", a problem of "skills" only, linked to "know-how" (and nothing more); a complex issue to be faced by staking everything on speed and simulation.

If we fail to rethink/re-imagine education and, even more so, thinking concerning education, modifying in this sense the choices and strategies regarding both teaching and training (continuous and systematic, though partially flexible and modular) and of all the figures involved, at various levels, including those in charge of making decisions-making, we will not go very far and we will continue to try to control (manage) and/or surf the waves of change, their ambiguity and indeterminacy, resorting to the same old short-term logic. Playing it by ear. And everywhere we turn, we think we are only seeing…BLACK SWANS!

6.3 Culture and Knowledge Reduced "to One Dimension": That of Quantity

Study, research, culture, education all find themselves more and more trapped between the logic of the market and conformism.

From this point of view, today more than ever, there is a palpable urgency for a kind of education (not only digital), which must always be imagined and rethought, in every case, in the direction of the social and cultural construction (first) of the Person and (then) of the Citizen. Teaching complexity, the scientific method, critical and systemic thinking, nourishes and nurtures a kind of thinking that can only be *multidimensional*, profoundly aware that although data and empirical evidence are indubitably fundamental, once again, they fail to give a complete picture of phenomena, of reality, and of the hypercomplexity of the Social and the Vital.

On the contrary, what has been happening for some time now, in the various realms of intellectual and cultural production, of the arts and creativity and, even, methodologically and epistemologically speaking, within the more rigorous spheres of study and scientific research, is that everything has been changing, inevitably and irretrievably, proceeding in a disturbing downward direction (in any case, on a slippery slope) that makes it very difficult to control or try to halt; in other words, all the spheres, areas and dimensions of intellectual and cultural production, even scientific production and research, have been rendered functional and reduced "to one dimension"[14] (from the title itself I recall a famous essay by Herbert Marcuse), the apparently neutral and objective dimension of quantitative data/quantity/numbers. All the spheres of intellectual and cultural production, without exception, have been made functional, not only to profit, but also to the deceptive principle of utility (on which, as I have been repeating for many years, we have based the counter-reform of our educational and training institutions). What the effects of these dynamics and drifts have had on the (few) opportunities and calls for change and emancipation, which had already been hindered from the start by the "Asymmetrical Society" (2003), comes down to an excessive simplification, or better yet a weakening; in reality, we could indeed say, "a one-dimensional reduction."

Although numbers have their (undeniable) importance, a dimension in which "only numbers count", along with "data" and quantity, is a dimension that goes from being a "tool" or "means" of organizational and social action to becoming an "end", with innumerable consequences that also impact organisational, digital and social cultures.

Everything is reduced to quantities, more or less exact, and to quantitative data; everything is reduced to numbers, indicators, parameters (?), medians and criteria that indicate/should indicate—and to some extent, reassure us—"precision", objectivity, satisfaction, value, "quality" (which is a complex, multidimensional concept), originality and satisfaction, in the belief—derived from this view/approach—of

[14] The reference, as is obvious from the title of this section as well, is to the well-known essay by Herbert Marcuse.

being able to measure quality, transforming it into quantity, a definitive transformation of "qualitative" into "quantitative", moreover, which heads exclusively in the direction of gratifying and satisfying a need.

I spoke about this, in times before such things were common knowledge, as one of the "grand illusions of the hyper-technological and hyper-connected civilization": the illusion of *measurability,* which implies the equally risky, albeit necessary, illusion of *predictability*.

Nowadays, everything seems to be—or actually is (!)—traceable back to (and reduced to) careful, systematic, scrupulous, obsessive, maniacal observation, even of what is not "observable", to total acknowledgement of and uncritical adherence to all the values (?) and (quantitative) criteria relating to the measurement of every aspect/dimension/detail/characteristic, to values and criteria useful for the construction and maintenance of visibility, popularity (often confused with trust), reputation, fame, success (a certain idea/vision of success), of sales, of the ability to build consensus but also to polarize, without, however, any in-depth analysis; in short, everything reduced to quantity, calibrated, as said before, to degrees of satisfaction/approval(?).

In other words, somehow, everything has/seems to have changed since they started counting. Only in terms of numbers (e.g. of citations, copies sold, followers, hits, shares, media appearances, even of concepts that are simply fabricated. And again, fame, visibility, being *followed*, downloaded and shared, being able to *influence* and direct public debate (?) and, even more, being able to fuel the very "polarizations", "bubbles", and controversies, which in general, are criticized and denounced by (almost) everyone; many of whom, in any case, ultimately end up greatly benefitting from all points of view (publications, reviews, invitations, careers, assignments, awards etc. etc.). A gigantic mechanism that needs to be constantly and continuously powered. The "tyranny of concreteness" manifests itself in infinite forms and constraints that are conventional and hegemonic at the same time.

As usual, we witness the continuous proliferation of rankings of all kinds: newspapers, sites, blogs, *influencers* that, at all levels and in all sectors, tell us/suggest/indicate (?) what to read, watch, listen to, who is truly "deserving", who is the "best", what is "excellent", what the "best practices" are, etc. etc. etc. They do this—they are able to do this—on the basis of numbers, which are always evident and unequivocal, because, whatever we are talking about, if a "product" is mentioned, read about, discussed, heard of, suggested and appreciated by many, if not by all, if this "product" features in some ranking or other, it is considered valid regardless of any other evaluation and/or feedback.

What counts are numbers, popularity, the ability to trigger and fuel polarizations, attract listeners/views/comments/shares and move audiences and followers. This is also because—as I have been repeating for years and as is evident—there is always less and less time to study, verify, listen, look for what might be truly original and innovative.

At times, I cannot deny that I try to imagine what many of these people/groups/cultural centres/industries would do without these extraordinary sounding boards, the continuous, systematic marketing campaigns (bolstered by awards, "top" rankings full of praise for "excellence" and the "best", documents, posters, guidelines etc.). And in these last years, I have been wondering what they would do without the (strategic and scientifically studied) assiduous, real-time, in-plain-sight presence on social media, which- let us be clear—despite the well-known critical issues at stake, certainly provide extraordinary opportunities, still little understood, except very partially. (Education, training, research). Ultimately, among the many issues/dimensions to consider, there is once again confusion between the means and the ends.

Complex processes and dynamics which, as usual, have been a long time coming. By way of example, the birth of the so-called commercial television comes to mind, whose effects, including the introduction of changes in values and social behaviour, with particular reference to many of the social, political and cultural phenomena observed in recent years, have never been (and never will be) sufficiently investigated. From the first moments of the advent of this kind of television, (nearly) everyone was highly critical—in my opinion, appropriately so—of the logic(s) thus introduced and, at times, radicalised: logics and strategies that have placed, at the centre of every choice, the dominance of ratings/audience/fame/audience numbers and the value of success at all costs.

Choices and strategies that have since been translated into resources (publicity, consensus at all levels and in all professional and organizational fields, and beyond), at times impacting the possibility of survival and/or numerous other advantages.

The same logics, which have always been of an industrial nature, have found multiple operational translations and methods of affirmation and hegemony also in all the other spheres/areas of study, intellectual, artistic and cultural production and, even, as mentioned before, in scientific research.

The underlying view (one of the "grand illusions" of the hyper-technological civilisation) is that of always being able to measure everything, in quantitative terms, orienting all choices and decisions towards the illusory certainty of being able to predict everything; because, ultimately—this is the common, widespread perception—these numbers also convey, or bring to light, the concrete and effective "value" of "things" (?). Because, in other words, such numbers, mistakenly considered "self-evident", almost automatically reveal/unveil, from this point of view/approach/paradigm, a great deal about the quality of the "products" of whatever kind they may be, even those generated by scientific research (the decision to refer to results obtained in the field of scientific and university research with the term "products" is certainly no coincidence.

It has thus come about that, while subscribing to the idea that building a culture of evaluation is a necessary objective (one that we are however very far from reaching), that pursuing and defining more impartial, reliable (?) criteria for judging and evaluating intellectual, scientific, and cultural production is imperative, many of us, even those who study and carry out research, in every field and at various levels, might entertain the (illusory) certainty of having achieved these aspirations. And for these reasons, perhaps—and I underline *perhaps*—most of us have also ceased, to some extent, to study, to be truly curious, to delve deeper, research, investigate, compare, and instead have begun to deem satisfactory solely and exclusively those indicators and parameters (that are presented and narrated as objective), those lists of rankings, the ones that "everyone reads, everyone quotes, everyone sees, everyone listens to". We have thus ended up being content to discuss and debate only what everyone discusses and debates, to read (and review) the books that everyone reads (and reviews), always on the basis of the same rankings and opinions of influencers who make suggestions in an apparently spontaneous and disinterested way. We have ended up being content to listen to the music everyone listens to; we have ended up being content to quote and listen only to what everyone (?) quotes and listens to. We have ended up by settling for labels, memberships, fame and popularity, visibility and followers, reviews, and in the world of research as well, the number of citations and/or the prestige of authors and/or publishers (these last phenomena complete with virtual price-tags for purchase and sale). These are, arguably, very important factors/variables, but they cannot be "*the*" criterion, the only criterion, if not the actual distinguishing variable that somehow frees us from the responsibility of verifying, studying, examining, comparing, and making rigorous evaluations that are at the same time specific and complex.

And all of this in a continuous and unstoppable rush to define and potentially become part of the rankings and listings, which are constantly pinned on everyone and everything, on and beyond the classification of the best and *the excellent* (with all its relative devastating rhetoric). As always, there would be much to write on this topic.

Ultimately, these are the same criteria and logics previously introduced by the so-called cultural industry and radicalised, in particular, by commercial television. Logics which, even then, were harshly criticized and recognized as the main cause of a certain decline in quality and of cultural impoverishment: well, over time, these same logics—all of them!—have become dominant/hegemonic in all areas of public, social and cultural life.

All of which ends up extending to the personal sphere, involving the People themselves and individual behaviour. To be even clearer, at times it seems that the above observations are being unanimously sustained, that absolutely everyone "has it in for" the so-called influencers ("new" or same-old opinion leaders as they may be), criticizing the "hegemony of numbers and quantity" as well as criticizing the "industries" that build and determine their success. However, it turns out that those who criticize them for the objectives they set themselves, the more or less sophisticated strategies and tools they adopt to achieve them, and for the audiences and the logics that support them, are in reality criticizing exactly what they themselves support and nourish. What they are actually working toward, what they are truly committed to, is precisely to move in the quite opposite direction; that is, to become exactly like those they are criticizing, to be recognized as "someone" themselves, and to remain so forever, despite declaring very different objectives and purposes.

Because, at the end of the day, those who disdain those objectives and that type of success and consensus, based so firmly on "objective" and (apparently) indisputable evidence (numbers), with the far-flung fame and visibility they offer, are truly very few…unless it is on others that they are being bestowed. What wouldn't people do, in fact—what wouldn't they continue to do—to achieve a bit of success, to maintain their visibility and notoriety, in the media, online, and not only.

Because at the end of the day, when we take a closer look, it does seem that in People, in Culture as an open, shared and common good/heritage, or in what we call Society (and its state of well-being), there is essentially very, very little interest.

Chapter 7
To Cross or Inhabit the Frontiers of the Hypercomplex Society? Rethinking Education and the General Architecture of the Fields of Knowledge

Our aptitude for solving problems can end up by being sterilized by their own successes: thus, a successful strategy turns into a programmed recipe of knowledge and the mind loses the ability to face novelty and invent it.

Edgar Morin (2007, 123).

Abstract School is the crucial level for educating Montaigne's well-made heads, aware that the significance of life and practice is innervated by errors and relationships. *Education is citizenship, participation, and inclusion:* fundamental questions that have now assumed even more strategic centrality, as schools and universities have once again become "agencies of selection" and no longer "agencies of emancipation". Moreover, the crucial question of radically *rethinking education* continues to be limited to *digital education,* seen solely as skills and "know-how". In spite of the exponential growth of connective opportunities, in which information and knowledge have become the main resources, society is as yet unable to guarantee that these networks generate fully *communicative* relationships, based on symmetrical relationships and authentic sharing. The sheer acceleration of processes, furthermore, has left little or no time for reflection, critical thinking, imagination or creativity, which are nourished precisely by contamination(s) between fields of knowledge, experiences, and socio-cultural landscapes. The (complex) organizations where young people will be employed are *adaptive* social systems, actual "organisms"; consequently, students must be educated (not *indoctrinated*!) to be capable of inhabiting and interacting with complexity, trained *in unpredictability, open-minded thinking, imagining, and ultimately, in coping with the unexpected and the indeterminate.*

P. Dominici, *Beyond Black Swans: Inhabiting Indeterminacy*, Philosophical Studies Series 160, https://doi.org/10.1007/978-3-032-09029-4_7

Keywords Rethinking education · Selection Agencies · Emancipation Agencies · Socio-cultural Landscapes · The unexpected · The indeterminate · Contamination(s) · Symmetrical relationships · Authentic Sharing

It is education, along with educational processes, that have always been the strategic level capable of conditioning not only the evolution and survival of social systems and the quality of democratic regimes, but the metabolization of and/or resistance to innovation processes and change as well. Among the protagonists of these complex and ambivalent processes we find, above all, the School, along with other agencies of socialisation, albeit within a general framework of an educational polycentrism which, in recent decades, has seen the media, networks and peer groups devour relational and communicative spaces. School is the crucial level where it is possible to educate and train well-made heads (Montaigne), aware that life and practice assume significance precisely because they are innervated by errors and relationships, rather than producing well-filled heads, an issue currently of great relevance in the so-called society of information and knowledge. However, it is also the tactical level upon which to try (at least) to cultivate and practice empathy, respect, pluralism and recognition of the value of diversity and conflict in order to build *open and truly inclusive* societies, founded on a culture of legality, prevention, responsibility, safety, respect, non-discrimination and on a "new culture of communication and complexity" (1996). It is the vital level upon which to nurture a continuous dialectic between liberty and equality, teaching and educating the appreciation of heterogeneity and deviance, the meaning of error and the importance of being able to make mistakes, and the variability and vulnerability of people and systems. All this because—as often argued in the past—"Democracy is complexity". Finally, it is the crucial level where the socio-cultural conditions of a never-fully achieved emancipation can be determined, which also passes necessarily through a reduction in the hegemony of individualistic and selfish values, which have significantly contributed to the weakening of social bonds and of belonging. From this perspective, the close correlation between education and equality of starting conditions and inclusion, as well as that between education and citizenship appear evident yet again.

Even more, I repeat: *Education is citizenship, education makes participation possible, education is inclusion.* It must be clear that these are topics and questions that have always been fundamental but which, as compared to the past, have now assumed an even more strategic value and centrality, as schools and universities have once again become "agencies of selection" and no longer "agencies of emancipation".

Starting from these "solid" premises, I shall now strive, without any claims of being exhaustive, to address these crucial questions concerning the urgency of radically rethinking education and training. I shall try to develop the relevant arguments, in an attempt to answer some important questions: (a) why should we rethink education and training radically? Can it not suffice to modify and update the contents of the various subjects/teachings/disciplines? (b) Is it not enough, for this hyper-technological and hyperconnected civilization, to adapt and/or extend education to technological change and the practices that it entails? Are digital literacy[1] and digital education[2] not sufficient? And finally, how could/should we try to rethink/re-invent/re-imagine/transform education and, with it, training and research?

Evidently, I am profoundly convinced that education and training need to be rethought in depth also, and above all, but not only, in light of the digital revolution, whose scope and implications (not only epistemological) continue to be underestimated. We have to acknowledge now that the (complex) problem no longer concerns the level of contents and applications only. It is no longer a question of "what" to teach (although this is in itself an important topic) or of certain multimedia practices or "special effects" to be developed. There is, furthermore, a series of ineluctable problems, far more profound and complex, linked to the architectures and spaces of knowledge, and to the dialogue between the fields of knowledge, associated with methods and an approach (a systemic approach to complexity) that call into question the urgency of actively practising multidisciplinarity, interdisciplinarity, and transdisciplinarity (Fig. 7.1), openly hindered, at the moment, within educational and training institutions, increasing awareness of the numerous epistemological and methodologies related, in the first place, to the paradigm shift and the advent of the hyperconnected digital civilization.

As can be seen above, in order to clarify the distinctions between these three approaches in connecting disciplinary fields, it is best to begin by first defining multidisciplinarity, in which diverse fields undertake to analyze the same topic of study, without, however, collaborating actively together: in fact, each one works in parallel without further integration, whereas interdisciplinarity, on the other hand, requires an integration between disciplines, along with the establishment of a methodological connection, aiming to create genuine relations between both the fields themselves and the experts who represent them, who interact and share hypotheses, concepts and methods. Carrying the concept much further, one achieves transdisciplinarity, an even more integrated approach, which not only connects the fields of study, but transcends the borders between them toward new forms of knowledge from a systemic, holistic perspective. In other words, transdisciplinarity implicates

---

[1] "Digital Literacy" is currently the object of great national and international debate.

[2] For years, there has been much debate, often carried out in a sterile and aporetic manner, on "digital education" or "digital civic education": the formulas change, but not the approaches, the methodologies, the epistemologies, or the cultures. See: Dominici P., L'ipercomplessità, l'educazione e la condizione dei saperi nella Società Interconnessa/iperconnessa, *Il Nodo. Per una pedagogia della Persona, Anno XXI, n° 47, pp. 81–104;* Dominici P., For an Inclusive Innovation. Healing the fracture between the human and the technological, *European Journal of Future Research, 2017;* Dominici P., Dentro la "civiltà senza errore". Sintesi complesse: l'Umano, il legame sociale e la civiltà ipertecnologica, in A. De Cesaris (Ed.), *Vite digitali. Essere umani nella società del XXI Secolo,* Franco Angeli, Milano 2020.

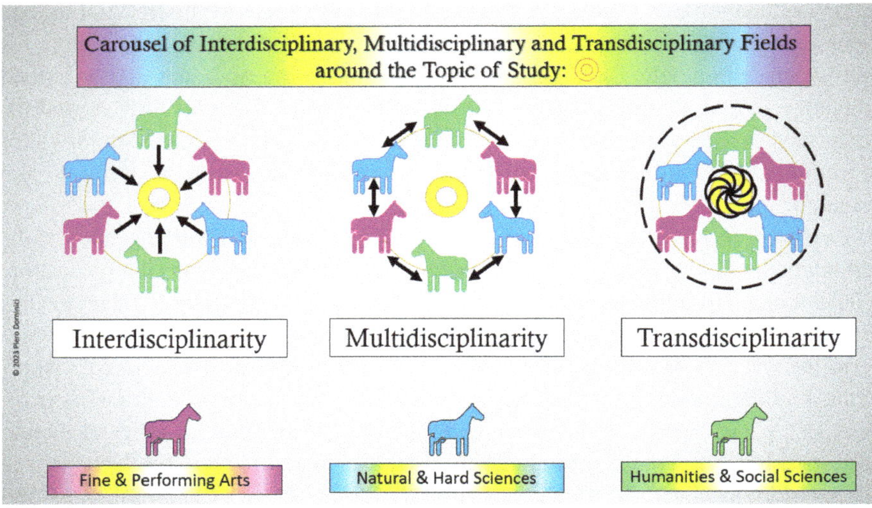

Fig. 7.1 A multidisciplinary, interdisciplinary, transdisciplinary merry-go-round

a profound reorganization of the complex architecture of the disciplinary fields, a new definition of the processes of constructing such fields, and above all, a *new epistemology*.

Despite this, the anthropological transformation[3] determined by scientific discoveries, connectivity technologies[4] and digital evolution, continues to be explained and addressed by the systematic use of traditional *linear models*, as well as through various (even "new") forms of reductionism and determinism, which continue to reveal all their weaknesses and inadequacies when it comes to the challenges, dilemmas, fluctuations, and tensions that mark the ongoing change, but also, and above all, in relation to the complexity, irreversibility, radical variety, uniqueness and heterogeneity of the Human, the Social and of Life. To put it another way, we continue to address—and try to manage (?) and/or, better still, understand—the (hyper)complexity of complex (adaptive) systems as if they were complicated systems (mechanical, predictable, linear, closed, that can be broken down into parts etc.), with little awareness—and here I repeat myself, once again—of how certain "boundaries" and *false dichotomies* have now been completely done away with.[5]

---

[3] See Dominici P, (1996), *Per un'etica dei new-media,* Firenze Libri Ed., Florence 1998. For further information, see also later works, among which Dominici, P. (2016c), *L'utopia Post-Umanista e la ricerca di un Nuovo Umanesimo per la Società Ipercomplessa*, in "Comunicazioni Sociali", n°3/2016, Vita & Pensiero, Milan.

[4] *Ibidem.*

[5] See in particular: Bateson, G. (1972), *Verso un'ecologia della mente*, Adelphi, Milan 1976; Bertalanffy von, L. (1968), *Teoria generale dei sistemi*, Isedi, Milan1975; Foerster von, H. (1981), *Sistemi che osservano*, Astrolabio, Rome 1987; Maturana H., Varela F. (1972), *Autopoiesi e cognizione. La realizzazione del vivente*, Venezia, Marsilio, 1985; Gandolfi, A. (2008), *Formicai, imperi, cervelli. Introduzione alla scienza della complessità*, Bollati Boringhieri, Turin; Gell-

In recent years, among the most harmful forms of reductionism and reductionist approaches, the so-called *digital education and/or digital civic education* have laid bare all the weaknesses and inadequacies of the approaches and views that support them. By way of confirmation of what is being said, moreover, we can see how the concepts and relative operational definitions are periodically updated, although the radical, necessary change of direction that the paradigm shift continues to require is never put into practice. We thus find ourselves still having to deal with a series of urgent matters, with a great need to:

(a) overcome the age-old linear and cumulative models which continue to impact deeply on the delivery and organization of the fields of knowledge, both in schools and universities;
(b) go beyond the logics of the separation and confinement of the fields of knowledge which, in fact, confine educational and training processes within individualistic dynamics that only permit the transmission of the fields of knowledge, not their communication and sharing;
(c) surpass the traditional idea/view of learning, understood as a process of accumulation of fields of knowledge, in favour of increasingly complex and multifaceted learning processes, above all oriented increasingly towards cooperation[6] and collaboration;
(d) reorganise, not only the teaching and training pathways by encouraging interdisciplinarity and multidisciplinarity, but also by rethinking the entire thought system and increasing knowledge of knowledge (Morin), augmenting the relative awareness.[7]

Mann M. (1994), *Il quark e il giaguaro. Avventura nel semplice e nel complesso*, Bollati Boringhieri, Turin 1996–2017; Dominici p. (1996), *op.cit.*; Dominici, P. (2018), *For an Inclusive Innovation. Healing the fracture between the human and the technological*, in, European Journal of Future Research, Springer. Again with reference to these issues, see some very interesting works, belonging to different eras, disciplinary fields and approaches, which reaffirm or introduce different elements and ideas for analysis and reflection: Haraway D. J. (1991), *Simians, Cyborg and Women: The Reinvention of Nature*, Routledge, New York, (Italian translation, di L. Borghi), Italian translation, *Manifesto cyborg. Donne, tecnologie e biopolitiche del corpo*, Feltrinelli, Milan 1995; Kohn E. (2013), *How Forests Think: Toward an Anthropology Beyond The Human*, Italian translation, *Come pensano le foreste*, Nottetempo, Milan 2021; Zimmer C., *Life's Edge: The Search for What it Means to Be Alive*, Palgrave Macmillan, 2021.

[6] See in particular, on the complex dimension of cooperating and cooperation: Axelrod R. (1984), *The Evolution of Cooperation*, Italian translation, *Giochi di reciprocità. L'insorgenza della cooperazione*, Feltrinelli, Milan 1985; Axelrod R., *The Complexity of Cooperation. Agent-based Models of Competition and Collaboration*, Princeton University Press, Princeton 1997; Gardner H. (1991), *The Unschooled Mind. How Children think and How Schools should Teach*, Italian translation, *Educare al comprendere. Stereotipi infantile e apprendimento scolastico*, Feltrinelli, Milan 1993. Among more recent publications, see Bearzi F.—Colazzo S., *New WebQuest. Apprendimento cooperativo, comunità creative di ricerca e complex learning nella scuola di oggi*, FrancoAngeli, Milan 2016; Gopnik A. (2016), *The Gardener and the Carpenter. What the New Science of Child Development Tell Us about the Relationship Between Parents and Children*, Italian translation: *Essere genitori non è un mestiere*, Bollati Boringhieri, Turin 2017.

[7] See Dominici P, (1996–2018), *op.cit.*

The crucial question of radically *rethinking education* has continued, for many years, to be declined only in terms of *digital education* and/or the so-called new media (a definition no longer valid). It continues to be presented and seen, in many cases, solely as a matter of skills that need to be expressed above all in terms of "know-how": knowing "how to do" and "how to use", a concept which is broadened and extended each time it is expressed in public discourse, more often than not for the purpose of showing that one's vision is the most original, despite the fact that experiential and practical evidence shows that what is important is not just a question of skills, of "knowing how to do and how to use" to provide quick solutions.

Beyond this dimension, digital education continues to be seen and presented as the "tool" for protecting new generations from the risks and dangers determined by the digital revolution and, specifically, by the advent of "new" connective technologies (1995–1996).

Although for almost thirty years now I have been posing the question of the need to radically rethink/re-imagine—not the so-called "digital and/or media education"—but education itself, not many were speaking in these terms at the time, on the contrary *(verba volant...).* It was necessary, and has become increasingly so, to consider the "epistemological and ethical implications associated with the advent of digitality, rethink/re-invent/re-imagine education, evidently, in a systemic perspective, always and inevitably "along with" teaching, training, and research".[8]

Instead, thinking (?) and reasoning only in terms of the *mere adaptation and extension of education to the nature of technological transformation,* that is, focusing on digital education and/or on the media and/or on digital literacy (or,

[8]As repeated many times, before these things were generally spoken of (1995–1996), the fundamental objectives of digital and, more generally, technological education are, in my opinion, different and concern multiple levels of analysis and intervention which I shall try to summarize in a few points: (1) so-called *digital education* must (should), at least, increase awareness (there is now a fair amount of consensus on this) regarding the multiple variables at stake; (2) digital education must (should) define and create the conditions for a truly critical and *systemic approach* to the transformation taking place at present: on this issue, however, there is truly much work to be done, since certain concepts continue to be used essentially as *slogans;* (3) digital education should enable People (and Citizens) to address and manage the dynamics and processes that arise not only from technological innovation, but from numerous other factors (economic, social, political, cultural) that distinguish the new ecosystem, in such a way that young people (and adults) are able, not only to defend themselves from the "dark side" of the digital (of social reality), to "know how to use" the tools and inhabit new environments, but also, and above all, be able to exploit the advantages and enormous potential both for sharing information and knowledge and for the construction/strengthening/intensification of relational networks (communication vs connection). And I add, again with reference to the issue of "digital education", which continues to be at the centre of the international political agenda of Politics and Technocracy, this cannot and must not be addressed only in terms of control and protection, otherwise the risk of an exclusively repressive, reductionist, deterministic approach, built on fear and non-knowledge (connected variables), reduced to means and instruments (including networks and social media) will return. While, once again, attention should be paid concretely, to People, the system of relationships, the educational and cultural context and vital worlds (!), in a perspective which—I will continue to reiterate—can only be systemic, multidisciplinary, interdisciplinary and transdisciplinary. See, in particular Dominici P. (1995–1996) (2005), op.cit. and following.

worse still, focusing only on "digital skills", necessary but never sufficient) which is—as I have reiterated for quite some time—a limiting and limited, misleading and counterproductive perspective (which has already lasted almost three decades!); yet another reduction/simplification of something that cannot be simplified! Unfortunately, these views and approaches have actually become the hegemonic paradigm, left unchallenged even at international level; a paradigm which, for some years now, has been merged completely with the studies known as STEM (Science, Technology, Engineering & Mathematics—which I have defined the STEM "Doctrine"). This *hegemonic doctrine* has produced and determined, among the numerous consequences (some of which not evaluable), the affirmation of further new forms of *reductionism* and *determinism* and the consequent, inevitable inadequacy in dealing with the emerging change that is in progress; that is, with *nature* and other interdependent/interconnected dimensions of the new systemic, global emergencies that "disjunctive thinking" and the new specialisms are unable to analyse or understand (Fig. 7.2).

However, this is the global perspective, promoted by large transnational organisations, which—I repeat—continues/will continue to condemn us to a state of "cultural backwardness" and substantial inadequacy regarding the complex process of building a "culture of complexity and unpredictability". To *inhabit the (hyper)complexity of digital and new ecosystems* along with all the dimensions and implications related to the advent of artificial intelligence, which I have called a "new epistemological fracture"[9] referring in particular to the distinction between "complicated systems" and "complex systems", it is necessary to go to the root of the issues and … start over, concretely, *from below*.

7.1 Education and Hypercomplexity. Contextual Variables and Factors

It is precisely in the attempt to adopt a systemic perspective that is attentive to complexity, that we are also obliged to deal with the characteristics of the new global ecosystem. We need to come to terms with an Interconnected and Hyperconnected Society which "is a hypercomplex society where the treatment and processing of information and knowledge have now become the main resources. This is a type of society where the exponential growth of connective opportunities and the transmission of information, which constitute fundamental factors of economic and social development, do not correspond yet to an equivalent increase in opportunities for communication, which I define as the *social process of sharing knowledge* (1995–1996), which implicates equality and reciprocity (inclusion). Technology, social networks and, more generally, the digital revolution, despite having brought

[9] On the question of what I had defined a "new epistemological fracture", long before it came into vogue, see, I particular: Dominici P., (1995–1996) (2003), op. cit. and following.

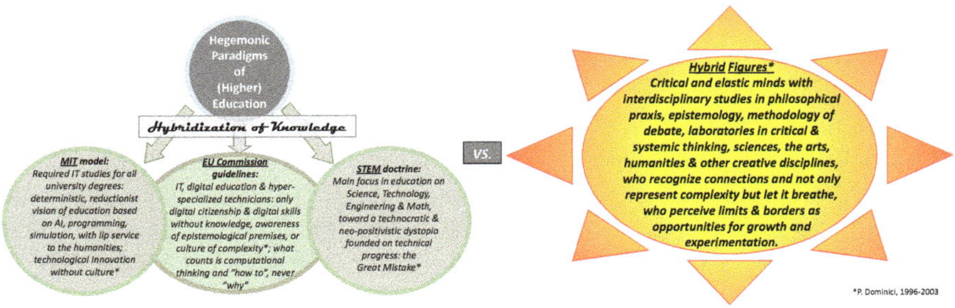

Fig. 7.2 Hegemonic paradigms of higher education

about a *paradigm shift*, creating the structural conditions for the interdependence (and efficiency) of systems and organizations and intensifying the immaterial flows between social actors, are as yet unable to guarantee that the networks thus created generate fully *communicative* relationships, that is, based on symmetrical relationships and authentic sharing. In other words, although the Web has created a new communications ecosystem (1996), which is redefining the space of knowledge, it cannot guarantee, in and of itself, horizontality or more symmetrical relationships. Once again, it is the people and the uses that they make of technology which make the difference, beyond the many interests at stake. For these same reasons, I use the term "connection technologies" and not "communication technologies". A substantial distinction which—as I have been repeating for many years—materialises/translates into a series of dilemmas—at times, *open dialectics*—typical of human societies, and into the consequent definition/adoption of social, political and cultural choices/strategies, which have profound systemic repercussions. Dilemmas (and choices) whose implications and dynamics have become further radicalized in the hyper-technological civilization without finding even a minimal re-composition of fractures and distances, which are, once again, social and cultural (→education, training, research, organizational cultures etc.): control vs cooperation; surveillance vs *privacy*; security vs liberty; simulation/automation vs autonomy/responsibility; automation (= management and predictability) vs error (and unpredictability); Technological vs Human".[10] With reference to the issues of horizontality (without adequate knowledge and skills, it is merely "simulated" horizontality) and the symmetry of social relations, it appears evident that we have been trapped, for some time, in the illusion of having a less asymmetrical relationship with the State and with the "elites". It is precisely in this perspective that I have proposed, in the past, the concept (and operational definition) of an "Asymmetrical Society": a society that "guarantees citizenship and inclusion only from the point of view of legal recognition"; a society (people, languages, systems and forms of relationships,

[10] See Dominici P., *Dentro la Società Interconnessa. Prospettive etiche per un nuovo ecosistema della comunicazione*, FrancoAngeli, Milan 2014, p. 9.

information and energy flows, connections, levels of interconnection, ecosystems etc.) where the rules of engagement relating to citizenship and inclusion have profoundly changed. On the other hand, those same rules of engagement have long been defined, produced and reproduced within the very educational and training institutions that should have influenced and modified, over time, certain "starting conditions".[11]

The Asymmetrical Society places the unavoidable urgency of a "new social contract[12]" on the agenda. In actual fact, the conditions and variables that constitute and characterise complex phenomena and processes like those linked to freedom, equality, citizenship, inclusion, democracy have changed profoundly.[13] For some time now, we have been going through a crisis that is only partially economic; it is, above all, cultural and civilizational. This cannot be repeated too often: all in all, what is being pondered here are not only the conditions that can render complex processes like those concerning inclusion and citizenship effective, but also the opportunity and necessity, from within a systemic and network-logics perspective, to work on the definition and construction of a culture of citizenship and inclusion. The very growth of any Nation-State, which is a crucial issue that cannot be explained or managed using an economistic paradigm alone (globalization has amply demonstrated this), would benefit enormously from these. The educational and cultural question, the social and cultural factors,[14] are absolutely decisive, well beyond any legal, regulatory and ethical-professional framework. This is a complex issue of

---

[11] *Ibidem.*

[12] See Dominici P, (1996–2018), *op.cit.*

[13] Within a scholarly literature that is, to say the least, endless, we suggest some pathways: Gramsci, A. (1948–1951), *Quaderni del carcere*, Einaudi, Turin 1975, 4 vol.; Bobbio, N. (1984), *Il futuro della democrazia*, Einaudi, Turin 1995; Dahl, R. A. (1998), *On Democracy*, Italian translation, *Sulla democrazia*, Laterza, Rome -Bari, 2000; Dahrendorf, R. (2001), *Dopo la democrazia*, Laterza, Rome -Bari; Dominici, P. (2008), *Sfera pubblica e società della conoscenza*, in Morcellini M., Mazza, B., (Eds.) *Oltre l'individualismo. Comunicazione, nuovi diritti e capitale sociale*, Franco Angeli, Milan; Dominici, P. (2014a), *Dentro la società interconnessa. Prospettive etiche per un nuovo ecosistema della comunicazione*, FrancoAngeli, Milan; Dominici, P. (2014b), *La modernità complessa tra istanze di emancipazione e derive dell'individualismo*, in "Studi di Sociologia", n°3/2014, Vita & Pensiero, Milan; Dominici, P. (2016a), *L'anello debole e le reti "fuori" dalla Rete: ripensare la cittadinanza nella Società Interconnessa*, in Pietrafesa E., Flavia Marzano F., Medici T. (Eds.), *La Rete e il fattore C. Cultura, complessità, collaborazione*, Stati Generali dell'Innovazione, Rome; Dominici, P. (2016b), *La filosofia come "dispositivo" di risposta alla società asimmetrica e ipercomplessa*, in Candiotti L., Gambetti F. (Eds.), *Il diritto alla filosofia. Quale filosofia nel terzo millennio?*, Diogene Multimedia, Bologna; Foucault, M. (1975), *Sorvegliare e punire. Nascita della prigione*, Einaudi, Turin 1976; Marshall, T.H. (1950), *Citizenship and Social Class and Other Essays*, Cambridge University Press, Cambridge 2002; Nussbaum M.C. (2010), *Non per profitto*, Il Mulino, Bologna 2011; Rawls, J. (1971), *Una teoria della giustizia*, Feltrinelli, Milan 1982; Rodotà, S. (1997), *Tecnopolitica. La democrazia e le nuove tecnologie della comunicazione*, Laterza, Bari; Minogue K. (2010), *The Servile Mind. How Democracy Erodes the Moral Life*, Italian translation, *La mente servile. La vita morale nell'era della democrazia*, IBL Libri, Turin 2012.

[14] Beyond the substantial scholarly literature on these issues, I also refer to the interesting, and always relevant, conference held in 1893 by Thomas Henry Huxley, at Oxford University: Huxley T.H. (1894), *Evolution and Ethics*, Italian translation, *Evoluzione ed etica*, Bollati Boringhieri, Turin 1995.

crucial importance which, beyond public discourse and certain media spectacularization, linked in particular to events broadcast on the news, continues to be underestimated. Moreover, it is a question which calls into question the freedom and, with it, the responsibility (relational concepts) of People, citizens and, more generally, of social, individual and collective agents. *Civility*, teaching citizenship, a shared ethos and strong cultural and identity models—always, it goes without saying, within open communities—are all "devices" fundamental to the evolution of social and organizational systems, as well as, evidently, for the very survival of democracies. The clear impression is that we find ourselves on a slippery slope which—I repeat—if we do not seriously, profoundly, and rigorously address instruction, education, training and research, will risk leading us towards a slow but inexorable decline, in terms of the quality of democracy and of "living together" as well. Ethics and morality cannot be imposed—it has never been possible to impose them—and need to be built socially and culturally, starting, above all, from educational processes. Furthermore, the overall social situation, both local and global, has become even more fragile: the innovation(s) we are witnessing are accessible just to the few, while the inclusion we hear discussed so frequently in the media is but little put into practice.

For some time now, citizenship has no longer been an (exclusively) legal question or, at least, it is only minimally so! I have often underlined, in this sense, the risk of technological innovation without culture, the risk of a citizenship without citizens[15] and of a decline, common to the most "advanced" countries, which stems precisely from the Schools and Universities, whose vital functions have been stripped away or at best weakened: precisely those functions that are vital for a full-blown form of democracy based on citizens and not subjects,[16] on tangible, effective, rather than simulated, participation.[17]

We are facing—as mentioned several times above- an "Asymmetrical Society", characterised increasingly by inequalities and new asymmetries and by technological innovation bereft of culture. A hyper-technological and hyperconnected civilization in which—as already mentioned—the rules of engagement of citizenship have no longer been written by the Legislator for some time now. Schools and universities, and not only from today, in addition to failing to perform their essential, strategic function as "social elevators", have returned to being "agencies of selection", after a few seasons during which they carried out an essential function as "agencies of emancipation".[18] As regards this point, with reference to Italy, it is interesting to

[15] See Dominici P, (1996–2018), *op.cit.*

[16] See in particular: de La Boétie, É. (1549–1576), *Discorso della servitù volontaria*, Feltrinelli, Milan 2014; Dominici, P. (2017a), *Oltre la libertà ...di "essere sudditi"*, in F. Varanini (ed.)*Corpi, menti, macchine per pensare*, Casa della Cultura, Anno 2, numero 4, Milan.

[17] *Ibidem.*

[18] See for further ideas and insights on these issues, I refer, once again, to some contributions written for the "Il Sole 24 Ore" newspaper precisely on the "asymmetric society": *La "società asim-*

read the latest ISTAT Reports (reports from the national institute of statistics) regarding knowledge,[19] where, among the many issues, the empirical evidence

---

metrica" e la centralità della questione culturale http://pierodominici.nova100.ilsole24ore.com/2015/09/23/la-societa-asimmetrica-e-la-centralita-della-questione-culturale-le-resistenze-al-cambiamento-e-le-leve-per-innescarlo/; *Oltre lo storytelling: il Sociale, il digitale e "la Persona al centro"*– Il Sole 24 Ore http://pierodominici.nova100.ilsole24ore.com/2016/05/07/oltre-lo-storytelling-il-sociale-il-digitale-e-la-persona-al-centro/

[19] I invite readers, once again, to consult the interesting "*Rapporto Istat sulla Conoscenza 2018. Economia e società*", of which I quote for obvious reasons, only some excerpts "The knowledge capital deriving from education expands personal employment and income opportunities (see 6.2) and the performance of businesses (see 5.1, 5.2). It is also reflected in an increased ability to grasp cultural stimuli (see 4.1), to use technology effectively (see 4.2) and, more generally, to decode the complexity of information (!), allowing for greater awareness and, often, better health and a longer life, all other conditions being equal. *Italy presents a historical delay in education compared to more advanced countries.* In 2016, the percentage of people between 25 and 64 years of age with at least an upper secondary education qualification reached 60.1%, with a slight prevalence of females (62 against 58). Despite an increase of 8 points compared to 2007, this share remains 16.8 percentage points lower than the European average. Similarly, people between the ages of 25 and 64 with a third-level qualification total 17.7%, just a little above half of the respective European value (30.7%). The Italian delay is due largely, but not exclusively, to the poor education of the older cohorts. Among people between the ages of 25 and 34, 73.9% have at least an upper secondary education qualification, while in the EU they stand at 83.4%, a difference of 9.5 points. For third-level qualifications the differential is 12.6 points (25.6% and 38.2% respectively; see 3.2). The educational levels of the adult population vary greatly across the territory (...)" (p. 44.). And again: "[...] *Italy is a high-income but anomalous industrial economy because it is characterised, compared with the other major European economies, by modest, albeit growing, levels of education and skills (see 3.1, 3.2, 3.7). Mirroring these characteristics are the lower incidence in the employment of professionals and technicians and, in particular, of personnel with university degrees in these areas.* Italy, with its low educational levels and the low intensity of research and development as well as its modest patenting activity (see 2.1–2.2; 2.4), has based an important part of its well-being on types of productions with a relatively limited content of specialized knowledge, easily replicable at lower costs elsewhere". (p. 11) "[...] In this regard, the educational level of individuals represents a variable that permits us to embrace the majority of the topics covered in the Thematic frameworks, because it appears to be the element that influences behaviour and performance most in a variety of areas, from the most obvious one of basic skills (see 3.7), to involvement in creative (see 2.9) and cultural activities (see. 4.1, 6.4, 6.6) even later in life, to digital skills and activities carried out online (see 4.2, 4.3): for example, the diffusion of Internet, primarily associated with age, is greater among well-educated elderly people and young people between the ages of 16 and 24. Furthermore, in the area of small businesses, the educational levels of entrepreneurs and employees are associated mostly with survival, placement on the market, the tendency to innovate and adopt information technologies (see Chapter 5)". (p. 12) "[...] On the other hand, the frameworks presented in the Report also show, from different perspectives, how the origins of individuals (their country of birth, territory, socio-economic characteristics) influence their access to education, qualifications and the results achieved: in 2016, over 30% of young Italians aged between 25 and 34 obtained a university degree, while among foreign residents the incidence was 10%, only 6% among males. In vocational schools, where the children of less well-educated parents mainly enrol, the linguistic and numerical skills of fifteen-year-olds are dramatically lower than those of their lyceum-attending peers. In 2016/17 the rates for the transition to university of the former stopped short at 11.3%, compared to 73.8% from the latter. According to the results of the Invalsi survey, in the South of Italy even students attending lyceums have much lower-level skills than their peers in the Central-Northern regions". Finally: "The European countries that have

presented highlights: (a) "a historical delay in education compared to more advanced countries"; (b) the importance, influence and impact of individuals' geographical, social, economic, cultural origins on their access to education, their qualifications and the results achieved; (c) once again, the educational level of individuals represents a variable that turns out to be, without fail, the element that most influences behaviour and performance in a variety of areas; (d) the extremely concrete risk is that Italy will find itself falling back economically—as has occurred already in the past—unless it enhances people's opportunities by "reducing the area of social exclusion determined by knowledge", by guaranteeing access to information and culture (common good), not to mention the profound implications of exclusion from the so-called digital revolution.

In other words, it is a matter of rethinking the very conditions of the "social contract". Speaking in an extremely simple and straightforward manner, if by "social contract" we mean that *system of rules, agreements and conventions—socially and culturally negotiated and co-built*—which makes "Society" possible (WE, the system of relationships, networks and social mechanisms etc.), taking the place of an age-old "state of nature", from which Society itself can/must arise. And if by using the word "Society"[20] we mean a mega-system of social actors, linked, to some

achieved better results in terms of the spread of university education have invested in orientation, ongoing support and opportunities, with lower percentages of students enrolled in vocational institutions and a system of scholarships with extensive coverage. This shows how the improvement of a population's level of education—the increase in attendance and success, particularly at university, and the quality of the service provided, at all levels—continues to remain the privileged ambit when it comes to intervention in favour of knowledge. This to avoid that Italy may fall behind economically—as during the past crisis—while at the same time, broadening people's opportunities by reducing the area of social exclusion determined by knowledge, in particular digital knowledge" (p. 13). And we are obliged to go on repeating Italy's current reality is a far cry from inclusion, a far cry from digital citizenship, a far cry from a knowledge society. What is taking form is an innovation meant more and more for the few, which concretely risks translating into a "society of ignorance" (Dominici, 1996, 2009 and following)—evidently, "ignorance" intended as a lack of education and absence of knowledge (also with reference to its creation, transmission, access, sharing, use, etc.)—where very many are excluded (also) from the extraordinary opportunities offered by hyper-technological and hyper-connected civilization.

[20] As is well-known, the operational definitions, like the approaches and *paradigms* of reference in the study and analysis of Society (positivism, structural-functionalism, conflictism, constructivism, symbolic interactionism; another distinction accepted in scholarly literature is that between the paradigms of order, conflict, structure and action which are multiple and can be traced back to a vast and complex scientific literature. I shall simply indicate, among the great "classics": Marx K. (1867 e 1885–94), *Das Kapital. Kritik der politischen Ökonomie*, 3 voll., Italian translation, *Il capitale. Critica dell'economia politica*, Ed.: D. Cantimori, R. Panzieri, M.L. Boggeri, Editori Riuniti, Rome 1951–56; Tönnies F. (1887), *Gemeinschaft und Gesellschaft*, Italian translation, *Comunità e società*, Comunità, Milan 1963; Durkheim É. (1893), *De la division du travail social*, trad. it., *La divisione del lavoro sociale*, Milan 1971; Weber M. (1922b), *Wirthschaft und Gesellschaft. Grundriß der verstehenden Soziologie*, Italian translation, *Economia e società*, Vol. II, Milan, Comunità, 1961; Mead G.H. (1934). *Mind, Self and Society*, Italian translation, *Mente, Sè e Società*, Barbera, Florence 1966; Dahrendorf R. (1959), *Class and Class Conflict in Industrial Society*, Italian translation, *Classi e conflitto di classe nella società industriale*, Laterza, Bari 1963; Parsons T. (1951), *The Social System*, Italian translation, *Il sistema sociale*, (intr. di L. Gallino),

extent, to one another by a network of relationships and interactions of various kinds, where exchange, cooperation, collaboration and the division of roles and tasks are established to ensure the survival, cohesion and reproduction of the whole/system itself and the members that constitute it,[21] in that case, we cannot help but note how, precisely within the so-called *knowledge society*, increasingly (hyper) complex and hyperconnected, and based more and more on "limited rationality" (contrary to what we are told by the dominant narratives and storytelling), on the brink of paradox (if one thinks of the infinite availability of data and information), the relationship between the offer and creation of a "new social contract[22] and the educational question is, to say the least, intimate and very close-knit. Precisely because this system of agreements and conventions, which makes "society"

Comunità, Milan 1965; Goffman E. (1959), *The Presentation of Self in Everyday Life*, Italian translation *La vita quotidiana come rappresentazione*, Il Mulino, Bologna 1969; Elias N. (1987), *Die Gesellschaft der Individuen*, Frankfurt, Suhrkamp, Italian translation *La società degli individui*, Il Mulino, Bologna 1990; Habermas J. (1981), *Theorie des kommunikativen Handelns*, Bd. I *Handlungsrationalität und gesellschaftliche Rationalisierung*, Bd. II *Zur Kritik der funktionalistischen Vernunft*, Frankfurt am Main, Suhrkamp, Italian translation *Teoria dell'agire comunicativo*, Vol. I *Razionalità nell'azione e razionalizzazione sociale*, Vol. II *Critica della ragione funzionalistica*, Il Mulino, Bologna 1986; Luhmann N. (1984), *Soziale Systeme*, Suhrkamp, Frankfurt 1984, Italian translation, *Sistemi sociali. Fondamenti di una teoria generale,* Il Mulino, Bologna 1990; Among the significant works to be considered are Gallino L. (1978), *Dizionario di Sociologia*, Utet, Milan; Ferrarotti F. (1991), *Trattato di sociologia*, UTET, Turin; Crespi F., *Manuale di Sociologia della cultura*, Laterza, Rome -Bari 1996; Izzo A. (1992), *Storia del pensiero sociologico*, Bologna, Il Mulino; Ferrara A., *Pensare la società. L'idea di una filosofia sociale*, Carocci, Rome-Bari 2001; Hedström P. (2005), *Dissecting the Social. On the Principles of Analytical Sociology*, Italian translation, *Anatomia del sociale. Sui principi della sociologia analitica*, Bruno Mondadori, Milan 2006; Cipolla C. (2015), *Dalla relazione alla connessione nella web society*, FrancoAngeli, Milan; Rainie L., Wellman B. (2012), *Networked: The New Social Operating System*, Italian translation, *Networked. Il nuovo sistema operativo sociale*, Milan: Guerini 2012; Cipolla C. (2019), *Per una scienza sociale eclettica. Riflessioni aforistiche e algoritmiche*, FrancoAngeli, Milan; Latour B. (2005), *Reasssembling the Social*, Italian translation, *Riassemblare il Sociale*, Meltemi, Milan 2022.

[21] For further information, see the historic *Dizionario di Sociologia* (Utet, Turin 1978 e 1993), ed. by Luciano Gallino, the great sociologist and social scientist.

[22] See Dominici P, (1996–2018), *op.cit.* Speaking of the "new social contract", in the month of November (2021), I took part, also as a delegate of the World Academy of Art and Science, at the UNESCO International Conference in Paris where there was much talk of the "Future of "Education", evoking the formula of the "new social contract": this is actually the title of UNESCO's new global action, as well as that of an old scientific publication of mine. From the point of view of the public debate and the commitment of a grand transnational organization, these are certainly important signals. Finally, (perhaps) we are starting to become aware and act concretely, starting from the assumption that education—well aware of the thin line between education and indoctrination—is not a simple "tool" capable of accompanying change and social transformation; education—the quality (complex concept) of educational and training processes—is the fundamental infrastructure of every process of innovation and change. In other words, we may be starting to realize how social and cultural are as important as economic factors, if not more so. This is the link to the UNESCO global action website: https://uil.unesco.org/lifelong-learning/new-social-contract-education; see also: https://news.un.org/en/story/2021/11/1105492. On the "Future of Education" action, UNESCO see https://en.unesco.org/futuresofeducation/

possible"[23] and helps to define the "forms" of interaction and social stratification, both local and global, is increasingly marked, shaped, and structured by the quality (itself a complex concept, to be explored) of education and educational processes, by the opportunities, sometimes denied even from the start, to access adequate education, information, knowledge and culture. They have always been the true agents of citizenship and democratization. As always, a non-egalitarian and non-"quality" school is the precondition and the circumstance most likely to define, nourish and reproduce an unequal and, indeed, asymmetrical society.

The impression, at times, is that of moving slowly yet inexorably towards a *Società dell'ignoranza* [Society of Ignorance] (2009)– which I have rendered in English as a *No-Knowledge Society*—based on a *feudal*, corporate social model, which stakes everything on technocracy and the possibility of *simplifying* everything, even democracy itself. This ideal type of Society is founded on criteria of exclusivity rather than inclusiveness because it provides for exclusively horizontal social mobility and the progressive marginalization of any deviance and/or unpredictability. These issues add to the continuing lack of awareness of the well-known correlation between education and innovation, between education and inclusion, between education and democracy. With all the risks and opportunities that the hypertechnological and hyperconnected civilization brings with it: the hegemony of new forms of reductionism and determinism, the affirmation of a "technological-juridical solutionism" built on the mechanism of granting "carte blanche" to technology for questions that are absolutely vital for social systems and organizations, regarding thinking, control, rationality, protection, security, trust and social bonds. In developing the aforementioned points, we cannot help but consider a whole series of factors and critical issues that intercept diverse and, above all, no longer negligible levels of analysis and intervention:

- the absence of a thought system and of a systemic view and the contemporary underestimation of the importance of research into thought and education[24]—"disjunctive thinking is still hegemonic";
- the absence of (long-term) policies relating to education, training, and research;
- the inconsistency and inadequacy of investments in education, training and research (UNESCO, OECD, ISTAT etc.);
- education, instruction, training and research are still struggling to be (genuinely) acknowledged and perceived as the true infrastructure of change and innovation that cannot be just for the few;
- the School and University continue to be thought of, imagined and designed as separate entities, while they are a single ecosystem (1996);
- the absence of policies of orientation, delegated totally to marketing practices;

[23] I repeat: "Society" is not a merely theoretical, abstract construct, as some would have us believe. It is a system of relationships and interactions, characterized also by signs, symbols, beliefs, representations, behavioural and cultural models, social mechanisms of trust and cooperation/competition, systems of value and cognitive orientation etc.

[24] *Ibidem.*

- the triumph of the deceptive principle of "useful knowledge" and "useful fields of knowledge";
- the mistake of continuing to chase after markets and business, in an era of rapid, very rapid obsolescence of all knowledge, skills, training and professional profiles;
- the dominance and hegemony of a culture of standardization[25] that pervades the entire culture of evaluation and communication.

7.2 *Broadening One's View*. The Strong Bond: Education and Sociocultural Change

Change is *the non-linear evolution of social complexity*, that is to say, change is *social complexity*.[26] It is a very particular kind of complexity, typical of living systems, which are complex adaptive systems (CAS), irreversible in their non-linear evolution, marked by non-linearity, an intrinsic dynamism, and an unpredictability which, in the first place, is associated with the *non-observability* of all the constituent dimensions and combined with their ability to adapt and self-organise. In this sense, there is a close, very close correlation between education and change, between education and cultural evolution, that strategic type of evolution that allows you to try, if not to govern (as many would like), to *inhabit* the subjective, ethical, social, systemic, linguistic, communicative hypercomplexity that characterizes the new interconnected ecosystems.

However, more and more frequently, change, unlike what one might believe, lurks in the zones of *tension* and *conflict*, in our weaknesses and inadequacies, in the anomalies, deviances, the fluctuations and dilemmas that characterize knowledge, social action, complex (adaptive) systems, organizational and social practice. Change even lurks inside our *incompleteness* which, in any case, allows us to be creative and dip into our imagination, seeking alternative pathways, abandoning, if necessary, the paths already trodden. More and more frequently, change lurks in the moments of uncertainty, in those errors, fragilities and vulnerabilities that we often ignore and/or try not to see. These complex dimensions are those we keep striving to remove, and/or at least marginalise, precisely within educational and training processes, "communicating" the illusion of a *pervasive rationality*, of total control and predictability[27] which, moreover, are combined with,—indeed, I might say, are practically welded onto another well-rooted illusion, both of the individual and of

[25] See, in particular: Robinson, K. (2015), *Scuola Creativa. Manifesto per una nuova educazione,* Erickson, Trento 2016; Hammersley M. (2013), *Il mito dell'evidence-based,* Raffaello Cortina Ed., Milan 2016.

[26] See Dominici P, (1996–2022), *op.cit.*

[27] *Ibidem.*

the collective imaginary: that of being able to know and understand complex phenomena with great precision and *explanatory depth.*[28]

Change (and innovation) which risk, however, being opportunities "for a select few". As I often repeat, it is necessary to question the fields of knowledge, the borders between them, and the consolidated practices, reconsidering the strategic force of emotions and of individual and collective imagination(s). In other terms, as also recalled at the beginning of this work, it is also necessary to have the courage to shatter equilibriums, *breaking the chains of tradition*, abandoning the certain for the uncertain, choosing, at least temporarily, to run *the risk of being vulnerable.* Inhabiting the borderlines, the unexplored territories, overstepping those limits and those logics of separation (typical of educational and training institutions) that prevent us from grasping the deepest meaning of the vital, the social, the relational and from understanding their complexity and ambivalence. Dimensions, in effect, that are complex, never reducible/attributable to mathematical formulas and/or sequences of data.

It is time to rethink the Human and its somehow ambiguous interaction with the technical[29] and the technological: an interaction from which only a complex synthesis can arise and of which we are as yet unable to evaluate the prospects, developments and implications. Amid "new" utopias and dystopias. Amid forces of interdependence and those of fragmentation. Amid inclusiveness and exclusivity, within asymmetries that travel along discontinuous trajectories.

Because, it is precisely in this era of programming, of the illusion of total control and of the simulation of all processes and dynamics, of the dominance (apparent, at least) of automation and the progressive marginalization of the Human and of responsibility (vs. speed), precisely in the era of Big Data and the infinite availability of data and information, that social systems and complex organizations appear to be dominated increasingly by profound uncertainties and vulnerabilities, increasingly chaotic and disorderly, even within the contradictions and ambivalence of productive, economic and cultural processes. This hypercomplexity—it is worth clarifying, once again—is cognitive, social, subjective, ethical, and as it affects every area of life and praxis, it requires us to rethink the categories, paradigms, education and the "forms" of citizenship that arise from it.

The false and misleading contrast between the (hyper)specialization of the fields of knowledge and their complexity and interdisciplinarity makes the inadequacy of

---

[28] See in particular, the interesting article by Rozenblit L., Keil F, *The misunderstood limits of folk science: an illusion of explanatory depth*, in "Cognitive Science", 26 (2002) 521–562, where the question of the "illusion of explanatory depth" is addressed. To further broaden one's view, see also Sloman S., Fernbach P. (2017), *The Knowledge Illusion. Why We Never Think Alone*, Italian translation, *L'illusione della conoscenza,* Milan: Raffaello Cortina Editore, 2018; Boorstin D. J. (1994), *Cleopatra's Nose. Essays on the Unexpected,* Vintage Books, New York; Boorstin D. J. (1998), *The Seekers. The Story of man's continuing quest to understand the world*, Italian translation, *L'avventura della ricerca*, Raffaello Cortina Ed., Milan 2002; Hustvedt S. (2016), *The Delusions of Certainty*, Italian translation, *Le illusioni della certezza*, Einaudi, Turin 2018.

[29] See, in particular Mumford, L. (1934), *Tecnica e cultura,* Il Saggiatore, Milan 1961; see also: Foucault M., *Tecnologie del Sé. Un seminario con Michel Foucault*, Bollati Boringhieri, Turin 1992.

schools and universities even more evident when faced with the complexity and ambivalence of life, of the irruption, in some ways prepotent, of chaos and disorder; when faced with processes and dynamics (and rationalizations) that continue to evolve in an increasingly swift, non-linear manner. The "old" boundaries between scientific and humanistic education have become unrecognizable in the presence of dilemmas that require us, first of all, to question fields of knowledge and consolidated practices, individual and collective imaginaries (risks vs opportunities). As a result, a reformulation of thinking and of the fields of knowledge that involve both schools and universities directly, becomes even more urgent (but resources are required!). Unfortunately, schools and universities are still thought of and organized as separate "entities", whose policies should be reviewed and re-designed in systemic terms. We need to become aware, definitively, that the real strategic "factor", at the basis of change and innovation processes, is the cultural "factor".

The "fact" is that we are not ready to face the challenges and dilemmas of complexity and of the new (global) ecosystem, not so much in terms of research methodology(ies) or of the increasingly refined tools of detection and analysis—as the theoretical-interpretative models that must (should) guide/orient the empirical, not only scientific, observations of phenomena and processes. Yet, education and training in complexity, in the scientific method and a renewed awareness of the need for an interdisciplinary and multidisciplinary approach to this reality of phenomena are needed; an approach which brings with it a redefinition of the space allotted to the fields of knowledge and the overturning of the logics of power and control which, at all levels, have sanctioned its fragmentation and confinement within the narrow "borders" of disciplines.

For some time now, we have no longer been able to look at/observe the whole, the system, the entirety, the globality, nor the system of relationships and/or interactions that characterize them. In other words, we have difficulty recognizing connections, correlations, and causal links precisely because we have been educated and trained (at best) to describe and record "regularities", the "hows" and not the "whys". We have been educated and trained to seek and be content with simple and/or pre-codified answers (in any case, obtained in a short time), to seek—I always say—*simple solutions to (hyper)complex problems.*

The interconnected economy and the hyper-technological civilization require strategic choices and a new ethical sensitivity towards the problems concerning the social agents, the system of relationships and the spaces of knowledge. What is needed is a "new culture of communication" oriented towards a kind of sharing and understanding, that is capable of impacting the social mechanisms of trust and cooperation.[30] Putting the Person, the sphere of relations and educational processes back at the centre. Fully aware, in agreement with Maturana, that every time we accept an idea, a concept or a stated "fact" without reflecting, simply for the reason that it represents something that is universally understood, we are donning what

[30] See, in particular, Coleman, J.S. (1990), *Fondamenti di teoria sociale*, Il Mulino, Bologna 2005; Putnam R.D. (2000), *Capitale sociale e individualismo. Crisi e rinascita della cultura civica in America*, Bologna: Il Mulino 2004.

Maturana terms "blinkers". For example, insisting that human beings are by definition rational is a *blinker*, because it does not allow us to perceive emotions, which are dismissed as being something pertaining to animals or else negations of rationality...the two-way links between reason and emotion that make up our ever day existence as human beings become invisible to us, thus we are unaware of an important truth: that every rational system is based on emotion.

Cognitive sphere, emotional sphere and—I would add—social sphere. It is time to heal some of the fractures that characterize not only the fields of knowledge and skills, but knowledge itself, and become aware of the intrinsically collective and collaborative nature of knowledge.[31] The fractures in question also mark individual existences, reality and our views of reality. They are important fractures, rooted in organizational and even in scientific cultures; fractures that condition not only the evolution of the fields of knowledge and knowledge itself, but also our abilities and capacity to inhabit hypercomplexity and respond—through cultural models, for one—to the instances of uncertainty, as well as to the anomalies of the living and of reality. Above all, these fractures are also elements that condition our existences, our social and cultural experiences, and the very way we conceive of life and existence, relationships, of the encounter with the Other from Ourselves, and of our thoughts and actions concerning what is forever ungovernable, unpredictable and at times unknown. "Inside" and "outside": it is time to inhabit the boundaries and tensions that this hypercomplexity entails.

7.3 The Urgency of a Radical Change

The question of all questions: education, which has always been strategic for social cohesion itself, and even more vital for inclusion and the *quality* of democracy, has become overwhelmingly relevant once again, this time, even more evidently, on a global plane. And this has come about in the train of the latest universal, systemic emergency (the umpteenth universal, systemic emergency) to impact the world system. With it, the questions (re)formulated at the beginning have also become

[31] See Sloman, S., Fernbach P. (2017), *L'illusione della conoscenza. Perché non pensiamo mai da soli*, Raffello Cortina Ed., Milan 2018; Morin, E. (2015), *7 lezioni sul Pensiero globale*, Raffaello Cortina Ed., Milan 2016; Morin, E. (2017), *op.cit.*; Dominici, P. (1996), *op.cit.*; Dominici, P. (2005), *op.cit.*; Dominici, P. (2008), *Sfera pubblica e società della conoscenza*, in Morcellini M., Mazza B., (Eds.) *Oltre l'individualismo. Comunicazione, nuovi diritti e capitale sociale*, Franco Angeli, Milan; Dominici, P. (2014a), *Dentro la società interconnessa. Prospettive etiche per un nuovo ecosistema della comunicazione*, FrancoAngeli, Milan; Dominici, P. (2014b), *La modernità complessa tra istanze di emancipazione e derive dell'individualismo*, in "Studi di Sociologia", n°3/2014, Vita & Pensiero, Milan; Dominici, P. (2015a), *Communication and Social Production of Knowledge. A new contract for the Society of Individuals*, in "Comunicazioni Sociali", n°1/2015, Vita & Pensiero, Milan; Dominici, P. (2016a), *L'anello debole e le reti "fuori" dalla Rete: rethink la cittadinanza nella Società Interconnessa*, in Pietrafesa E., Marzano F., Medici T. (Eds.) *La Rete e il fattore C. Cultura, complessità, collaborazione*, Stati Generali dell'Innovazione, Rome.

relevant again: why on earth do we need to radically rethink educational processes? Cannot *digital literacy*, digital education and/or media education suffice? We need to again ask ourselves (see the long-standing and, at this point, repetitive debate regarding digital literacy) whether technologies and digitality are/will ultimately be able to eliminate any critical issue or vulnerability typical of social networks and organizational systems that may arise, in terms of social bonds, trust, protection, security, inclusion, etc. This is, in reality, the still-hegemonic *narrative*, which fits the continuous and systematic reference to the "STEM Doctrine" perfectly: a narrative which, despite the public declarations, along with a certain literature (scientific and other) and institutional documents that are now, without exception, calling for "critical thinking" and interdisciplinarity, and even speaking about complexity (sic!), nevertheless continues to push us in a certain direction, due in no small part to the considerable economic and political interests at stake.

This is a narrative/storytelling, one that did not begin today, which assumes concrete form in strategies and practices, not only involving educational and training, which takes shape, in my opinion, in the most erroneous and misleading framework and approach.[32] All this at a time when, on the contrary, we need a radical change of direction. Because we are faced with a hypercomplexity—a challenge that is evidently epistemological and ethical—which has expanded to such an extent as to make any attempt to provide/formulate schemes for its reduction extremely difficult and complicated. This is a type of (hyper)complexity that is further increased by the more and more strategic relevance that communication and technological innovation have assumed, not only in processes of education and socialization,[33] but also and above all in the representation and perception of evolutionary systemic dynamics and processes which also closely concern the production of fields of knowledge, of "tools", of scientific knowledge[34]; a production functional precisely to the analysis and management of this hypercomplexity and also for the creation of social and cultural conditions capable of counteracting the uncertainty and unpredictability that characterize organizational and social systems.[35] Fundamental, decisive issues, of strategic importance for both organizations and democracies, which are also marked by a profound crisis. And yet in the "hyper-complex society" (2003), all this risks becoming no longer sufficient. It is increasingly crucial to know how to communicate this (hyper)complexity, bringing the question of knowledge and skills to the foreground (as if there were still the need to do so), as well as the urgency of overcoming, once and for all, the "false dichotomies".[36]

---

[32] See Dominici P, (1996–2018), *op.cit.*

[33] On the complex issue of socialization and the sociological approach to education, see in particular: Benadusi L., *Scuola, riproduzione, mutamento. Sociologie dell'educazione a confronto*, La Nuova Italia, Florence 1984; Benadusi L., Censi A., Faletti V., *Educazione e socializzazione*, FrancoAngeli, Milan 2004.

[34] See Losee J. (1972), *A Historical Introduction to the Philosophy of Science*, Italian translation, *Filosofia della Scienza. Un'introduzione*, Il Saggiatore, Milan 2001.

[35] *Ibidem.*

[36] *Ibidem.*

7.4 The Paradigm Shift and a Cultural Evolution That Is Now Conditioning Biological Evolution

For these reasons, I have returned several times to the formula I had proposed, which is, precisely: "objects as systems".[37] In the hypertechnological and hyperconnected civilization, it is necessary that we teach/learn to see, observe, watch, recognize objects as systems and not vice versa, because we must always measure ourselves with complex adaptive systems and complex issues that involve numerous different disciplinary fields, which draw our attention to the importance of a systemic, interdisciplinary and multidisciplinary perspective[38] that escapes traditional categories and definitions; a hypercomplexity, characterized by increasingly imperceptible borders between nature and culture, natural and artificial, human and non-human and which cannot find simple answers and solutions to problems that are evidently complex. In this sense, it is necessary to start again precisely from an overall rethinking (1) of knowledge as shared and transdisciplinary knowledge, (2) of the space between the fields of knowledge and, at a second level, the space between skills; but at the same time, an overall rethinking of relational space is also necessary (freedom is responsibility—the centrality of educational processes) which places the Person at the core of the complex process of the change in progress within the framework of a renewed interaction with Technique and technologies. By recomposing what, precisely within the hypertechnological and hyper-modern

[37] See Dominici P, (1996–2018), *op.cit.*

[38] Within an endless array of scholarly literature, besides those attributable to different approaches and disciplinary fields, see in particular: Wiener, N. (1948), *La cibernetica*, Il Saggiatore, Milan 1968; Wiener, N. (1950), *Introduzione alla cibernetica. L'uso umano degli esseri umani*, Bollati Boringhieri, Turin 1966; Arendt, H. (1958), *op.cit.*; Ashby, W.R., (1956), *op.cit.*; Simon, H.A. (1962), *The Architecture of Complexity*, in "Proceedings of the American Philosophical Society", 106, pp. 467–82; Bertalanffy von, L. (1968), *op.cit.*; Bateson, G. (1972), *op.cit.*; Lakatos I.—Musgrave A. (1970), *Critica e crescita della conoscenza*, Feltrinelli, Milan 1976; Foerster von, H. (1981), *op. cit.*; Maturana H., Varela F. (1972), *op.cit.*; Maturana H.R., Varela F.J. (1985), *L'albero della conoscenza*, Garzanti, Milan 1987; Bocchi G. e Ceruti M. (Eds.) (1985), *La sfida della complessità*, Bruno Mondadori, Milan 2007; Luhmann, N. (1984), *op.cit.*; Gleick J. (1987), *Caos*, Rizzoli, Milan 1989; Gallino, L. (1992), *L'incerta alleanza*, Einaudi, Turin; Kauffman, S.A. (1993), *op.cit.*; Gell-Mann M. (1994), *op.cit.*; Prigogine I. (1996), *op.cit.*; Diamond, J. (1997), *op.cit.*; Diamond, J. (2005), *op.cit.*; Morin, E., Ciurana, É.-R., Motta, D.R. (2003), *op.cit.*; Morin, E. (1973), *op.cit.*; Morin E. (1977–2004), *op.cit.*; Morin, E. (1990), *Introduzione al pensiero complesso*, Sperling & Kupfer, Milan 1993; Morin, E. (2015), *op.cit.*; Morin, E. (2017), *op.cit.*, Milan 2018; Capra, F. (1975), *Il Tao della fisica*, Adelphi, Milan 1982; Capra, F. (1996), *op.cit.*; Emery, F.E. (Ed.) (2001), *La teoria dei sistemi*, FrancoAngeli, Milan; Barabási A.L. (2002), *Link. La scienza delle reti*, Einaudi, Turin 2004; Israel, G., *The Science of Complexity. Epistemological Problems and Perspectives*, in "Science in Context", 18, Anno 2005, pp. 1–31; Gandolfi, A. (2008), *op.cit.*; Dominici P., *op.cit.* (1996–2017); Dominici, P. (2018), *op.cit.*; Taleb, N.N. (2012), *op.cit.*; Tegmark M. (2017), *Vita 3.0.*, Raffaello Cortina Ed., Milan 2018; Blastland M. (2019), *The Hidden Half. How the World Conceals its Secrets*, Italian translation, *La metà nascosta. Le forze invisibili che influenzano ogni cosa*, Bollati Boringhieri, Turin 2021; Al-Khalili J., McFadden J. (2014), *Life on the Edge. The Coming of Age of Quantum Biology*, Italian translation, *La fisica della vita. La nuova scienza della biologia quantistica*, Bollati Boringhieri, Turin 2015.

civilization, is configured, as I have defined it, as a "huge fracture" (albeit not the only one that Modernity has produced): the fracture between culture and technology/technologies, between I and WE, between I/WE and the environment, between ecosystems, between I/WE and the world of objects, between the "inside" and the "outside" of Subjectivity/Subjectivities (another dialogic pair that has been completely done away with). A fracture that arises directly from the rift between humanistic and scientific education that has also determined others, for which we are paying, moreover, a very heavy "price", concerning the world of scientific research in every one of its facets, as well as the multiple universes of intellectual and creative production, also because we are faced with a type of hypercomplexity which—as repeated over and over again in the past—bears witness to a cultural evolution now capable of influencing biological evolution. All this requires a paradigm shift which finds its fulcro, its fundamental "lever", in the urgent need to redefine (or, finally, dismantle!) the boundaries between *the natural and the artificial*, between the human and the non-human, between the mind (individual/collective) and the environment, between systems and new ecosystem, etc.

As argued many times in the past, we find ourselves "hurled into hypercomplexity", obliged to inhabit it and interact with it, without having the "forma mentis" and the tools necessary to do so. Thus, the urgent need to radically rethink education, instruction and training: educational and training institutions have long shown their inadequacy in the face of such hypercomplexity, in the face of indeterminacy and the ambivalence of the ongoing metamorphosis, in the face of the extension of all political, social and cultural processes on a global scale.

7.5 Speed vs. Critical Thinking. Questions of a "Non-Technical" Nature

The "new" speed of the digital, within the ambit of its complex interaction with the "human factor" and the system of social relations, maintains the original ambivalence of any factor of change and of any type of social and cultural process; this ambivalence, in addition to providing an extraordinary opportunity, highlights more than ever our limits, our inadequacies (socially and culturally "constructed" within educational and training institutions), our fragilities and our inefficiencies—in personal, organizational and social terms—but, above all, it leaves us little time for reflection, critical analysis and attempts at "viewing the overall picture". When acknowledging these inadequacies, and the irreversibility of these processes and dynamics, we realise that there is an authentic risk of focusing exclusively on the "technical" and "technological" dimensions (and, consequently, on the need to provide training exclusively in technical and hyper-specialised fields) and, more generally, in applicative areas, underestimating, once again, the spheres concerning People (and their creativity), systems of relationships, educational and cultural contexts, the lifeworlds[39] and new asymmetries. In more general but essential terms (!), we are focusing only on the "hows" and not on the "whys", on the "solutions" (?) and not on the "problems", on "confirmations" and not on "errors", only on the "answers" and not on the "questions". Leaving no room for the fundamental dimensions of imagination and creativity, which are nourished precisely by contaminations between disciplines and fields of knowledge, between experience and experiences, between social and cultural landscapes. By focusing totally on "mechanical thinking", only partly on "analytical" thinking but, above all, completely neglecting "systemic thinking" exactly when we most need a multidimensional type of thinking, capable of identifying (more or less complex) connections and correlations. Errors and questions are the true condiment of cognitive and innovative processes, as well as of scientific research; for many years, perhaps too many, we have continued to be lulled (also with regards to learning, education and training) by the illusions of the hypertechnological civilization, supported by narratives and economic interests. Hence confusing methodology and technology, methodology and digitality. In this sense, education and critical training in complexity and responsibility are configured as the complex "instruments" of the social construction of the Person (in the first place) and the Citizen (subsequently); instruments capable of defining the rules of engagement of the "new" forms of (global) citizenship and inclusion, related to the advent of the so-called knowledge society.[40] This because it is not now, nor will it ever be, technology and/or digitality to determine emancipation, citizenship or inclusion.

From this perspective, therefore, the social and cultural construction of the Person, (as said above, first and foremost), and (then) of the citizen, are also complex processes that must (should) be activated/triggered/accompanied from the earliest years of life and should not be postponed in time. These processes are fundamental pre-requisites that, at the same time, are functional to the attempt to reconstitute/strengthen an extremely weakened social fabric by effectively creating—we might say—the "empirical" conditions capable of counteracting that weakening of social bonds and the void of ethos and meaning which, beyond their representations on the media and relative emotional outbursts, seem to be spreading more and more, not only among the new generations. They represent the "germs" of the well-known "cultural question" which constitute the real obstacles to the

[39] See Habermas J. (1981), *Teoria dell'agire comunicativo*, Vol. I *Razionalità nell'azione e razionalizzazione sociale*, Vol. II *Critica della ragione funzionalistica*, Il Mulino, Bologna 1986; see also: Habermas J. (1968), *Conoscenza e interesse*, Laterza, Bari 1970.

[40] See Castells M. (1996–1998), *The Information Age, Economy, Society and Culture* (vol. III), Oxford: Blackwell Publishers; Castells M. (2009); Castells M., *Comunicazione e potere*, EGEA-Università Bocconi, Milan 2009; Himanen P. (2001) *L'etica hacker e lo spirito dell'età dell'informazione*, Milan: Feltrinelli, 2001; Benkler Y. (2006), *La ricchezza della Rete. La produzione sociale trasforma il mercato e aumenta le libertà*, Università Bocconi Ed., Milan 2007. Hess C., Ostrom E. (2007) *La conoscenza come bene comune. Dalla teoria alla pratica*, Bruno Mondadori, Milan 2009; Rainie L., Wellman B. (2012), *Networked. Il nuovo sistema operativo sociale*, Guerini, Milan 2012; Rifkin J. (2000), *L'era dell'accesso. La rivoluzione della new economy*, Mondatori, Milan 2000.

affirmation of true (social and cultural) innovation and of more open, inclusive social systems.

Questions and problems that have profound implications regarding the very conception/design/definition of any model or practice of citizenship and participation. Because this is the crucial level of cultural change that is capable, in the long term, of triggering and accompanying economic, political and social change. And, as we continue repeating, there is no room for improvisation and/or shortcuts. The strategic level is that which concerns educational processes and relational and communicative space.

The route cannot continue being that of "simple solutions to complex problems",[41] an extremely dangerous pathway, even for the destiny of modern democracies because—as stated long before it became a buzzword- "democracy is complexity" (1998) and the value of simplification—though we are scarcely aware of it—is not absolute.[42]

7.6 Teaching Unpredictability and Training "Hybrid Figures"

With respect to the various issues addressed, we cannot fail to consider another "factor": the cultural delay in terms of "communication culture", at various levels of analysis and action. The new virality of communication, in effect, is one of the elements that has determined the transition from complexity to hypercomplexity. A (hyper)complexity that is not an option; it is a "fact of life": the real problem is that we have not been educated and trained to recognize it, or in any case, not by using our own heads. In reality, it is not only as of today that "technology has become part of the synthesis of new values and new evaluation criteria,[43] rendering even more evident the centrality and the strategic function of kind of evolution that is above all cultural, which has placed itself alongside biological evolution, conditioning it profoundly and determining processes of dynamics and feedback (here the technological progress linked to artificial intelligence, robotics, information technology, nanotechnologies, genomics etc. comes to mind). This progressive appropriation of the levers of one's own evolution calls radically into question traditional models and categories, obliging us (?) even to re-evaluate/reformulate concepts and operational definitions.

Yet, we are still bogged down in false dichotomies, in the opposition between studies in the humanities vs. studies in scientific fields, in rankings (of all kinds) and in discussions regarding the usefulness of fields of knowledge and knowledge itself (deceptive principle) or, much more banally, in "all we need are technicians" vs

[41] See Dominici P. (1996–2017), op.cit.

[42] *Ibidem.*

[43] See Dominici P. (1995–1996), op. cit.

"humanists are better", "we need more graduates in..." vs "too many graduates in..." during a phase of change so swift and unpredictable as to accelerate the obsolescence of knowledge and skills, of curricular and professional profiles even more rapidly.[44] As a matter of fact, keeping in mind the "long term" above all, it is precisely during this very delicate phase of transition and complex evolution—in the midst of a paradigm shift and an anthropological transformation—that we ought to rethink education, education, training and research thoroughly, moving toward (hyper)complexity, multidisciplinarity and interdisciplinarity, with a view to educating and training "hybrid figures",[45] that is, forging open and elastic minds capable of inhabiting the tensions and distances born of the fusion, not of the separation/fracture, of the "two cultures" (Snow), a fusion generated precisely by overcoming "false dichotomies",[46] as I have been saying for years, long before it came into fashion (1996)—a social, relational, organizational complexity that cannot be objectified in any mathematical formula and/or sequence of data and which is capable of escaping any process of reduction. We cannot/must not be satisfied any longer with defining/establishing "what to teach"; the challenge goes far deeper and is much more compelling.

The (complex) organizations where young people are employed—or will be employed—are *adaptive* social systems, actual "organisms"; consequently, what we have to/will have to do is to educate (not *indoctrinate*!), train, and update these future workers with this in mind, so that they will be thoroughly capable of inhabiting and interacting with this complexity, which will never be controllable or predictable to the core. We will have to *teach and train them in unpredictability, to be open-minded in thinking, imagining, and ultimately confronting the unexpected and the indetermined.* And instead, on the contrary, besides continuing to educate and train them staking everything on control, total rationality and simulation, in terms of public discourse, we continue to talk/write/debate solely in terms of the "usefulness" and "functionality" of the fields of knowledge and skills, we continue exclusively to talk about how much (only) engineers, graduates in the hard sciences, and certain figures rather than others, are needed. We are still thinking—as hard as it is to believe—within the logics of separation and confinement of the fields of knowledge, on the basis of the false dichotomy created between a humanistic and a scientific education, yet we must (at least try to) overcome these dichotomies. Continuing to believe that for this hypertechnological civilization, what are needed are solely figures who are highly trained in "knowing how to do" and "knowing how to

---

[44] The "World Economic Forum" itself has long continued to highlight these dynamics—soon destined to become radicalized further among robotics, big data and artificial intelligence—underlining that 65% of children currently attending primary school will, "when they grow up" have jobs that we are unable to even imagine at present.

[45] See Dominici P. (1996), *op. cit.* and following.

[46] It is useful to point out that the definition "complexity manager" was also used in the past for the sake of brevity and convenience, to express awareness in all publications and works that the *management of complexity is practically an oxymoron*, certainly, a *contradiction in terms* even more so if referred to social, relational and human complexity (Dominici P., 1995–2022).

use"—within a purely technical and technological dimension—corresponds to a short-sighted approach that will keep us trapped within a condition of perennial cultural backwardness. For nearly twenty years now, everyone has been talking about skills (in particular, "digital competences") and "know-how", but "know-how" without "knowledge", especially without a critical and systemic approach, leads/will lead us to *"do things as they have always been done"*, because doing them that way has always worked so far, which for many reasons continues to reassure us in the face of the unpredictability and indeterminacy of reality.[47] And so it seems that we shall go on telling ourselves that *technology moves faster than culture*, as if the former were something "external" to the latter. In conclusion, it is well worth repeating how urgent it is to overcome the "false dichotomies" and the "great mistake", freeing ourselves from certain logics and reasonings that are the result of deceptive and misleading *polarisations* that have been of considerable use to us insofar as we are incomplete and limited beings,

but from which we must absolutely *free ourselves*. The hypertechnological and hyperconnected civilisation—and the so-called knowledge society—will have more and more need of hybrid figures (hybrid figures, but not "experts in everything") and complexity managers (I use this formula for convenience and for the sake of brevity) who know how to see opportunities in what we today define and recognize as risks, vulnerabilities, dangerously disorderly variables, capable of making systems and social life even more unstable and insecure. "Hybrid figures", *complexity managers*,[48] who will prove to be even more indispensable within complex systems and organisations that stake everything on automation and on the total simulation of all processes and dynamics. Creativity and imagination—evidently "grafted" onto an epistemological, methodological and transdisciplinary preparation, as well as on continuous updating—will be the variables, increasingly capable of making a difference, also and above all in the era of "intelligent (?) machines". It is high time to recompose "something" that has been mistakenly kept separated. It is high time to facilitate the construction of *bridges between the fields of knowledge*, between skills, between the natural and the artificial (whose boundaries have been done away with), between the fields of knowledge and life, between the Human and the technological.

---

[47] *Ibidem.*

[48] These are an operational concept and a definition that I have always used for the sake of clarity and explanatory effectiveness, since, as I have striven to show throughout this work, complexity, for many reasons, cannot be managed.

Chapter 8
Thought and Thinking in the Hypertechnological Civilization of Automation and Simulation: Artificial Intelligence as a "New Epistemological Fracture"

> *Men fear thought as they fear nothing else on earth – more than ruin, more even than death. Thought is subversive and revolutionary, destructive and terrible, thought is merciless to privilege, established institutions, and comfortable habits; thought is anarchic and lawless, indifferent to authority, careless of the well-tried wisdom of the ages. Thought looks into the pit of hell and is not afraid. But if thought is to become the possession of many, not the privilege of the few, we must have done with fear.*
>
> *Bertrand Russell*

Abstract As yet incapable of grasping the complex, qualitative(!), unobservable dimensions and nuances, the ambiguities and contradictions of the Human and the Vital, we avoid thinking about (or "taking care of") the socio-cultural construction of the Person and the Citizen, about a truly systemic and long-term change which can only originate and flow *from below*. The issue, as always, is educational and cultural: we have not learned to relinquish our decrepit myths, rooted in the erroneous view that still confuses complicated systems with complex systems, mechanisms with organisms. Thought, which requires time and a capacity for *abstraction*, makes us slow and inefficient, whereas with the aim of becoming "perfect" (machines?)—we are staking everything on (hyper)velocity and on processes of simulation and *automation*. The objectives, openly declared, are to "instruct" (aha!) and train people to be hyper-specialised *technicians* and *mere executors of functions and rules*. Having severed thought, we fall into Arendt's vision of humanity enslaved by its own know-how, at the mercy of any technologically possible device, even the most ferocious. Meanwhile, with the advent and progressive tuning of artificial intelligence, a complex synthesis between man and machine, whose implications we cannot foresee, may create a new epistemological fracture.

P. Dominici, *Beyond Black Swans: Inhabiting Indeterminacy*, Philosophical Studies Series 160, https://doi.org/10.1007/978-3-032-09029-4_8

Keywords Epistemological fracture · Qualitative dimensions · From below · Hypervelocity · Automation · Simulation · Mere executors · Hyper-specialization · Complex synthesis · Humanity enslaved

In the hyper-technological civilization of automation and simulation, we humans are trying to eliminate unpredictability and any possibility of error, to physically control and connect everything, delegating all strategic functions and action, including that of thought, to intelligent machines (?), artificial intelligence systems and robots, scarcely aware that thought (and not only) with all its complexity and non-observability—cannot be simulated, emulated, reproduced, or replicated. Artificial Intelligence itself (albeit an extraordinary opportunity)—which I understand and define as a "new epistemological fracture"—is oriented increasingly—both in terms of research and applicability—towards the crucial objective of emulating/simulating intelligence (which, as I always insist, too many confuse with the ability to calculate and with so-called *problem solving*), human minds[1] and their non-replicable/non-reproducible, "complex product": thought.

And indeed... thought, thinking, "thought on thinking", "research on thought"... despite the attempts, doomed to failure, to reproduce it, emulate it, simulate it in all its complexity and *indeterminacy*, thought is a *dimension*, always essential and of vital importance; and yet, a dimension that is (often) deliberately ignored, devalued, belittled, considered useless, so little relevant for the civilization of "error-free" automation, to the point of believing that it can be assigned to artificial and mechanical systems (however sophisticated) and/or be reproduced/replicated by intelligent machines[2] and, more in general, by complicated systems, considered almost infallible, in any case, constantly perfectible (→*carte blanche* assigned to technology → technological solutionism). The dimension of thought is truly considered of minimal relevance or ascendance by contemporary civilization's hegemonic paradigm, which is founded upon a precise, functional "system logic" and, specifically, built upon what I have long called, before it came into fashion, "the grand illusions of the hypertechnological and hyperconnected civilization" (rationality, control, measurement, predictability, elimination of error),[3] upon the progressive automa-

[1] See in particular: Damasio A.R. (2021), *Feeling & Knowing. Making Minds Conscious*, Italian translation, *Sentire e conoscere*, Adelphi, Milan 2022.

[2] See J. Dorrier, *New Study Finds a Single Neuron Is a Surprisingly Complex Little Computer*, in, "Singularity Hub". Then the following link: https://singularityhub.com/2021/09/12/new-study-finds-a-single-neuron-is-a-surprisingly-complex-little-computer/; S. Fan, *How AI Is Deepening Our Understanding of the Brain*, in, "Singularity Hub", 2021. And then the following link: https://singularityhub.com/2021/11/23/how-ai-is-deepening-our-understanding-of-the-brain/

[3] We have often returned, over time and in various publications (Dominici P., 1995–1996, 2005–2021), to the strategic relevance of error, to its vital value, not only in cognitive and learning processes, but also, and above all, in research and scientific knowledge; not to mention the importance of "building a culture of error", from the very early years of school, bringing it to the centre of processes of education and training as well as of scientific research itself. Yet, errors continue to be stigmatized and defined/seen/acknowledged as something to be removed and avoided, at all

tion of any type of mechanism (the aspiration being to "manage", direct and/or predetermine even social mechanisms), as well as upon continuous and systematic

---

costs. Quite a serious…error, indeed. In the history of philosophy (and not only), for example, error can be traced back in general to the sphere of "judgment", and not to that of statement, in a context of its intimate correlation with the quest for "truth". In several classical theories, error has to do with the failure to apply logical rules and/or knowledge and also requires/would appear to require a valid criterion of judgment. An error can be an outcome of the imperfection of human nature, of an act of will and/or feelings occurring in various circumstances; at the same time, in the social sciences it can be considered a product of/from social and cultural processes of negotiation, labelling and, more generally, social construction. Social and cultural processes—which are also related to the refusal and/or failure to adhere to the system of norms, values, behavioural models—where errors encounter labels/stigmas/concepts such as deviance/abnormality/irrationality and the dimensions of the predictability of (expected) behaviour—always marked by arbitrariness and convention such as codes, languages, semantic and communicative processes. The Treccani Encyclopedia defines error in the following terms and with the following specifications, also associated with the so-called "exact sciences": erróre [Derivative of the Latin error -oris, from errare "to make a mistake"] In numerical calculation, the difference (positive or negative) between the calculated value of a number and its exact value, which is found when one limits oneself to a certain degree of approximation (by excess or by default, respectively) deemed satisfactory for a given purpose, or when an approximation is made necessary due to the very nature of the calculation. ◆ An experimental datum that does not correspond to the true one and also, with a quantitative meaning, the deviation of such a datum or result from the true or presumed value or result: see also: ◆ Accidental E: see also: measurement E. ◆ Apparent E.: in the Gaussian theory of error., the same as absolute e: see also: measurement E. ◆ Casual E.: the same as accidental e.: see also: measurement E. ◆ derivative E.: see physical measurements: IV 51 a. ◆ image focus E.: see geometric optics: IV 388 d. ◆ measurement E.: generic., the uncertainty that must be considered associated with the determination of the value of a quantity in a given procedure of measurement, on the basis of the observation that, by repeating the measurement of the same quantity several times with the same procedure, different results are generally obtained; this uncertainty is specifically expressed by the absolute difference with sign, between the measurement and the true value (or assumed such) or, equivalently, from the relative e., ratio, possibly percentage, between the absolute measurement e. and the true value; as for their nature, we speak of systematic e., that is, one which always occurs and with the same entity, and accidental e., by nature random, both in terms of occurrence and entity: see physical measurements: IV 48 b. ◆ Observation E. of: the same as measurement e.. Then, not only in terms of definitions, also by virtue and inconsistency of the disciplinary fields of reference, there exists—as is known—a complex typology of errors: "standard error", analytical, systematic, relative, random, absolute, algorithmic, accidental, topics and issues on which, once again, there is an endless and multidisciplinary literature. In addition to recalling the other parts of the monograph in which the issue is addressed, I shall simply point out, among the "classics" on these topics: Canguilhem G. (1966), *Le normal et le pathologique*, Italian translation, *Il normale e il patologico*, Einaudi, Turin 1998; Oppenheimer J.R, *Scienza e Pensiero comune*, Bollati Boringhieri, Turin 1965: another "classic", Oppenheimer, the great physicist and scientist speaks of "science" as the "discovery of errors" (pp. 61–63). Among the most recent publications, furthermore, introductory to the topic and the numerous related implications: Baldini M., *Epistemologia e pedagogia dell'errore*, La Scuola, Brescia 1986; Benes R. et al., *Per una pedagogia dell'errore*, Asterios, Trieste 2017; Dominici P., *Dentro la "civiltà senza errore". Sintesi complesse: l'Umano, il legame sociale e la civiltà ipertecnologica,* in A. De Cesaris (ed.) *Vite digitali. Essere umani nella società del XXI Secolo*, FrancoAngeli, Milan 2020. Among the "classics", I would also like to point out: Bruner J.S., *The Process of Education*, Harvard University Press, Cambridge 1961, a work in which, among the numerous issues addressed, error is defined and recognized as a constitutive element of learning.

simulation, upon the frequently mentioned marginalization of the Human, considered the "carrier" and potential creator/producer/executor of error, of errors that the systems cannot "reset". As for *error* itself, what it has always been considered is not a "source of knowledge and learning", but rather an element to be eliminated and removed, even from educational and training processes. A disastrous prospect, to say the least, as I have been repeating for almost thirty years now.

As I was saying, the only "holder" of the possibility of making mistakes, of committing "errors", even knowingly and voluntarily, is "the Human (being)", who emerges from unpredictability as "the Human factor": a "complex variable" of continuous, systemic and emergent unpredictability which, moreover, implies/involves/requires—necessarily—taking on responsibility.[4] These are precisely the elements—according to the abovementioned paradigm—which would need to be eliminated/removed in order to build perfectly functional, efficient, predictable, manageable, pre-determinable social systems, even in terms of single individual and collective instances of social behaviour.

According to this mechanistic and technocratic point of view, societies and organizations should be designed and imagined (?) as perfect (therefore, not intelligent) "machines", or better still, as "mechanisms", rather than the "organisms"—the hypercomplex systems—which, in reality, they are, that is: a "new nature" of complex human ecosystems. If we imagine the global ecosystem to be an immense *reticular nervous tissue*, then the aim/function of the systems of artificial intelligence and processes of automation, which have been designed and created by human beings, is to progressively replace human and relational networks and mechanisms. The ambiguity and ambivalence, the multidimensionality and unpredictability of human beings do not combine well with the axioms of the civilization of automation and algorithms. Those same algorithms do not allow any ambivalence and ambiguity. And so…

And so, and it could not be otherwise, in the civilization of automation and of the "tyranny of concreteness", "facts"—or what we label as such—and "data", presented as self-evident "facts of life"—dominate unchallenged. According, at least to certain misleading, deceptive and more than probably compromised interpretations/narratives, in perfect line of continuity, as well, with the "grand illusions of the hyper-technological and hyper-connected civilization", it would seem that neither human logical, epistemological and methodological preparation nor intervention, nor capacity for non-linear/causal systemic thinking is required any longer. According to these same narratives and interpretative perspectives, well supported by new forms of reductionism and determinism, functional to the creation of technocratic and dystopian societies, all that is necessary and sufficient is training in know-how, in how to use software, in artificial intelligence systems and/or, specifically, in *machine learning.* The latter being seen and presented by the computer science professor Pedro Domingos as "the scientific method on steroids" *(sic!)* and metaphorically compared by Domingos (but also by other scholars and experts) to

[4] Dominici P., op.cit., 1995–1996, 2005 and following.

the "sword with which to slay the monster of complexity". These theses and arguments contained in Domingos's work—but also in other publications that have enjoyed significant editorial success—reveal unequivocally, however, a certain confusion, not only in terminology and in terms of operational definition, as regards the study of and research on complexity and complex systems. This confusion which also appears in the analogy proposed and which reaffirms, in no uncertain terms, what I have called the "ultimate error", that is to say, the gigantic and, in some ways, atavistic misunderstanding between complicated systems and complex systems, between the world of life and organic aggregates and the technological and artificial world:

> The power of machine learning is perhaps best explained by a low-tech analogy: farming. In an industrial society, goods are made in factories, which means that engineers have to figure out exactly how to assemble them from their parts, how to make those parts, and so on – all the way to raw materials. It's a lot of work. Computers are the most complex goods ever invented [note by this author: *sic!*], and designing them, the factories that make them, and the programs that run on them is a ton of work. But there's another, much older way in which we can get some of the things we need: by letting nature make them. In farming, we plant the seeds, make sure they have enough water and nutrients, and reap the grown crops. Why can't technology be more like this? It can, and that's the promise of machine learning. Learning algorithms are the seeds, data is the soil, and the learned programs are the grown plants. The machine-learning expert is like a farmer, sowing the seeds, irrigating and fertilizing the soil, and keeping an eye on the health of the crop but otherwise staying out of the way. [Note by this author: this point is also significant… also with regard to the *marginalization of the Human*!]
>
> Once we look at machine learning this way, two things immediately jump out. The first is that the more data we have, the more we can learn. No data? Nothing to learn. Big data? Lots to learn. That's why machine learning has been turning up everywhere, driven by exponentially growing mountains of data. If machine learning was something you bought in the supermarket, its carton would say: "Just add data."
>
> The second thing is that machine learning is a sword with which to slay the complexity monster. Given enough data, a learning program that's only a few hundred lines long can easily generate a program with millions of lines, and it can do this again and again for different problems. The reduction in complexity for the programmer is phenomenal. Of course, like the Hydra, the complexity monster sprouts new heads as soon as we cut off the old ones, but they start off smaller and take a while to grow, so we still get a big leg up.[5]

Some of the statements expounded here are, to some extent, surreal, since they deny the value of error, unpredictability, and deviance, true constitutive elements of the Social, the Human, of Life.

Once again, "data" and "learning programmes" return to reduce complexity and seek to understand/know it in a "phenomenal way". As if this infinite availability of data (*big data*) were, in itself, "knowledge", with a clear underestimation/non-consideration of the processes of signification and of logical, methodological, epistemological and, more generally, theoretical mediation. These "data" are, without any doubt, fundamental (decisive) but, as I always repeat, are erroneously presented

---

[5] See Domingos P. (2015). *op.cit.*, pp. 29–30. On these issues see also: Finn E. (2017), *What algorithms want: immagination in the age of computing*, Italian translation, *Che cosa vogliono gli algoritmi. L'immaginazione nell'era dei computer*, Einaudi, Turin 2018.

as "facts". Yet, as we were taught, precisely in our research methodology and epistemology courses in the olden days, "the data can never speak for themselves"; as a result, alongside the paradigm of the hyper-technological civilization, they cannot but dominate unchallenged—at least, always according to that same hegemonic paradigm—counting only skills, digital skills, and "know-how". Necessary dimensions but, as I have always repeated, insufficient, especially in terms of approaches and new epistemologies.

And, in this regard, remembering the words of the extraordinary Hannah Arendt: "If it should turn out to be true that knowledge (in the modern sense of know-how) and thought have parted company for good, then we would indeed become the helpless slaves, not so much of our machines as of our know-how, thoughtless creatures at the mercy of every gadget which is technically possible, no matter how murderous it is."[6]

And, historically speaking, the absence/lack of thought can even prove dangerous.[7]

Essentially, we continue to avoid thinking about the social and cultural construction of the Person and the Citizen; we continue to fail to think about "taking care of Society"; in other words, we never think about the long term, about a true "systemic change" which can only originate and flow *from below*. Even more worryingly, we continue not to think/not to be aware of the urgency of "rethinking how we think", of thinking itself, and, at the same time, of the thought system. "Know-how", automatisms, the ideology of concreteness, the obsession with concreteness dominate unchallenged and, moreover, combine well with the ideology and obsession with *simplification*. An obsession, that of the concreteness and measurement/quantification of every entity, phenomenon, process and even emotion which, moreover, and not only from today, undermines any possibility of innovation and profound change; the *uncritical exaltation of speed* and *fashionable technological solutionism* dominates unchallenged, in every dimension of the Social and the Vital.[8] In other words, the "tyranny of concreteness" dominates unchallenged. *The Tyranny of Concreteness* (upon which I have worked extensively) is a cognitive and cultural dictatorship, which, supported by the current *overall architecture of fields of knowledge and skills*, has brought with it a totally inadequate culture of evaluation, incapable of grasping the complex and qualitative (!) unobservable dimensions, the nuances, the ambiguities, the contradictions, the conflicts, the ambivalences, the coexistences, typical of the Human, the Social, the Vital.

---

[6] How can we forget the words of Hannah Arendt, "I was struck by a manifest shallowness in the doer which made it impossible to trace the incontestable evil of his deeds to any deeper level of roots or motives. The deeds were monstrous, but the doer—at least the very effective one now on trial—was quite ordinary, commonplace, and neither demonic nor monstrous. There was no sign in him of firm ideological convictions or of specific evil motives and the only notable characteristic one could detect … was something entirely negative; it was not stupidity but thoughtlessness."

[7] See Arendt H. (1958), *The Human Condition*, Italian translation, *Vita activa. La condizione umana*, Bompiani, Milan 1964.

[8] Ibidem.

Above all, it now seems that "thinking" (but also thought, thoughts, the thought system, thought systems) which, as we know, requires "time" and a capacity for *abstraction* (among the many dimensions), causes us to move slowly and, as a result, not very "efficiently" according to certain longstanding myths of quite a few organizational cultures. These are age-old myths which, beyond certain narratives, and despite claims to the contrary on the part of the majority of experts, along with public discourse itself, have never waned, quite the contrary! I cannot help but repeat myself, yet again, that it is always an "educational and cultural issue"! Decrepit myths of designing, creating and managing organizations and social systems, which are rooted precisely in the erroneous overall view that still leads to confusing complicated systems with complex systems, mechanisms and organisms.

Yes, indeed, *Thinking* and *Thought*: what good are they, what "use" can they be for the human race if, ultimately, bottom line, with the aim of becoming ever more infallible, almost "perfect like machines" (?), increasingly efficient and effective, we—as organizations and social systems—are staking everything on (hyper)velocity and on processes of simulation and of *automation*. In this sense, the objective and intention—which are, moreover, openly declared—of "instructing" (aha!), educating (?), training, practically *drilling* people to be exclusively hyper-specialised[9] *technicians* who are fully functional and adequate and/or, worse still, as said before, "*mere executors of functions and rule*s".[10] This is why, beyond other contextual factors to be considered, we have remained so behind in terms of thinking, thought systems, educational and training processes, of the (re)planning/rethinking of teaching and, even, of research itself. This because we continue to educate and (de)form "disjunctive thinking" that *enables us* to separate, divide and isolate what is (profoundly) united, and not just correlated and interdependent, belittling and devaluing, moreover, all and every capacity for abstraction and imagination. Breaking down, isolating and simplifying (even what cannot be simplified) have always constituted our (simple) answers, the illusory devices that reassure us—now, more and more, in the name of *technological solutionism* and the inevitable reductionist, deterministic explanations—with respect to the anthropological transformation that regards the (hyper)complexity of life and society and our responses to this (hyper) complexity.

Thus, by identifying, acknowledging and representing as separable, isolatable and decomposable, the parts/objects/entities, which are always "relationships" and "systems" that constitute the phenomena of reality, we do not hesitate even for a moment to close, separate and segregate the fields of knowledge, knowledge itself, experience and experiences. This operation is certainly not simple, but complex, something which has always marked the evolution of social action, organizations, systems, of LIFE itself. Isolating, selecting, distinguishing/separating, secluding, but also recomposing, re-uniting, re-connecting, are, in other words, all strategic

---

[9] See also what I have called the "STEM Doctrine", "new reductionist paradigm", perfectly functional to the neoliberal vision of educational and training institutions. See, in particular Dominici P., 1996–2021, op.cit.

[10] Dominici P., 1995–1996, 2005, 2014, 2021, op.cit., and following.

functions carried out precisely by educational and cultural models and by the very *forms of symbolic mediation* (Cassirer) (codes, languages, culture(s), the law, communication, religion, art etc.) which, essentially, render our limits and incompleteness less evident. These processes and dynamics highlight, once again, the substantial inadequacy of our educational and training institutions and, more generally, of our thought system and the paradigms that, *up until now*, have helped/guided us to make the world, the reality, the world (apparently) "outside of US" readable and interpretable.

It is paradoxical, to say the least, to educate and train in disjunctive thinking, to boost specialisms, to create closed *fields of knowledge* that prevent any dialogue or contamination (without even reaching trans-disciplinarity) during this very sensitive phase of transition and paradigm shifts, a phase, as mentioned before, of "complex synthesis" and progressive dissolution of all boundaries and/or limits.[11] Precisely in the light of what has just been said, and from this perspective, among the various areas/fields of study, research and action, reflection and research concerning *man-machine* relationships/interaction/synthesis are assuming particular importance, a type of relationship and interaction considered, not only in the past, also in terms of the "subalternity" of the human being (the new Prometheus) with respect to the universe of machines that the same human being has designed and created. A human being who, in an era when the Machine has (nearly) become a dominant historical subject, rediscovers that he is "antiquated" and reduced, like all the rest of the phenomena of reality, to the status of *raw material* to be manipulated.[12] The risks are considerable and difficult to assess, but the objective, in my opinion, is/needs to be/ remain that of co-building the conditions favouring a "complex synthesis" that succeeds, finally, in accompanying us beyond the devastating separation between *culture and technology,* with a view to "recomposing the fracture between the Human and the Technological".

As already stated, this progressive taking possession of the levers of one's own evolution forces us, therefore, to rethink the Human and its, in some ways ambiguous, interaction with technology and technological devices, specifically with intelligent machines (?), robots, artificial intelligence and the new ecosystems as they are being structured: a *complex interaction* from which a *complex synthesis* cannot but arise and of which we are as yet unable to evaluate the prospects, developments and implications. However, again with regard to the issues of "natural versus

---

[11] On the topic of "limits", see in particular: Bodei R., *Limite*, Il Mulino, Bologna 2016.

[12] In particular, on the complex, and always current, question of the "subalternity" of human beings to machines, I refer to a great "classic", Anders G. (1956), *Die Antiquiertheit des Menschen, I: Über die Seele im Zeitalter der zweiten industriellen Revolution*, Italian translation, *L'uomo è antiquato. Vol. I. Considerazioni sull'anima nell'epoca della seconda rivoluzione industriale*, Bollati Boringhieri, Turin 2003; Anders G.(1980), *Die Antiquiertheit des Menschen, II: Über die Zerstörung des Lebens im Zeitalter der dritten industriellen Revolution,* Italian translation, *L'uomo è antiquato. Vol. I. Sulla distruzione della vita nell'epoca della terza rivoluzione industriale*, Bollati Boringhieri, Turin 2003. See also Brynjolfsson E., McAfee A. (2014), *The Second Machine Age*, Italian translation, *La nuova rivoluzione delle macchine*, Feltrinelli, Milan, 2015; Cingolani R, Metta G., *Umani e umanoidi*, Il Mulino, Bologna 2015.

artificial" ("false dichotomy"), "Human versus Technological" and, specifically, with reference to "artificial intelligence", I confess that I have always been extremely critical of the very concept of "artificial intelligence", since it is based on the erroneous assumption that the very idea/concept of intelligence (as well as of thought) coincides with the ability to calculate and find frequently repetitive, systematic and/or modelled solutions to problems (*problem solving*).

We may well ask ourselves, therefore, in what direction we are going. Sometimes, the impression—but, evidently, it is more than an impression—is that of continuing to "play things by ear" in a condition of *increasingly limited rationality* and unmanageable dynamics/processes. Despite the extraordinary scientific discoveries and technological innovations, awareness of our limits and incompleteness as "human beings" has not increased in any way. It is a matter, when all comes to all, of the grand paradoxes of the hypertechnological and hyperconnected civilization: a civilization which, as has been said several times, feeds on knowledge but also on illusions of knowledge.[13] It feeds on empirical evidence (which is, in any case, decidedly important yet not sufficient), increasingly defined/acknowledged as "objective" and self-evident. At the same time, it feeds on dilemmas and illusory certainties, without the indispensable culture of complexity and unpredictability required to face and metabolise them. Finally, it feeds on scientific paradigms and models and, at the same time, on increasingly hegemonic forms of reductionism and determinism, in the face of a growing (hyper)complexity which, moreover, continues to clearly reveal its limits and inadequacies. In my opinion, the words of Giorgio Israel are enlightening:

> The idea of rationality is something much broader than what is preached by the various reductionist conceptions. It should be clear that the battle in defence of rationality – which we should cherish no less than some of its frenzied lip-service-providing supporters – has absolutely nothing to do with the all-out defence of a materialistic type of metaphysics. On the contrary, we are convinced that the affirmation of an authentically rationalistic, scientific view has to pass through the assumption of a phenomenological approach and a refusal to present the various metaphysical ideologies, whether explicit or latent, as objective scientific results. This point of view is undoubtedly difficult – because an ideological attitude is much more convenient, especially if masquerading as scientific objectivism -, but it represents the only effective way of defending rationality. At the heart of materialist metaphysics, which aims at taking science and rationality hostage, there is the metaphor of the living machine, understood in a twofold sense: as an expression of the idea that man is *nothing more than a machine*, and as an affirmation of the possibility of *creating a machine that becomes human*. Perhaps it is precisely in this that the mystification consisting in passing off as "positive" scientific results that are simply beliefs, aspirations or mythologies manifests itself, reaches its most extreme peak. […] However, we must not forget that science is

---

[13] On issues pertaining to "knowledge" and illusions of knowledge see Sloman S., Fernbach P. (2017), *The Knowledge Illusion. Why We Never Think Alone*, Italian translation, *L'illusione della conoscenza,* Milan: Raffaello Cortina Editore, 2018; Boorstin D. J. (1994), *Cleopatra's Nose. Essays on the Unespected,* Vintage Books, New York; Boorstin D. J. (1998), *The Seekers. The Story of man's continuing quest to understand the world*, Italian translation, *L'avventura della ricerca*, Raffaello Cortina Ed., Milan 2002; Hustvedt S. (2016), *The Delusions of Certainty*, Italian translation, *Le illusioni della certezza*, Einaudi, Turin 2018.

nourished not only by objectivity, but also by metaphysics and mythologies, at least as sources of inspiration.[14]

As we have seen, the hypertechnological civilization brings with it not only extraordinary opportunities and innovative potential but also a series of illusions, which can be traced back to the idea that we "human beings" will be increasingly capable—and more and more will we do so—of delegating our choices and—I add—our responsibility to technological systems and devices. In other words, the dimensions of the technologically controlled will continue to increase and, with it, the illusory certainty that the human, social and relational "factor" will be increasingly less important, (precisely) because, in the meantime, we shall progressively *delegate* all decisions and choices to technologies, algorithms or robots. We are becoming ever more convinced that the digital and technological will be able to solve everything and to protect us from all dangers, while ignoring one important detail: the human factor which is and will always be decisive since it lies *behind* every process, *behind* every mechanism, *behind* every algorithm. The question that arises is whether decisions in the future will be made by AI or by those human beings who have maintained their capacity for creative (and critical) thinking (Fig. 8.1).

Fig. 8.1 Decisional processes: AI or creative human minds?

[14] See Israel G. (2004), *op.cit.*, pp. 10–11 On these dimensions, see also the interesting analysis contained in: Enriques F. (1906), *Problemi della scienza*, Bologna: Zanichelli 1985. The mathematician and historian of science Federigo Enriques, cited by Israel himself, also speaks of a "system of images" that stimulates "new associations" functional to the development of science. By the same author see also: Enriques F., *Il significato della storia del pensiero scientifico*, Zanichelli, Bologna 1932.

Yet it is useful to reiterate that faced with the challenges of hypercomplexity, we cannot but take note of the substantial inadequacy of the *architectures* that characterize our fields of knowledge and, with them, our Schools and our Universities.

This issue is so central, seeing that every process of technological innovation and change has always determined, and will always determine, elements of stress and vulnerability in systems. Unless we try to correct the "false dichotomies" and the substantial inadequacy of the current educational and teaching-training processes (in addition to issues concerning the *governance* of schools and universities), in another twenty years we will find ourselves still talking about how technology accelerates and how culture cannot keep up with it. In a critical condition of perennial "cultural delay".

How can we try, then, to counteract this *systemic*, by now structural, *cultural delay*? I continue to be convinced, as I have always been, that we can only do so by working thoroughly and radically on education, educational processes, educational institutions, thinking in the long term, and overcoming what I have always called the "great mistake", as well as the "false dichotomies", "disjunctive thinking", and certain historical logics of separation. These are topics and questions of enormous importance also, and above all, in terms of power and the negotiation of processes and algorithms (produced by humans!)—increasingly powerful and pervasive—but also, and above all, in terms of citizenship, inclusion and democracy.

Chapter 9
Rethinking the Human and Re-mediating the Social: Because "Democracy Is Complexity"

> *What is the destiny of democracy at a time when information and communication technologies redesign the places of politics, break down borders, deny the very constraints of space and time, erase ancient subjects and create new subjectivities? If these are the effects of change, then, it is not only a particular form of politics that is at play. It is the whole of society that, day after day, discovers that it is changing continually. And with it, the meanings of rights and languages change; the manner in which personality is constructed changes. Those same technologies, engaged in an incessant transformation of reality, create a favourable terrain for positive and negative utopias*
>
> *Stefano Rodotà, Tecnopolitica, 1997*

Abstract Citizenship must be redefined, not simply *adapted to technological practice*, as an open field where freedom and equality face each other in search of a practicable equilibrium, along with the awareness that (systemic) social change cannot be imposed from above, nor can it be achieved exclusively through technological innovation or by adherence to a regulatory framework—which is by no means a definition of democracy. Democracy is complexity: it is open dialectics, plurality of opinions, thoughts, cultures, and visions of society. Democracy is conflict, openness, order and chaos. "Connected citizens" will not suffice: what we need are citizens—not vassals—educated in critical thinking, complexity, liberty, and responsibility. Currently, the risk is that of building an illusory form of *citizenship without citizens,* through processes of "*simulation of participation*" in which social actors represent the *weakest link:* a simulation of participation that leads inevitably to a simulation of citizenship and ultimately, to a simulation of democracy. The very concepts of *participation* and *citizenship* call into question a fundamental issue: the urgency of rethinking the "social contract" and, consequently, through education, of redefining the rules of engagement regarding citizenship and inclusion, because, once again, "democracy is complexity".

P. Dominici, *Beyond Black Swans: Inhabiting Indeterminacy*, Philosophical Studies Series 160, https://doi.org/10.1007/978-3-032-09029-4_9

Keywords Simulated participation · Simulated citizenship · Simulated democracy · Weak link · Social Contract · Hyperconnected

The extremely difficult phase of transition and re-adaptation of the last years—which in a certain sense, still appears to be in progress—those months and months of social distancing and quarantine, or perhaps, to put it more appropriately, of self-isolation, as well as having obliged us to ponder many of our vulnerabilities and, more in general, the enormous fragility that has always characterised social systems, societies, and humans (complex adaptive systems), had the effect of placing fundamental questions regarding our private and public lives on the agenda once again. Furthermore, that period of crisis forced us to reflect on the strategic centrality of knowledge, of "shared knowledge"[1] and the importance of a systemic approach and a multi/inter/transdisciplinary perspective. This lengthy, problematic phase was also marked by the absence of values/dimensions/aspects/factors which we now tend to take *more and more for granted* and consider acquired: above all, that of "freedom", a complex concept with numerous nuances and meanings, frequently abused and mistreated, along with the term "relationship". Freedom and relationships are key concepts and essential values which also make us reflect profoundly on how much the *pandemic*, in its global dimension as a social phenomenon, changed/is still changing our lifestyles (not values, I think) as well as our expectations and way of thinking and relating to the Other than ourselves. In this regard, as I have been arguing for some time now, I do not in the least believe that, after those difficult years—and we are still, so to speak, in the thick of it all—we have all awakened to find ourselves "better", more generous, open and supportive (so many narratives!), prepared to understand the difficulties of the Other. *A society built on individualistic and utilitarian values*, on competition devoid of rules, marked by particularisms so strongly rooted in all the social strata, a society divided into elites and corporations (the feudal social model),[2] an "asymmetrical society"—built upon the logics of closure and conservation—cannot change or modify itself in depth, even in the presence of pandemics and/or global emergencies, even if they are systemic and viral. On the contrary, at most, – as actually occurred—it displays a radicalization of contradictions and asymmetries, of dynamics that enhance/emphasize conflict and inequality. There are several precedents, regarding other disasters, catastrophes and emergencies of all kinds. The fact is that we always return to the same exclusive and closed-off logics and, above all, to the traditional hegemonies, with the aggravating circumstance of finding ourselves *more alone* than before, victims, at times, of measures and manoeuvres due more to the

[1] Dominici P., (1995–1996). (2003), (2008), op.cit.

[2] *Ibidem.*

"shock doctrine"[3] than to a *strategic vision*, and accompanied by the psychological, cognitive, relational and emotional damage left behind by that isolation and fear of the Other which presented itself/presents itself, from hour to hour, in the guise of multiple forms and metaphors.[4]

---

[3] See Klein N., *The Shock Doctrine,* Knopf Canada, 2007.

[4] Here, I limit myself to indicating a text from Internazionale, where reference is made to various issues—of fundamental importance—such as the first research centres and study groups to work on the psychological and social effects of the pandemic, the therapies for these effects and access to care: Camilli A., *Gli effetti psicologici della pandemia*, in, "Internazionale" (10th December 2020).: https://www.internazionale.it/reportage/annalisa-camilli/2020/12/10/effetti-psicologici-pandemia. As we know, national and international centres and funds have been created for post-covid psychological support, although—I repeat—many effects will only be assessable in the long term. As regards the study of and research regarding psychological and social "systemic effects", on people and relationships, on sociality itself, which are difficult to evaluate even today, however profound and complex the analyses may be, I would like to point out and refer to some scientific articles (also relating to different contexts, categories, age groups and social classes): Talevi D. et al., *Mental health outcomes of the CoViD-19 pandemic*, in "Riv. Psichiatr." 55, 137–144 (2020); Mergel E. & Schützwohl M., *A longitudinal study on the COVID-19 pandemic and its divergent effects on social participation and mental health across different study groups with and without mental disorders*, in, "Soc. Psychiatry Psychiatr. Epidemiol." (2021) doi:10.1007/s00127-021-02025-9 PM—33569650; Brooks, S. K. et al., The psychological impact of quarantine and how to reduce it: rapid review of the evidence, in, *Lancet (London, England)* 395, 912–920 (2020); Rubin, G. J. & Wessely, S., *The psychological effects of quarantining a city*, in, *BMJ* 368, m313 (2020); Sher, L., *The impact of the COVID-19 pandemic on suicide rates*. *QJM* 113, 707–712 (2020); Sher, L. COVID-19, anxiety, sleep disturbances and suicide. *Sleep Med.* 70, 124 (2020); Sharma, A. et al. Impact of isolation precautions on quality of life: a meta-analysis. *J. Hosp. Infect.* 105, 35–42 (2020); Purssell, E., Gould, D. & Chudleigh, J., *Impact of isolation on hospitalised patients who are infectious: systematic review with meta-analysis*, in, *BMJ Open* 10, e030371 (2020); Zhang Y. & Ma Z. F., *Impact of the COVID-19 Pandemic on Mental Health and Quality of Life among Local Residents in Liaoning Province, China: A Cross-Sectional Study*, in, *Int. J. Environ. Res. Public Health* 17, (2020); Cao, W. et al. The psychological impact of the COVID-19 epidemic on college students in China. *Psychiatry Res.* 287, 112934 (2020); Rossi, R. et al. COVID-19 Pandemic and Lockdown Measures Impact on Mental Health Among the General Population in Italy. *Front. Psychiatry*, 11, 790 (2020); WHO, *Mental health and psychosocial considerations during the COVID-19 outbreak.* https://www.who.int/docs/default-source/coronaviruse/mental-health-considerations.pdf?sfvrsn=6d3578af_8 (2020); Yuan Y. et al., *Prevalence of Post-traumatic Stress Symptoms and Its Associations with Quality of Life, Demographic and Clinical Characteristics in COVID-19 Survivors During the Post-COVID-19 Era*, in, "Front. Psychiatry", 12, 665507 (2021); Pancani L., Marinucci M., Aureli N. & Riva P., *Forced Social Isolation and Mental Health: A Study on 1,006 Italians Under COVID-19 Lockdown*, in, *Front. Psychol.*, 12, 663799 (2021); Tyler, C. M. et al. A Study of Older Adults' Mental Health across 33 Countries during the COVID-19 Pandemic. *Int. J. Environ. Res. Public Health* 18, (2021); CDC, *Mental Health and Coping During COVID-19.* https://www.cdc.gov/coronavirus/2019-ncov/daily-life-coping/managing-stress-anxiety.html (2021); Ganesan B. et al., *Impact of Coronavirus Disease 2019 (COVID-19) Outbreak Quarantine, Isolation, and Lockdown Policies on Mental Health and Suicide*, in, *Front. Psychiatry*, 12, 565190 (2021); Feng, J. et al., *Psychological Impact During the First Outbreak of COVID-19 on Frontline Health Care Workers in Shanghai*, in, "Front. public Heal", **9**, 646780 (2021); Bo H.-X. et al., *Posttraumatic stress symptoms and attitude toward crisis mental health services among clinically stable patients with COVID-19 in China*, in, Psychol. Med., 51, 1052–1053 (2021).

Among the many paradoxes of the hypertechnological civilization which, as mentioned several times before, aspires to become a (completely) transparent, super-efficient civilization, "devoid of error" and of "deviance(s)", where everything is "predictable", there is also that of an increasingly more evident fragmentation of processes and social bonds in the face of an exponential growth in the opportunities provided by (hyper)connection. In these past years, we have never had so many opportunities (?) of communicating and exercising/practising "proximity". We have never been so hyperconnected and interconnected, and yet, never as in this phase, have we felt the absence of the Other, of Others, of the social bond, of (true) sharing and of immediacy. Never before have we felt the great *distance* existing between us and the Other, the Others, despite "mediated proximity" and real time. We have perceived *the absence of a complex relationship that can only be simulated*, because meeting, even just encountering People, exchanging glances, talking and interacting also with those one does not know well, *breathing* the situations and circumstances in which one comes to find oneself, but also listening to the sounds and noises of public places, is something truly irreplaceable and difficult to simulate/replicate.

Noises and silences, glances and thoughts, pauses and accelerations, moments of debate and confrontation, the breath and movement of bodies: an educational, relational, communicative, visual, perceptive, cognitive, linguistic, cultural space, a systemic, shared space, characterized by infinite dimensions; a systemic, vital, emotive and emotional space which, on my part, despite my having always believed (also) in the primordial utopia of the interconnected knowledge society, in remote learning as well, (which for me, should be combined with traditional teaching and not used only during emergencies), in the use of platforms and the opportunities provided by hyperconnected ecosystems, will never be able to replace the social, the vital and the fascinating nor the unpredictable beauty and complexity of relationships devoid of the *mediation* and filters/constraints imposed by connective technologies. Admittedly extraordinary technologies, as I wrote many years ago, whose (ethical and epistemological) potential and implications have not yet been fully understood, but which, in any case, allow only "simulation of communication" and social relations, at times even emptying them of certain fundamental dimensions. We too are "complex systems", we are "relationships", and we desperately need Others, relationships with Others, because we ourselves, in fact, are "relationships", complex, open, adaptive, dissipative systems. We need Others, and their proximity, just like plants need sunlight.

Let us try, therefore, to avoid making the mistake, for the umpteenth time, of thinking (and also acting as if it were so) that disasters and emergencies regenerate, almost magically, all the social bonds of trust, cooperation and solidarity, the "social mechanisms" that are of vital importance to the very existence of groups, organizations, of what we call "society", of a democracy that can neither be reduced to a few procedures nor simplified in all its vital and contradictory complexity. These are complex bonds and mechanisms, which require much stronger "efforts" and "energies" to be reactivated (education, training and research, in this case as well.). Disasters and emergencies carry out a function solely in the very moment when they occur, and, in any case, in the short term.

Although there is a substantial amount of scholarly literature on these topics,[5] highlighting that in the presence of disasters and emergencies, the *social mechanisms of trust, cooperation and solidarity* begin to strengthen again—at least in proximity to the events themselves—there has been a tendency to misinterpret these data, too often mistakenly based on a partial reading of these same studies and research reports which, despite undoubtedly having had the merit of "photographing" situations and contexts synchronically, have been taken to uphold the belief/thought that emergencies and disasters could/can (actually) recreate trust, cooperation, solidarity among people, groups, even among nation-states. Arguably, in certain contexts/historical periods this has in fact occurred (albeit occasionally and partially), but the conditions were totally different. In the mid/long term, however, the dynamics that have been witnessed are of the opposite sign.

As soon as the moment/phase (even a long one) of emergency, emotion and *solidarity induced by fear*, fueled by media and social-media coverage is over, we all return, inevitably, to our daily lives and existences, to a general state of disinterest, (*individualism* always hegemonic) locked within our particular interests, distant from others and mistrustful of everything. Which brings us back inevitably to the point I keep harping on, that of the "cultural and educational question". We really need to thoroughly rethink/re-mediate and act, thinking concretely about *the long term*, in the direction of a rediscovery of the (strong) reasons that keep social actors, People, Humans—to use an old sociological metaphor—*around a hearth*. The rationales for social bonds and the social mechanisms of trust and cooperation need to be found elsewhere, not in the existence and/or strengthening of interconnected infrastructures, networks and environments.

It is precisely from this perspective, looking back on the lengthy "pandemic period"—although I prefer to speak of an "emergency ecosystem" -, that we need to indicate, among the many paradoxes and paradoxical issues of those years, a detail that I consider particularly important: it was precisely that long interval of isolation and social (and emotional) distancing that probably made us (re)discover, albeit very slowly, on the one hand, (1) the extraordinary potential of digital and connective technologies (still unexplored), with digital cultures still to be co-built (and we are very far from understanding its ethical and epistemological implications), on the other hand, (2) how important, fundamental, vital, generative, relationships and the "physical" closeness of other people are, even of those we do not know too well. Networks, ties of various "natures", social bonds, vitality and energies. Because

[5] See in particular, including the bibliography, Dominici P. (2009), *La società dell'irresponsabilità. L'Aquila, i nuovi rischi, la carta stampata, le Scienze Sociali*, FrancoAngeli, Milan 2010.

sociality, as Todorov says, *is neither an accident nor a contingency; it is the very definition of the human condition.*

However, as already mentioned, what the hypertechnological and hyperconnected civilization is showing us, besides the paradoxes and dilemmas that innervate it, are the numerous risks that we are running today by carrying forward the idea/project/perspective of *delegating* everything to technology /to technologies, to systems of artificial intelligence, marginalizing the Human (autonomy vs automation) and Politics itself (hegemony of technicians and *technocracy*). It is a question of increasingly insidious and concrete risks—and their associated implications: simulation of *participation*, of *communication*, of *proximity* and of *citizenship*, and, closely related to these, simulation of *democracy*. While it can be said that these epistemological challenges are certainly extremely fascinating to research, at the same time, they require great awareness and a "culture of complexity"—which, for the moment, appear to be merely "feigned and flaunted"—in addition to the fundamental awareness that—this cannot be overly reiterated—not everything can be simplified/condensed (for example learning, language, communication, education, risk, emergency, and democracy) and that "simplification" is not an absolute value.[6]

More and more, phenomenal reality as a whole is being characterized as an "ecosystem of automation and simulation processes": everything is presented and narrated as "possible", achievable, translatable into ideal and idealized actions and environments; and, thus, one might "think" of replicating/reproducing—through simulation—thought, intelligence (as if this coincided only with the ability to carry out calculations and find data, simple solutions to complex problems, perhaps); to reproduce and simulate learning, imagination, creativity, emotions, communication, even consciousness (in whatever way we fool ourselves that this can be done), social relationships, or freedom. However, as I have been repeating for over twenty years now, thought is one thing, the simulation of thought another; intelligence is one thing, the simulation of intelligence another; empathy is one thing, the simulation of empathy another (many other examples might be given); relationships and communication (complexity) are one thing, the relative simulations are another, however sophisticated and technically perfect they may be. Finally, enlarging our gaze to embrace even more complex spheres: consciousness and Life are one thing, the "simulation of consciousness" and, even more so, the "simulation of Life" are something quite different.

[6] See Dominici P. (1995–1996, 2019), *op.cit.*

9.1 Democracy as a Complex Organism/Ecosystem of Communication

Communication is citizenship. It is on this powerful assertion that my analysis and approach are based, an assertion I have been making for years, long before this kind of expression became the talk of the town. On the other hand, it has long been known that there is an extremely close, intimate correlation between communication—which I have defined as the *social process of knowledge sharing* (1995–1996)—and citizenship (not *hetero-directed* but participating in the common good), between communication and democracy (= *complexity*); but also, between democracy and the visibility/publicity/accessibility of power. In other words, communication is not a superstructural element/variable of organizations and social systems; on the contrary, it is one of its constitutive, connotative and, moreover, ontological elements. Furthermore, the *culture of democracy* of a country and/or of a given historical-cultural context is, in my opinion, closely related to the "culture of communication" which characterizes not only the institutions but civil society itself, as it moulds dynamics and processes which, when all comes to all, are simply processes of continuous negotiation, conflict and mediation. In developing such an approach, aware of finding ourselves on the "shoulders of giants", we cannot but refer to Norberto Bobbio when he defines the government of democracy as the *government of public power in public,*[7] effectively acknowledging "making public"—as opposed to "secrecy"—as one of the fundamental cornerstones of democracy. However, despite their acknowledged, as well as basic, importance, the principles of *visibility* and *public communication*—and, currently, the discussion needs to be extended also to those concerning *access* and *transparency*—are required to guarantee (at least they should…) "information" from the Public Administration/the State/the Sovereign to the citizens. Yet, despite their fundamental importance, they do not contemplate the opportunity of "communication" (as I have defined it) for/with them (→ symmetry—reciprocity). Whereas communication—as I shall strive to argue—is a complex social process that implies, first of all, the *parity* and *reciprocity* of the subjects and elements involved (knowledge, skills, means, tools, etc.), but also possibility of access, transparency, sharing, involvement, *engagement*. However, a point that must be reiterated, so that (at least) the conditions of the above-listed principles/values may be achieved (although effectuating an operational translation of these values becomes even more complicated and complex) is the necessity, above all, that the process of communication (and the complexity that characterizes it) relating to complex adaptive systems—both in terms of interpersonal communication and organizational communication and social systems (in this case from the State to citizens → public sphere)—must encounter and involve people and citizens who are endowed with awareness and with Montaigne's *well-made heads,* (→ which again brings us to the strategic role of the school, instruction and education), aware, informed and competent (not only from a "technical" point of view)

[7] See Bobbio N. (1984), *op.cit.*, p. 86.

because—though this may look like a mere *slogan*, for this author, at least, it has never been such—*it is possible to be "vassals" even in a democracy*, even more so in an era so marked by insecurity, precariousness and vulnerability, an era identifiable within an overall framework of general downsizing of the sphere of rights, both at a local and global level. Ours is *a hypercomplex* and hyperconnected *society* where, for some time now, the *rules of engagement* of the citizenry are no longer being defined/created, as in the past, by the so-called Legislator: the rules of engagement of citizenship today are defined, constructed and reproduced precisely within the educational and training institutions, which in the past were the active protagonists, in charge of and fully responsible for the creation of new conditions of emancipation and inclusion. For this reason, in the midst of new asymmetries and inequalities, and with educational and training institutions having, in the meantime, become "agencies of selection", *one can, indeed, be a vassal in a democracy*. First of all, by not knowing one's rights/duties, by not knowing which tools to use and which channels to pursue (and not having access to them); by not being sufficiently literate, educated and (precisely) competent to pressure, dialogue and deal with any type of authority, but also to engage actively in the construction of an autonomous *public sphere*,[8] capable of bringing pressure to bear on politics itself and on the "Sovereign" (power) as well as of influencing the relative decision-making processes. *One can be a vassal in a democracy* out of *cultural habit* (De La Boétie), but also, and above all, by not having the *ability* and/or even the minimum possible guarantee permitting access to that *equality of starting conditions* which is, in fact, the fundamental pre-requisite for even simply addressing these issues.

Because—on this point, I wish to be perfectly clear, and not solely with reference to my approach—*not only can it be asserted that communication is complexity, but that democracy is complexity also,* (1995–1996) as well as a continuous process of communication. Democracy means/implies open dialectics. It is a system of systems in which, once again, the whole does not coincide with the sum of the parts involved, but is much greater; it is plurality of opinions and thoughts, of codes and languages, of cultures and utopias/visions of society. Democracy is conflict, openness: it is order and chaos at the same time etc. etc. The risks of simplification, of a certain kind of simplification (populism) have always been present as well as being very concrete, while, on the contrary, we desperately need to move from linearity to complexity, also with reference to what we consider/believe to be possible (simple) "solutions" to problems that are complex. Evidently, they closely concern *communication* (codes, languages, models, cultures etc.), forms of relationality, conflicts and the numerous connections as well as democracy itself, along with the culture of communication which distinguishes it historically. Unfortunately, there is still little awareness of how *extremely slender the boundary between citizenship and subjugation* is; it is a line that has become even narrower within the civilisation of hypertechnology and hyperconnectivity, *within which*, not only in terms of narratives/storytelling, there exists the *increasingly widespread illusion* that, without even

[8] See Dominici P., op.cit., 1995–1996 and following.

taking into consideration other variables and contributing causes, it is *precisely* the *digital and connective technologies* (1996) that can/will be able to create the conditions for a "true" democracy,(finally) active and *non-simulated* form of engagement and even strengthen the social bonds and mechanisms of trust and cooperation.

Attempting to understand and manage the (complex) processes and dynamics of this hypercomplexity necessarily requires a culture of complexity—intended also as a culture of responsibility and sustainability—and a view that can only be systemic. In this sense, it is useful to reiterate the strategic relevance of the social fabric, of the quality of the *social capital*, the networks and movements, the forms of cooperation and associationism, which, today, find in the Internet, despite it now having become a place of maximum control and maximum surveillance, the infrastructure and ecosystem fundamental to self-reproduction and intensification of bonds (→ see also the concept of *autopoiesis*). Furthermore, it is important to underline the fact that the recognition of the value of transparency (always as a means and not an end) which has now occurred, definitively, not only at a legislative level, pushes complex (public and private) organizations more and more to seek configuration as "open systems" (therefore, complex and uncomplicated systems), capable of managing (?) complexity in the best possible way, by opening up, precisely, to the environment and ecosystems: a very particular social and political complexity that is thoroughly and totally chaotic and unpredictable, conditioned by a lack of knowledge (or in any case, by poor management), knowledge that is marked and moulded, as always, by the renowned and ever-present "emergent properties". As a result, processes of this kind necessarily imply an overall rethinking of the organizational models and cultures on which they are based, of the very concept of communication and, more generally, a pressing need for what, several times in the past, we have evoked (precisely) as a "new culture of communication".[9] At the same time, what needs to grow ever more urgently and rapidly is the awareness that social and organizational change, in order to be "systemic", cannot be imposed from above, that it cannot be achieved exclusively through technological innovation nor by a better, rigorous and increasingly specific definition and/or adherence to a regulatory framework.

The modes, forms, languages, even the data, through which we communicate, represent and visualize *complex systems*, determining, in a non-linear but profound and effective manner, perceptions, emotions and, as a result, individual and collective behaviours (hypercomplexity), functional, in some way, for dealing with phenomenal reality. Old, age-old questions linked to an idea/view of efficiency and (economic) rationality which identifies the "legal" and the "technological factor" as necessary (which they are) and sufficient (which they are not) conditions for efficiency, innovation and the correct *functioning* of organizations. On the contrary, and quite trivially (often and precisely because of its apparent banality, this "principle" is scarcely considered), it is people, groups, communities, and related cultures/subcultures that are capable of hindering, in a more or less voluntary way, the external and internal changes of organizations (one need only think of the concept of

[9] *Ibidem.*

organizational climate) and social systems. This raises the issue of knowledge, monitoring, listening (tools and techniques of social research, subsequently borrowed by *marketing*), of knowing how to communicate, which requires the kinds of training, skills, listening skills, etc., which are a far cry from certain clichés/stereotypes (still listed on professional profiles) and from the still-persisting confusion between communication and communication media, between communication and marketing, between communication and connection. Because, as I have been repeating since the mid-nineties: *communication is complexity, communication is organization, communication is citizenship*. In the background, once again, the confusion, and not only terminological, between "complicated systems" and "complex systems", between organizations conceived and managed as if they were "mechanisms" and not "organisms".

9.2 The Hyperconnected Society and the Illusion of Citizenship

In line with this approach and analytical perspective, I would like recall that the concept of "citizenship" is, as is well known, a complex one that boasts an extremely complex scientific literature, not only in the juridical area.[10] This is a concept or, to phrase it better, a category of political and social thought which, like several of the other categories of Modernity and the fields of knowledge it has produced, urgently requires a redefinition and rethinking (1998) and not—as often emerges from the public debate too- a simple functional *extension/adaptation to technological practice*. We are talking about a territory and a province of meaning defined by new variables and contributing factors; a territory where freedom and equality face each other in an open field, in search of a possible, practicable form of equilibrium.

Endowed with a vast semantic field, it is a complex and multidimensional concept, which can be traced back to some extent to a WE which contrasts with a YOU and calls into question the equally fundamental concepts of identity, recognition, subjectivity, the (political) community, territory, social rights, culture, inclusion vs exclusion etc., which repeatedly confirm its historically determined and problematic nature. As a result, the rights of citizenship themselves need to be rethought if

[10] See Marshall T.H. (1950), *Citizenship and Social Class and Other Essays*, Cambridge University Press, Cambridge 2002; Veca S., *Cittadinanza. Riflessioni filosofiche sull'idea di emancipazione*, Feltrinelli, Milan 1990; Zolo D., *Complessità e democrazia. Per una ricostruzione della teoria democratica*, Giappichelli, Turin 1987; Dahrendorf R. (1959), *Class and Class Conflict in Industrial Society*, Italian translation, *Classi e conflitto di classe nella società industriale*, Laterza, Bari 1963; Dahrendorf R. (1988), *The Modern Social Conflict*, Italian translation, Il *conflitto sociale nella modernità: saggio sulla politica della libertà,* Laterza, Rome 1992; Dahrendorf R. *Dopo la democrazia*, Laterza, Rome -Bari 2001; Bellamy R., *Citizenship. A Very Short Introduction*, Oxford University Press, Oxford 2008; Norris P., *Democratic Deficits: Critical Citizens Revisited*, Cambridge University Press, Cambridge 2011; Balibar É., *Cittadinanza*, Bollati Boringhieri, Turin 2012.

only because we are now all *the social-system actors* in a society which, despite its dramatic conflicts and obvious asymmetries/inequalities, is—and will increasingly be—global and cosmopolitan. Topics and questions which, precisely in the era of globalization and the new ecosystem, of the political economy of insecurity and large migratory flows, assume an even more strategic centrality, despite the risk of being defined and rendered in a reductionist and/or, at least, simplistic manner. At the centre of every discussion there are/need to be People and Subjectivities, intended, however, as members of a political community, which confirms the complexity of an analysis that is evidently linked to a multiplicity of indicators and variables. Having said this, as I have been repeating for years, what we continue not to consider with due attention is the question of who these citizens/recipients of services, policies (?), and strategies actually are and, beyond the technologies, platforms and public declarations of intent, the fact that the services in question continue to be delivered, essentially, *top-down.* "Who are these citizens and what characteristics do they possess?" should be asked with reference not only to the so-called structural variables but also, and above all, to variables and indicators which can no longer be neglected, such as those due to functional illiteracy and educational poverty as well as schooling and training in complexity and critical thinking.

As I had been warning long before these concepts were taken up as buzz-words, *the risk continues to be that of building a form of citizenship without citizens,*[11] *as well as that of determining technological innovation without culture.* In other words, the risk continues to be that of promoting the engagement of social subjects/actors who have not been empowered to participate actively, in a "non-hetero-directed" manner, because they do not possess the "tools" (I am not referring, evidently, to technical and technological ones) that would permit them to participate, interact concretely, discuss, contest, or negotiate themes, issues and requests that concern them very closely. In actual fact, we are faced with processes that are *not* participatory as they are based on a type of *negotiation* that, apparently, always arrives downstream of the processes, strategies and policies themselves. In other words, we are confronted with processes of "*simulation of participation*" (1996–2017), within which the so-called civil society, the public sphere and the social bodies represent,

[11] On these issues, see, in particular: T.H. Marshall, *Citizenship and Social Class* del 1950; S. Veca, *Cittadinanza. Riflessioni filosofiche sull'idea di emancipazione*, Feltrinelli, Milan 1990; Zolo, 1994 e 2000; R. A. Dahl (1998), *On Democracy*, Italian translation, *Sulla democrazia*, Laterza, Rome -Bari, 2000; R. Dahrendorf *Dopo la democrazia*, Laterza, Rome -Bari 2001; R. Bellamy, *Citizenship. A Very Short Introduction*, Oxford University Press, Oxford 2008; Dominici, *op.cit.*, 2008, 2014–2018; Norris P., *Democratic Deficits: Critical Citizens Revisited*, Cambridge University Press, Cambridge 2011. Among the popular essays, I highlight some published on "Forum PA" and "Agenda Digitale". http://www.forumpa.it/riforma-pa/i-rischi-di-una-cittadinanza-senza-cittadini. However, with regard to the idea/vision/narrative, which has been dominant for years, of designing/building educational and training institutions on the market and on the specific requests of companies, please refer to the article "Rethink l'educazione nell'era della rapida obsolescenza" https://www.agendadigitale.eu/scuola-digitale/rethink-leducazione-nellera-della-rapida-obsolescenza/

undoubtedly, the *weakest link* in *that complex systemic chain that is democracy*[12] (a non-ideal mechanistic metaphor!).

Let us go back to underline an essential clarification that had previously been made. I intend communication as the "social process of knowledge sharing (power)" (1996), in which social "actors" are involved, *flesh-and-blood* people who, by virtue of the knowledge and skills they possess, their psychological profile, their system of relationships and the characteristics of the environment they dwell in, of their possibility/ability to access and process information and knowledge, define more or less symmetrical relationships among one another (power—asymmetries of information and cognition). As already stated in the past, if we consider the *knowledge = power* equation to be well founded, it follows that all the processes, dynamics and tools aimed at knowledge sharing can only determine a sharing of power or, in any case, a reconfiguration of power systems and hierarchies within organizations (in the long term). However, it will certainly not be the connective and digital technologies that guarantee the creation of these conditions and opportunities. Once again, it is a question of the "quality" of processes of education and training.

From this perspective, as reiterated several times, the new social and communicative ecosystem seemed/seems to be able to open up interesting prospects for processes of democratization of knowledge and increase the possibilities of access to information and the processing of knowledge; but, for this to happen, it is necessary to deal seriously not so much with the *digital divide* (an evidently important issue that, in all likelihood, will be resolved over time)—as with the *cultural divide*. This is an issue of vital importance—and not only for the *governance* of the Internet and the new global ecosystem. I have spoken several times about "rethinking citizenship" and the urgency of a "new social contract"[13]—which closely concerns schools and education, requiring, in general, an overall reform of thinking, and more specifically, of teaching and education (something I will never tire of repeating). But the redefinition of citizenship (and the quality of democracy) urgently calls for citizens—and first and foremost, simply People—who are aware and responsible, capable of evaluating and monitoring, of not accepting narratives and/or media representations or, worse again, information passively acquired "by hearsay".

Cultural change, from this point of view, is the complex "product", on the one hand, of systemic processes and social mechanisms that must start "from below", and on the other, the complex outcome of the action of that civil society and that public sphere, currently absorbed and engulfed by Politics, which has usurped their autonomy (a few years ago I spoke of the "public sphere, handmaiden of the power system"). We need (long-term) policies designed and implemented by way of a systemic perspective (the absent dimension). Otherwise, precious little can (could) be achieved even through those processes, platforms and dynamics of inclusion (concretely) built according to the logics of participation, and activated by a Public

[12] See in particular: Dominici P., *Oltre la libertà ...di "essere sudditi"*, in F. Varanini (ed.)*Corpi, menti, macchine per pensare*, Casa della Cultura, Anno 2, numero 4, Milan 2017.

[13] See Dominici P., (2008–2018), *op.cit.*

Administration/State/Sovereign that had, in the meantime, become (as one would hope and wish) ever more transparent and efficient. *The risk is the construction of a form of citizenship/democracy without citizens* that is able to include only those who possess the proper tools and are capable of producing/elaborating/sharing knowledge. Let us reaffirm a central concept: in the fundamental perspective of overcoming (always, and only, in the long term) the sterile—though always functional to certain logics of power—dichotomy/separation between humanistic and scientific education, it is of vital importance to educate in critical thinking, complexity, unpredictability and a systemic vision of phenomena and of Life, in an attempt to overcome "disjunctive thinking". Beyond "false dichotomies",[14] knowledge, strategic when trying to metabolize and *inhabit* hypercomplexity and the ongoing change, is based, necessarily, on a *systemic approach to complexity,*[15]

[14] See Dominici P., *op.cit.*, 1995–1996 and following.

[15] Within an endless and multidisciplinary literature featuring these topics and issues, we expressly recall Morin E. (1973), *Le paradigme perdu: la nature humaine*, Italian translation, *Il paradigma perduto. Che cos'è la natura umana?*, Feltrinelli, Milan 1974; Bertalanffy von L. (1968), *General System Theory: Foundations, Development, Applications*, Italian translation, *Teoria generale dei sistemi*, Isedi, Milan 1975; Monod J. (1970), *Le hazard et la nécessité*, trad. it., *Il caso e la necessità. Saggio sulla filosofia naturale della biologia contemporanea*, Milan, Aronoldo Mondadori Ed. 1997; Morin E. (1977), *La Méthode I. La Nature de la Nature*, Italian translation, *Il metodo 1. La natura della natura,* Raffaello Cortina Editore, Milan 2001; Morin E. (1980), *La Méthode II. La Vie de la Vie*, trad. it., *Il metodo 2. La vita della vita*, Raffaello Cortina Editore, Milan 2004; Prigogine I. (1980), *From Being to Becoming: Time and Complexity in the Physical Sciences*, Italian translation di Bocchi G. e Ceruti M., *Dall'essere al divenire,* Einaudi, Turin 1986; Miller J.G. (1971), *Living system*, tr. it., *Teoria generale dei sistemi viventi*, FrancoAngeli, Milan 1986[3]; Cramer F. (1988), *Chaos und Ordnung. Die Complexe Struktur des Lebendigen*, Italian translation, *Caos e Ordine. La complessa struttura del vivente*, Turin: Bollati Boringhieri 1994; Prigogine I. (1996), *La fin des certitudes. Temps, chaos et les lois de la nature*, Italian translation, *La fine delle certezze. Il tempo, il caos e le leggi della natura*, Turin: Bollati Boringhieri, 1997; Maturana H., Varela F. (1972), *Autopoiesis and Cognition. The Realization of the Living*, Reidel Italian translation, *Autopoiesi e cognizione. La realizzazione del vivente*, Venezia, Marsilio, 1985. Again on these topics and, specifically, on the concept of "autopoietic organization", see Maturana H.R., Varela F.J. (1985), *The Tree of Knowledge*, Italian translation, *L'albero della conoscenza*, Milan: Garzanti 1987; Prigogine I., Stengers I. (1984), *Order out of Caos*, New York: Bentham Books; Heisenberg W., *Physics and Philosophy: the Revolution in Modern Science*, Prometheus Books, Buffalo N.Y., 1958; Margulis L., *Symbiosis in Cell Evolution in the Archean and Proterozoic Eons*, Freeman, New York 1993; Nicolis G.—Nicolis C. (2007), *Foundations of Complex Systems*, Singapore: World Scientific; Goodwin B., *How the Leopard Changed Its Spots. The evolution of Complexity,* Princeton Science Library, Princeton 1994; Israel G., *La macchina vivente. Contro le visioni meccanicistiche dell'uomo*, Bollati Boringhieri, Turin 2004; Bertuglia C.S., Vaio F., *Non linearità, caos, complessità*, Bollati Boringhieri, Turin 2003; Israel G., *The Science of Complexity. Epistemological Problems and Perspectives*, in "Science in Context", 18, Anno 2005, pp. 1–31. See also, and with particular attention, the studies and research reports published by the physicist Giorgio Parisi, an extraordinary figure not only as a scientist, recently awarded the Nobel Prize for Physics, with reference both to complexity and the theory of "disorderly systems" Parisi G., *Field Theory, Disorder and Simulations*, World Scientific Pub. Co Inc., Singapore 1992; Parisi G., *La complessità da un punto di vista di un fisico*, in Rendiconti Accademia Nazionale delle Scienze—*Memorie di Scienze Fisiche e Naturali* 119° (2001), Vol. XXV, pp. 7–14; Parisi G., Urbani P., Zamponi F., *Theory of Simple Glasses*, Cambridge University Press 2020. See also, with a view to

aware of the ambivalence of the processes taking place and of the multiple levels of the connections between them, as well as between the variables and contributing causes involved; a systemic approach that can only be interdisciplinary and multidisciplinary at the same time, since the challenge is facing a dynamic and unpredictable, hybrid *hypercomplexity*, not constituted and not describable/representable in terms of numbers, data, empirical evidence only or by what is mathematically measurable. These are topics and issues that evidently concern our *culture of evaluation*, at all levels and in all areas of organizational and social practice, as well.

Without an attempt (at least) to pursue these objectives, there can be no "true" citizenship nor, much less, can there be any process of inclusion, only a kind of citizenship which, on the contrary, is already proving to be, and will prove to be, increasingly exclusive (inclusivity vs exclusivity). A dramatic illusion fueled and made socially acceptable thanks to media narratives and public discourse on the Internet and the new global ecosystem of communication and knowledge. Topics, questions and critical issues that we have addressed in the past (1996) and to which Stefano Rodotà, referring specifically to the "new media" (not as simple consumer goods), evidently posed a question, first of all, of *equality*, "As these are, in any case, indispensable tools for cultural, social, political and economic participation, the new inequalities become a real risk for democracy. If these problems do not become the object of conscious policies, instead of insisting on an abstract capacity of the market to solve them [this author's note: the Weberian *self-normativity* of the market], there is a real risk that the information highways will turn into *the umpteenth road without an exit in the pursuit of equality.* Abstract predictions about the universality of the service and the opening up of access to new technologies will not suffice. Active public policies of literacy, aimed precisely at eliminating the factors that are producing widespread and growing inequalities, are necessary if we wish to avoid phenomena of mass exclusion that impact directly on the democratic nature of a system: an exclusion that might also be determined by policies regarding tariffs and the localisation of services. If these risks are not avoided, democracy itself will be in question, given that it cannot be separated from a continuous, uninterrupted process of citizen inclusion. *Literacy* policies cannot be reduced, however, to the minimum notions necessary for the use of machines, thus legitimising any kind of utilization of the machines themselves (and this risk becomes even greater if we move in the direction of the pure and simple material availability of technical means according to the logics that inspired the proposal to provide every student with a

---

further information, the latest book by Parisi G., *In un volo di storni. Le meraviglie dei sistemi complessi*, Rizzoli, Milan 2021. Mentre, among recent publications, see: Al-Khalili J., McFadden J. (2014), *Life on the Edge. The Coming of Age of Quantum Biology*, Italian translation, *La fisica della vita. La nuova scienza della biologia quantistica*, Bollati Boringhieri, Turin 2015; Catania K. (2020), *Great Adaptations*, Italian translation, *Adattamenti meravigliosi. Sette irresistibili misteri dell'evoluzione*, Bollati Boringhieri, Turin 2021; On "variability" as a primary property of living entities see Kupiec J-J., *La concezione anarchica del vivente*, Milan: Eleuthera 2021; Zimmer C., *Life's Edge: The Search for What it Means to Be Alive*, Palgrave Macmillan, 2021; Josup M., Holme P., Kanazawa K. Et al., *Social Physics*, in Physics Reports, 948 (2022), 1–148, Elsevier.

personal computer)".[16] And so, even talk about *collective intelligence* (Lèvy) and/or *connective intelligence* (De Kerckhove) will always be fascinating, stimulating but misleading, even surreal: conformism and homologation will always be the hegemonic social *forces* acting within a general framework of reference, and overall, of "simulation of communication and participation".[17] The future of this hypertechnological and hyperconnected civilisation is foreshadowed by a new, worrying, globally extended "spiral of silence" (Elisabeth Noelle-Neumann), with only a few areas and networks, independent of the dominant systems (to use a modern *figure of speech*, the *provinces* of the empire), capable of producing innovation and (small) changes. The utopias of the interconnected and asymmetrical society: a technocratic "Mechanism-Society", which has chosen, as its principal systemic objective, the removal of conflict and deviance/abnormality from organizational and social praxes.

9.3 *Democracy Is Complexity.* The Construction of Democratic *Governance in* the Era of Technopolitics

Democracy is complexity (1995–1996). Democracy is, to all intents and purposes, a complex process. It is an eco-system of complex processes, with numerous connections and levels of connection involved, the reduction/simplification of which may involve risks. At the same time, as well as being definable/identifiable as complexity/a(hyper)complex system, it is a communication ecosystem.

A (hyper)complexity which, also in this case, can be intended and defined as:

- Transition from linearity to complexity (non-linearity).
- Transition from simplicity to complexity.
- Everything is "relationship"—"objects" are "relationships".
- Everything depends on everything else: Everything is interdependent.
- *Everything is inter-independent.*
- Reciprocity of sets and multiplicities.
- Intensification and extension of connections and interconnection levels.
- Intensification and extension of bonds of interdependence.
- Coexistence of antithetical, contradictory and/or dichotomous elements.
- Open dialectics.
- Recognition of the non (complete) "observability" and "measurability" of all the dimensions of complexity.
- Crucial relevance of the so-called "emergent properties".
- Extreme sensitivity to *disturbances* in the environment of reference and ecosystems.

[16] See Rodotà S., *Tecnopolitica. La democrazia e le nuove tecnologie della comunicazione*, Laterza, Bari 1997, pp. 91–92. See by the same author Rodotà S., *Iperdemocrazia. Come cambia la sovranità democratica con il web*, Laterza, Rome-Bari 2013.

[17] See Dominici P. (1998–2017), op.cit.

- The generative, *co-creative* and self-organizational capacity of systems and of the subsystems/subunits/units that constitute them.
- Processes and dynamics of "complex synthesis".
- Recognition of the "irreversibility" of the processes and dynamics that characterize complex systems (dissipative systems).
- The existence of complex systems that are constantly dynamic and far removed from equilibrium and stasis.
- Overcoming "false dichotomies" (1995).
- Healing the fracture between the Human and the technological.
- Complex interaction with machines and technologies.
- Excess information and data.
- Excess communication—virality and pervasiveness.
- The epistemology of interdependence (for the "hypercomplex society").
- The epistemology of uncertainty (Morin).
- The epistemology and the culture of error (Dominici).
- Organization of experiences and fields of knowledge.
- An "Other" look at reality and at what is real.
- A continuous dialectic between liberty and equality.
- Recognition of error as a producer of knowledge.
- A systemic and multi/inter/transdisciplinary approach to problems.
- Recognition of error and unpredictability as structural elements.
- Recognition and mediation of conflict.
- The simultaneous presence of order and chaos.
- Multiplicity of identities and subjectivities.
- An enhancement of multiplicity and diversity.
- Pluralism of principles, values and visions.
- An enhancement of heterogeneity and recognition of deviance.
- The unpredictability and vulnerability of people and systems.
- An intensification of systems of rules and procedures.
- Opening up to indeterminacy.

From this perspective, the co-construction of democratic governance, with the related processes of participation and involvement *(engagement)* of citizens, is an extremely complex process characterized by ambivalences and *open dialectics*; dialectics which (almost) never find their moment of (complex) synthesis. It is a process that calls into question numerous variables, approaches, and methods of analysis and detection but which also, and above all, requires *imagination,* the ability to "network" (and "create a system") and, at the same time, a long-term systemic view. A process that can find its effective translation only if supported by a *culture of complexity and sharing* that still struggles to establish itself inside and outside (public and private) organizations. Social and cultural factors, often underestimated, have always played a role of fundamental importance: as stated several times, there still seems to be little awareness that it will certainly not be digital and connective technologies (1995), and even less so new laws and regulations, to resolve the main criticalities and vulnerabilities (these are necessary, but not sufficient, conditions). On the contrary, the continuous and sudden accelerations, in determining new opportunities, trigger and radicalize problems of systems management and control, often related as well to an inadequate culture of communication. In the background, but not too far back, stands the risk of the concession of carte blanche to technology/technologies in an era, moreover, marked increasingly by the *hypertrophy* of the technologically controlled, with profound implications for autonomy, responsibility, privacy, and rights.

In a scenario as complex as this, within *the new global ecosystem* (1996) and the *Mechanism-Society*, it proves to be anything but simple to define and *build the social and cultural conditions of public participation* which is not, and cannot be, mere construction and acquisition of *consensus*—obtained perhaps by means of sophisticated marketing strategies– based on models, actions, practices, services which have, in any case, been defined, designed, and imposed top-down by expert and elitist knowledge. One thinks, in this sense, of an increasingly invasive *technocracy* that occupies, more and more every day, those social and political spaces (of the πολις) left empty by an increasingly marginalised Politics, especially when it has to measure itself with economic powers and by a civil society/public sphere increasingly suspended between hetero-direction and radical disillusionment/disaffection.

Participation, on the contrary, is "made" of continuous and constant processes of *negotiation,* which need to be meticulously detailed from the moment of their *conception* to that of their *decision making.* At this level it is no longer possible to continue not to deal with "real citizens" (if I may use this term) who, beyond the (rightly, much debated) issue of "digital skills" differ significantly from that almost *ideal-typical* figure of the "ideal citizen" (critical, informed, competent, capable of interacting on an equal basis with the public administration and, more in general, with power), often imagined and considered a reference by *decision-makers* themselves. At the same time, it is no longer possible to continue not to deal with worrying variables and critical issues such as *functional illiteracy,*[18] *educational poverty* and, more generally, the critical conditions faced by schools and universities which, for some time now, have no longer even been able to carry out their traditional functions as social elevators. The so-called "civil society" is destined to remain the weak link,[19] for a long time to come, within the complex and ambiguous dialectics of democratic practice.

The very concepts of *participation and citizenship* call into question an issue of a more general nature, but of fundamental importance: the urgency of rethinking the "social contract" (2003) and, consequently, of redefining the rules of engagement regarding citizenship and inclusion. On these grounds, we cannot fail to take note of a crucial cultural backwardness, forcefully reiterating an old formula of ours:

---

[18] Regarding "functional illiteracy" and other related (strategic) issues, see the OECD's *Programme for International Student Assessment* at the PISA website::https://www.oeso-calledorg/pisa/

[19] See Dominici P., (1995–1996), (2005), (2019), (2022°) and following *op.cit.*

"connected citizens" will not suffice; what we need are critically trained and informed citizens, educated in critical thinking and complexity, in citizenship and not subjugation, educated in liberty and responsibility. Educated in citizenship (the same goes for the social construction of a culture of *legality and/or of prevention*: *constructed* in school!) which—let this be clear—is made up not only of rights, which need to be known, but also of *duties*. In any case, it is necessary to act and intervene, with a certain urgency, right there where information and knowledge, the "truly" strategic resources of the new ecosystem, are produced, processed and distributed. With the centrality, once again, placed on educational and training processes, on *social capital* and on People who, in turn, must contribute actively to the co-construction of a social and communicative space capable of generating and distributing value and, why not, "trust"; a device that is truly fundamental to the very existence of systems, be they democratic or simply, first and foremost, social.

Freedom comes with significant responsibilities that we need not fear. And to attempt (at least) to achieve all this, exclusively in the long term, education and instruction need to be concerned with bridging that worrying gap, mentioned above, between the "ideal citizen" and the "real citizen", by training People and Citizens capable of taking advantage of the opportunities brought about by technological innovation and the interconnected/hyperconnected society, but also, and above all, People and Citizens capable of contributing to a social and cultural change that cannot be imposed/guided and which can no longer fail to deal with the famous "educational and cultural question" and the lack of a shared public ethics.

In this line of discussion, only the affirmation and widespread diffusion of the culture of communication (as knowledge sharing), throughout social systems in general and, in particular, inside and outside of public administrations and business systems (concept of organization as an "open social system") will be effectively able to create the conditions for the realization of those fundamental rights/duties of citizenship without which the citizens-users-consumers cannot evidently find any type of legitimation/recognition for their requests, finding themselves, in reality, in a condition of subjugation, within a public sphere which, although (hyper)connected, will turn out to be increasingly inconsistent and hetero-directed.

The profound belief is that, at the level of practice, *the categories of risk and conflict in social systems and complex organizations are strictly correlated with poor/ineffective management of knowledge or, worse still, with the impossibility of having access to these and making conscious and rational use of them.* From this point of view, I shall never tire of reiterating the crucial importance of a principle which, only in appearance, might seem banal and/or obvious: *whoever possesses more knowledge* in terms of control, possession, access and processing) has more power (similar to whoever controls more information), both in terms of interpersonal communication and organizational or macro-systemic communication. In other words, at every level of analysis and practice, those who possess more knowledge—currently, this even includes those who have greater possibilities of processing it—and have more skills (the strategic relevance of education, training and research for a Nation-system comes to mind) is undoubtedly better equipped to direct the evolution of the dynamics and processes that characterise social, economic and political relationships. That is, knowledge and skills are able to determine—and will increasingly be able to do so—the balance of power in every sphere of social, organizational and systemic life with clear consequences for citizens and democracies. "Age-old" questions, that are still crucial → knowledge = power.

Once we have become aware of the characteristics of the new social ecosystems and the hypercomplexity that characterizes them, I believe that the most important challenge is, once again, to *empower people, citizens* (not only as consumers, as is currently the case) to deal, in the most conscious and competent manner possible, with the processes and dynamics distinguishing the new ecosystem that concern them closely. In other words, it is of vital importance to create the "structural" conditions that enable them to *inhabit* these ecosystems, to know how to inhabit what is, in fact, not just *a new unlimited public space* –capable of defining "identities", "subjectivities" and the public communicative space in which they are realized and evolve—but also, and above all, a global Panopticon, within which the logics of total control and surveillance have always been—and will always be—dominant.[20]

9.4 Hypercomplexity and (Hyper)Connected Citizens

To cope with this hypercomplexity—as I have always repeated—digital education is /has never been sufficient, "digital skills" or so-called *digital literacy* do not suffice, and—above all—"connected citizens", who frequent the multimedia, and/or are simply aware of digital complexity—which would be an important first though insufficient level—need to be critically trained and informed, educated in citizenship and not in subjugation *out of cultural habit* (De La Boétie). Citizens need to possess knowledge and skills that are not only technical and/or digital but, above all, they need to be educated and trained in complexity and "critical thinking", within the general framework of a progressive recovery—in the long term—of the *complex dimensions of educational complexity* (1996), educated and trained to comprehend the importance of sharing and cooperation if they are to overcome the old logics of possession and control effectively: because sharing and cooperation are essential to the social and collective production of knowledge and culture, the true *agents*—some would say "engines"—of innovation. These are all people and citizens who also need to be educated and trained in the value of "shared knowledge", not so much because these assumptions—in my opinion strategic, vital—represent the

[20] See Foucault M., 1975; Lyon D. (1994), *The Electronic Eye. The Rise of Surveillance Society*, Italian translation, *L'occhio elettronico. Privacy e filosofia della sorveglianza*, Feltrinelli, Milan 1997; Lyon D. (2001), *Surveillance Society. Monitoring Everyday Life*, Italian translation, *La società sorvegliata. Tecnologie di controllo della vita quotidiana*, Feltrinelli, Milan 2002; Bauman Z., Lyon D. (2013), *Liquid Surveillance. A conversation*, Italian translation *Sesto potere. La sorveglianza nella modernità liquida*, Editori Laterza, Bari 2014; Byung-Chul H. (2012). *Transparenzgesellschaft*, Italian translation, *La società della trasparenza*, nottetempo, Rome 2014; Byung-Chul H. (2013), *Im Schwarm. Ansichten des Digitalien*, Italian translation, *Nello Sciame. Visioni del digitale*, Nottetempo, Rome 2015.

"new utopia" to be pursued, but because—and it is incredible how, at all levels, there is still no awareness and unity of purpose—it is precisely the sharing economy and the knowledge society that require increasingly higher levels of education and training, as well as *continuous updating*[21] in all workplaces and professional fields (data and research into functional illiteracy and educational poverty provide a picture that is anything but reassuring).

In this sense, a "true" citizenry, active and engaged in the common good and, more in general, profound cultural change are always the complex products, on the one hand, of social processes and mechanisms that must be initiated from below; on the other, of the action of that civil society and that public sphere, currently absorbed and swallowed up by a Politics that has taken away their autonomy. We need (long-term) policies which, in addition to being imagined from a global perspective, must be designed and implemented with a systemic perspective, to be then evaluated and monitored constantly according to their effects. Dimensions that have been completely ignored; the absence of a "real" industrial policy in countries like Italy, for example, comes to mind.

As I have been repeating for quite some time now, innovation is a complex process, or, rather, it is complexity itself: education, instruction, training, research, evidently, must be its "backbones", not a set of "tools") that arrives "downstream" of the processes of change. The risks remain those of technological innovation without culture and the illusion of citizenship: a kind of citizenship and engagement that is not negotiated and socially and culturally constructed within inclusive processes, but imposed, instead, from above without completely and concretely engaging the points of view of the recipients of these actions/strategies. In other words, without their being nourished constantly by the participation of those called upon to exercise citizenship and engagement. We shall be increasingly obliged to choose between the "liberty/responsibility of being citizens" and the "liberty/responsibility of being subjects".[22] In this case, no third way is available!

On the other hand, technological innovation has always been a strategic factor for changing social systems and organizations. However, without a culture of communication, a systemic view of complexity and, at the level of political decision-makers, social policies capable of triggering and supporting cultural change, innovation fails to be achieved.

Beyond our inadequacies and *incompleteness*—which are constitutive/ontological elements of our "being human", *complex adaptive systems* (CASs)—compared

[21] Here the recent transnational actions and initiatives undertaken by UNESCO and the United Nations should be kept in mind, specifically designed to make "lifelong learning" a widespread and systemic practice. See in particular, the Report "Making lifelong learning a reality: a handbook" UNESCO—Institute for Lifelong Learning, 2022—ISBN 978-92-820-1238-3 For further information, visit the Institute's official website for Lifelong Learning: https://uil.unesco.org/

[22] See Dominici P., *Oltre la libertà ...di "essere sudditi"*, in F. Varanini (ed.)*Corpi, menti, macchine per pensare*, Casa della Cultura, Anno 2, numero 4, Milan 2017.

to the (hyper)complexity with which we need to deal,[23] beyond the *limited rationality* (Simon) that characterizes reality, and precisely because of the profound epistemological implications associated with the advent of the digital civilization (still underestimated and/or not adequately considered) and the *new connective technologies* (1996), "the" question (Education) is no longer—and, in my opinion, never has been—that of simply rethinking "digital education" or "media education" with a view to building the new (global) digital citizenship and a new "culture of citizenship". Furthermore, all the adjectives and terms that, from time to time are added, confirm, paradoxically, if there were still any need, the incompleteness and inadequacy of certain approaches/settings/views that continue to dominate, not only in public discourse, but also throughout numerous educational projects and practices (not all of them, evidently: there are exceptions and generalizations are always wrong).

Defining and building the social and cultural conditions required to draw up a "new social contract" and create a new (global) kind of citizenship means, even more so in the so-called knowledge society, *thoroughly rethinking education and training* (complex and strictly interdependent dimensions), profoundly rethinking the School and University (a single ecosystem), *recovering the complex dimensions of educational complexity*—empathy, critical thinking, a systemic view of phenomena, education in communication, as well as dimensions that we have deliberately put aside, such as imagination and creativity—and, compared to the recent past, having the courage and authority to rethink the two levels of education "together" and not as two separate entities (also at stake is the old, very old question of the "training of trainers"). We need to thoroughly rethink educational processes, the relational and communicative spaces that characterizes educational and training institutions, not only in the short term, but, rather, by defining objectives and results that can only be achieved in the "long term" (the sore point of the problem), aware—I strongly reiterate—that technology and the digital will not be what succeeds in reducing the new asymmetries and inequalities that characterize the hyperconnected and hypertechnological civilization.

A "new social contract",[24] on the other hand, is conceivable only as the result of a systemic process of cultural transformation capable of triggering economic, political and social change as well. We need—it is important to repeat this once again—to think about the long term and about a "long-term" political project centred around

[23] See also my essay published in September 2021in "Media Studies" with the following title: *Tutto sotto controllo! La (iper)complessità tra realtà e rappresentazione*, Anno 2021, settembre, https://www.mstudies.it/2021/09/03/la-ipercomplessita-tra-realta-e-rappresentazione/. On the complex topic of "control" in "globalized modernity", see also: Eriksen T.H. (2016), *Overheating. An Anthropology of Accelerated Change*, Italian translation, *Fuori controllo. Un'antropologia del cambiamento accelerato*, Einaudi, Turin 2017. For further insights into reality and representations of the world, also with reference to that provided by Science, I recommend another great "classic" by a true philosopher-scientist: Schrödinger E. (1958), *The Physical Basis of Consciousness, Science and Religion, The Mystery of the Sensual Qualities*, Italian translation, *L'immagine del mondo*, Bollati Bopringhieri, Turin 1963 and 2017.

[24] See Dominici P., (1996–2019), *op.cit.*

"educational processes" and, above all, a strong idea of education and learning, based on "critical thinking", an approach to complexity, a "humanistic education" as well as a scientific one. Overcoming the traditional logics of separation and confinement and what I have called "false dichotomies".

Since, as already discussed, the complexity which regards social and organizational systems is of *a very particular type*, difficult to manage[25] and simplify: technology can certainly make an important contribution in terms of control, simulation and replicability of complex processes and dynamics; but, for this contribution to prove even more strategic, we need to invest decisively and concretely in education, training, continuous updating and research. There is no alternative, since we have to deal with a continuously evolving and chaotic hypercomplexity, which cannot be traced back to the sole application of mathematical formulae, data (albeit clearly fundamental) and empirical evidence that can be framed within a kind of "scientific objectivism" and, more generally, within a culture of standardization, which have again become hegemonic, contributing to fuelling a certain type of conformism on these issues.

Social (and organizational) (hyper)complexity, in all its particularity, has always been a matter of knowledge and, at the same time, a complex problem regarding *access to and the sharing and management of knowledge,*[26] too. These are cognitive possibilities that can be selected and realized, translated into choices and decisions—which cannot but remind us of the Weberian *finite segment of the meaningless infinity of the world process*. A (hyper)complexity, as I have understood it, that has increased further and, contrary to what one might think (in the so-called Interconnected/Hyperconnected Society → + information + data = + rationality in choices and decisions), has become even more unpredictable—despite the fact that the dimension of what is controlled technologically has increased exponentially—precisely by virtue of the enormous (infinite) quantities of data and information which not only *never speak for themselves*, but determine a permanent and constant condition of *limited rationality* at all levels, from organizational to social. Topics and issues that closely concern not only possibilities for the emancipation and inclusion of individuals (versus people), but also, and above all, the quality of democracy and our *living together*. However, at least for the moment, we are trapped in the illusion of a less asymmetrical relationship with power and a participation that is only "simulated".

Between new utopias and dystopias that appear, finally, to be within reach... I always repeat: simple solutions to complex problems. But Democracy is complexity, it is relational, social, human complexity (1998–2005). Order and chaos,

---

[25] In this author's opinion, "managing complexity" has always been, by virtue and as a consequence (not only) of what has also been argued in this monograph, is a "contradiction in terms" (1995–1996 and following). It would suffice to think about the crucial and strategic relevance to complex adaptive systems of "emergent properties".

[26] See Dominici, P., (1996–2021) op.cit.

balance and conflict, control and unpredictability, normality[27] and deviance, pluralism and conformism, cohabit and coexist within this (hyper)complexity, along trajectories that are only apparently predictable and controllable.

It is necessary, as stated several times, to *disseminate* a renewed awareness and sensitivity, which are still wanting, concerning an issue, which is (above all) a matter of *thought* (for now, however and always, "disjunctive") and of pedagogical, organizational and social cultures that this disjunctive thought has defined and produced: we continue to think of organizations and societies as if they were *machines/ mechanisms* and not *organisms*.

The crisis of *thought*—thought which is an "emergent property" of the brain, the most complex system of them all, perhaps[28]—and the absence of a thought system, are proving to be even more significant and evident, precisely now that we are coming to realise how complex, interdependent, interconnected, unpredictable and chaotic the nature of phenomenal reality is. These are all fundamental challenges that "*linear thinking*" and "*causal thinking*" have attempted to face, only to reveal all their inadequacy.

A crisis of *thinking* and thought so profound and systemic by now, so consolidated that, by continuing to underestimate/devalue/belittle its scope and implications—in individual and social terms—we are slowly (though not too slowly) heading towards a (global, systemic and hyperconnected) type of Society which, besides exalting the (dis)values of ignorance and *non-knowledge,*[29] is also increasing inequalities and asymmetries (as has already been happening for some decades now), which will deny to itself more and more and, consequently to the social actors who *inhabit* it, any possibility/opportunity for innovation, change, emancipation or democratization.

The hypertechnological civilization—this point must be reiterated—is attempting to eliminate error and unpredictability from our lives and our existences, granting *carte blanche* to technologies; something which actually translates into the

[27] On "normality", "deviance" and "exception", see the interesting work of S. Belardinelli, *La normalità e l'eccezione. Il ritorno della natura nella cultura contemporanea*, Rubbettino, Soveria Mannelli 2002.

[28] On these issues, see in particular: James W. (1878), *The Brain and the Mind*, Lowell lectures on 'The brain and the mind', in, Burkhardt F (ed.) *Manuscript Lectures. Original lectures delivered 1878*, Harvard University Press, Cambridge, MA, pp. 16–43, a publication in which James recognizes the nervous system as a highly elastic system, with a great degree of plasticity and adaptive capacity to interact with the environment, contracting new habits. Among the most recent "classics" see Minsky M. (1985), *The Society of Mind*, Italian translation, *La Società della Mente*, Adelphi, Milan 1989. See also the following articles: J. Dorrier, *New Study Finds a Single Neuron Is a Surprisingly Complex Little Computer*, in, "Singularity Hub". https://singularityhub.com/2021/09/12/new-study-finds-a-single-neuron-is-a-surprisingly-complex-little-computer/; S. Fan, *How AI Is Deepening Our Understanding of the Brain*, in, "Singularity Hub", 2021. https://singularityhub.com/2021/11/23/how-ai-is-deepening-our-understanding-of-the-brain/

[29] In international scientific publications, in the past, I have proposed and used the formula "*No-Knowledge Society*".

progressive de-responsibility of the People, supported apparently and at times even replaced by intelligent (?) machines and systems.

But this objective is nothing but a mere illusion: the illusion of control, the illusion of being able to *manage* complexity/hypercomplexity, of being able to direct, pre-determine, even "eliminate" it. I feel truly sorry for all the aspiring "complexity managers", but (hyper)complexity is not manageable/predictable or, at least never thoroughly manageable. As mentioned before, (hyper)complexity (such as interconnection, interdependence, inter-independence, continuous and systemic interaction of the variables, and of the ecosystem elements) is systemically and systematically marked in depth, by dynamism, entropy, error, unpredictability, variability, diversity, heterogeneity, instability, etc. It is a constitutive, structural, ontological element of the Social, the Human and of everything that is and belongs to the Vital. *We can only strive to inhabit it.*

But—in the light of the research pathways addressed in this work—how can we try to inhabit hypercomplexity and, even more so, *open ourselves to* indeterminacy?

- Rethinking educational and training processes, from a perspective that can only be socio-emotional and systemic.
- Teaching unpredictability
- Building/creating/co-creating a "culture of error" (from the earliest years of school).
- Teaching and practising an *epistemology of error*, doubt and uncertainty within institutions and research groups.
- Educating and learning to "see objects as systems".
- Overcoming the "false dichotomies", beyond the "Great Mistake".
- Adopting all possible measures and strategies required to overcome, in the long term, cognitive cages, fractures and logics of the confinement of knowledge and the various *monodisciplinarisms* that are still hegemonic.
- Educating and training "Hybrid Figures" (who are not "all-knowing experts" or experts in everything), and not (only) technicians and/or (hyper)specialists (1995 and following).

Long, very long-term actions and strategies with the fundamental objective of trying to *re-mediate*, that is, of trying to reconstruct mediation and re-activate the social mechanisms of trust and cooperation. Because... "democracy is complexity" (1995).

9.5 *The Ultimate (Systemic) Error:* Treating Complex Systems as If They Were Complicated Systems

In a civilization based increasingly on programming, automation and (hyper) simulation of processes and dynamics—the digital and hyperconnected civilization of Artificial Intelligence—whose extraordinary technological innovations give us a convincing illusion of total control and predictability,[30] an illusion, or to be precise, a series of illusions, which consolidate in the progressive marginalization of the Human within processes, systems and ecosystems which are evidently complex, hypercomplex, that is to say, constitutively unstable, dynamic and unpredictable, impossible to manage and govern (emergent properties, extreme sensitivity to external disturbances, capacity for self-organization etc.).

Hence, we are obliged to go back to square one. Hurled into hypercomplexity, we try to inhabit the hypertechnological and hyperconnected civilization of intelligent machines (?), based on programming and total (hyper)simulation of processes and actions; a hypercomplexity marked by a progressive, as well as exponential, growth in the dimension of the *technologically controlled* (a sort of "new Nature") which, in reality, narrows down the space of responsibility even further, along with (at least apparently) that of the unexpected and of error (the latter being a fundamental prerequisite of any kind of knowledge and of freedom itself).

Within the hyperconnected civilization, the governance of information and communication—and the complexity that characterizes them—assumes, as a result, an increasingly strategic relevance which produces the necessity of trying to interact with the variety, ambivalence and unpredictability of social, relational, informative and communicative processes. It signifies—even more so in the so-called *era of access* (Rifkin)—trying to bestow meaning and systematicity on that endless/infinite availability of data and information, which, however, will not necessarily contribute to the reduction/simplification of complexity. It also means having to deal—in the transition from simple to complicated, and from complex to hypercomplex—with a further increase, in terms of both quantity and quality, of the variables, the contributing causes and the parameters to be considered, in order to observe, describe, interpret and understand phenomena. In doing so, we need to be very careful not to fall into the use of rhetoric or the "grand" narratives on disintermediation, on a direct and simplified form of democracy, founded on a radical horizontality of social and cultural processes which, at least for the moment, can only be simulated and declared.[31] In the background, an idea/view of the very concept of simplification, seen as an absolute value and practice to be pursued at all costs and in every area of praxis.

The Interconnected and Hyperconnected Society—founded on *the interdependence of systems of limited rationality*—presents itself as the supporting *megastructure* of digital capitalism and the sharing economy (1996). As a new ecosystem capable of defining unprecedented opportunities of inclusion but also—as has happened—as the bearer of new asymmetries. Capable of creating the socio-cultural conditions for the affirmation of a *new transnational public sphere* while, at the same time, profoundly impacting (individual and collective) perceptions of fear and

[30] See Dominici P. (1995–1996), (2003), (2014) and following, *op.cit.*

[31] On these topics and issues, among the various national surveys and research projects carried out in recent years, see the latest Istat Reports on Knowledge; while, among the international Reports see OCSE "Education at Glance", 2021 https://www.oeso-calledorg/education/education-at-a-glance-19991487.htm/?refcode=20190209ig

insecurity, of dangers and *social acceptability,* and affecting the collective imaginary of public opinion, exposed more than ever to the emotional outbursts of the global media system. New environments of communication and of public opinion, more and more conditioned and malleable, which in turn are then capable of bringing pressure to bear on an increasingly weaker and more marginalized Politics, in the presence of economic and financial power. And yet it is Politics that is called upon to provide answers and make choices in a transnational key and rethink a system of rules conceived within a context widely differing from that of a strong Nation-State. The networks and ecosystems of complex connectivity constitute, today more than ever, the fundamental lever for bringing the conflict, which already presents completely different characteristics compared to those of the past, upon an increasingly strategic not to mention vital level of praxis, both in micro and macro terms: the level of viral and widespread communication in real time.

Managing information and communication in this kind of context calls even more for *managing* (?) complexity (which, as I have said before, is an oxymoron, which I use here for the sake of simplification), paying great attention, however, not to fall into the rhetorical traps of disintermediation and of an ideological vision of *simplification* intended as a principle and an absolute value. *Not everything can be simplified* (Fig. 9.1).

For some time now, there has been a great deal of talk about *direct democracy*, of an efficient, simplified form of democracy, of *an almost infallible sort of*

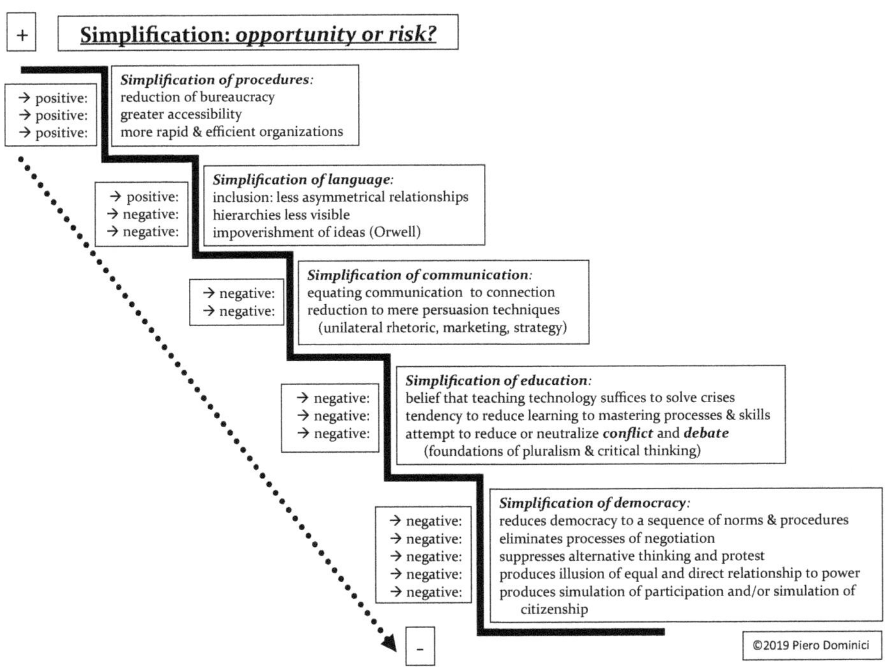

Fig. 9.1 Simplification: opportunity or risk?

mechanism (precisely), in perfect continuity with the hegemony of a technocratic vision of Society and the Social, along with the rhetoric of disintermediation and the great narratives regarding a kind of digital civilization represented as inclusive, supportive and open, by its very "nature", as it were. This is the reason that we had, in the past, used the definition: "asymmetrical society". We know that the variables and factors to be considered are numerous and that the crucial (and complex) issues to be addressed do not concern technological (opportunities), but rather education and training, that is, People and citizens. However—it is important to be clear on this—democracy is first and foremost complexity, while simplification cannot, in any way, be considered an absolute value. Democracy is mediation, negotiation, pluralism and plurality, convergence, listening, conflict, effort, failure, error, responsibility. *There are no simple solutions to complex problems*, even at an organizational level. Unless we educate and train the new generations in this wonderful and unpredictable complexity, the consequences will cost us very dearly.

The information and knowledge society lives and feeds on risks, illusions and paradoxes. Among these, the most obvious one concerns disintermediation and the decline of traditional processes and figures of mediation. In the background, stands a return to an almost neopositivist view regarding cognitive, informational, organisational, social and cultural processes. The extraordinary availability of data and information seems to give us—as mentioned before- the illusion of full control over any type of social and organizational process. Information and communication play an absolutely strategic role in this complex and ambiguous scenario.

From this perspective, what is needed is a far cry from dis-intermediation, a far cry from simplification (however important in certain areas/dimensions). It is high time to abandon, not only reductionist and deterministic explanations, but also the dominant, and extremely seductive, logic of "simple solutions to complex problems" (1995), at every level of organizational and social practice, above all, when dealing with the cornerstones of a society which, despite numerous critical issues and uncontrollable factors, intends to be truly open, inclusive, and democratic in its education, instruction, training, research, information, and communication. We can no longer afford to go on perpetuating *the ultimate error, that* of treating complex systems as if they were complicated systems (1995).

9.6 (Hyper)Complexity, Our Limits and Knowledge

It is an "expanded" complexity, rendered hypertrophic by our (ontological) limits, by our socially and culturally constructed inadequacies.

Our ontological limits and the inadequacies of our education, training and, more generally, culture urge, stimulate (→absence of an adequate thought system, capable of…), oblige us to keep "objects" (which are always "systems") and phenomena separate, thus placing us in a position of not knowing how/being able to recognise, not only the connections, but also the different levels, both in terms of quantity and quality, of interdependence and reaction (*feedback*- non- linearity- ability to adapt and *self-organize*, interactive agents).

As I always repeat, educated and trained in "disjunctive thinking" as we are, we continue to treat complex systems as if they were "complicated systems", with the aggravating circumstance of the exclusion of numerous fields of "expert knowledge" from attempts at analysis, explanation and management of processes and the contemporary use of reductionist and deterministic analyses and explanations. And the almost obsessive quest for *simple solutions to complex problems* which, moreover, not only as of today, has profound implications for social life and politics itself. At the level of public discourse, we all talk about complexity, paradigm shifts, *systemic thinking,*[32] etc. but, frequently, we continue to reproduce and fuel the logics of separation and restriction that innervate our educational and training systems (and processes) and, more generally, our organizations. Logics of separation and "false dichotomies" which, as we have verified (not only) in the course of the present work, constitute the obstacle most difficult to overcome with respect to our possibilities of adapting, managing, governing but, above all, *inhabiting* the process of anthropological transformation and global change underway (1995–96).

It is useful, necessary actually, to reiterate the point that the educational and cultural question is "*the*" question, the mother of all questions. At all levels of analysis and discourse and, among these, that concerning the "social contract", citizenship, inclusion, democracy, above all. From this point of view, there is still no awareness of the urgency of recovering/revitalizing the institutions and the *figures of mediation* (teachers, educators, journalists, communicators, various mediators etc.) by reaffirming their importance and strategic value precisely in the era of so-called disintermediation—and of the grand narratives (illusions, risks, paradoxes, dilemmas, between new utopias and dystopias) relating to the hyper-connected and hyper-technological civilization—an urgency which manifests itself in a very clear and evident way.

As things stand at present,[33] the horizontality of social and communicative processes, their inclusiveness and the opportunity of fostering a less asymmetrical relationship with the system of power and the elites, are little more than mere illusions, anchored to ambiguous and opaque processes that "simulate" the engagement and *dialectics* typical of democratic regimes. There cannot/will not be any citizenship and/or inclusion without first-quality education and training (a complex concept, to be resolved). There cannot/will be no "true" (social and cultural) innovation without profoundly and radically reforming education, schooling, training and research. Important policies and investments are needed. Today, truly everyone supports the idea, but it is regarding the approach, the "how" and "why", that the differences, the

[32] See, in particular, Capra F. (2021), *Patterns of Connection*, Italian translation, *Le relazioni nella vita. I percorsi del pensiero sistemico*, Aboca, Sansepolcro 2022.

[33] With reference to these issues, in various publications, I recall having proposed the following concepts and operational definitions: exclusive innovation, citizenship without citizens, risk of inclusion for a few, *technological innovation without culture*, simulation of communication, simulation of participation.

distances, even the inconsistencies between saying and doing emerge. I always repeat the well-known formula, "we stand on the shoulders of giants",[34] though, unfortunately, we are scarcely aware, recalling a great scientist (Prigogine), that only from the "end of certainties" will creativity and knowledge (collaboration, cooperation, sharing) arise. We are still scarcely aware that that *knowledge itself*, to paraphrase another great (philosopher) *is sown and takes root in the errors of life*. And, in an increasingly hyperconnected and interdependent world, where traditional limits and boundaries have broken down, to continue to keep "objects", phenomena, systems, dynamics and processes separate means following the pathways of an illusory simplification capable, indeed, of hindering any kind of innovation and change. Inside the *No-Knowledge Society*. After yet another "black swan", waiting for the next one to arrive. *Beyond Black Swans…to inhabit hypercomplexity*.

9.7 Epilogue. For a *Non-Mechanistic View* of Society and Life

9.7.1 Beyond the "New Epistemological Fracture"

The non-linear and unforeseeable dynamic that can spring from a new epistemological fracture is, I believe, a progressive evolution and transformation of all "complicated systems" into "complex systems" due to a *non-linear differentiation* accompanied by a re-found, and, to some extent, paradoxical centrality of the Human/the Natural and, as a result, of a re-found/renewed centrality of the error, of what is a non-compliant "derivative" and unpredictability itself. All this will occur inevitably, while awareness of the systemic and relational dimension of all the social and communicational processes, is still lacking. Thus, the Human, in other words, contrary to what is claimed and imagined by a considerable part of the scholarly and popular literature regarding these issues, returns to the centre, with new powers and responsibilities, unprecedented in the history of society and human relations.

My hope and wish are that all this may take place within a framework of the rediscovered centrality of "qualitative" dimensions and variables within educational and training processes, along with a *renewed systemic relationship* with machines, artificial/natural duality and the world outside of us ("inside" vs. "outside", another false dichotomy that, perhaps, we shall be able to rethink).

On the other hand, within the framework of certain recurring, dystopian utopias, such as the design and creation of logical machines/technological *automata, intelligent machines* (Turing)[35] capable of imitating and behaving—at least,

[34] Merton R.K. (1965), *On the Shoulders of Giants: A Shandean Postscript*; It. Trans., *Sulle spalle dei giganti*, il Mulino, Bologna 1991.

[35] See, in particular A.M. Turing (1950), *Computing Machinery and Intelligence*, in, *Mind: A Quarterly Review of Psicology and Philosophy*, 59, 1950, pp. 433–460; A.M. Turing (ed. Gabriele Lolli), Collected Worksof A.M. Turing. *Mechanical Intelligence*, Italian translation, *Intelligenza*

apparently—like human beings (replicating/reproducing) and/or, subsequently, creating hyper-technological and automated *environments/ecosystems*, even going so far as to co-construct "perfect societies"—according to some, authentic "algorithmic societies".

Once again, the question summed up in the concept and operational definition that I have proposed, that of the "Mechanism-Society" returns, that of a Society/Civilization entirely fashioned by *perfect, superefficient devices incapable of* making mistakes, an old, age-old idea/aspiration/view that has crossed/crosses different historical eras, infecting and marking all areas of the human and the social: from myth to religion, science/the sciences to art, literature to technique/technology itself, to arrive at all forms of social and cultural production.

Having reached the end of this anything but linear and reassuring route, which—I hope—may prove, once again, to be generative, I shall try to sum up some of the main issues treated. In particular, I cannot but point out, even here, how the paradigm of hyper-technological civilization has been, collectively and systematically, designed, built and co-built precisely upon the marginalization of the Human, as the "bearer" of error, unpredictability, uncertainty and assumption of responsibility.

With the advent of artificial intelligence—which I defined, before it became more widely used, as the "new epistemological fracture", following the fracture determined by Chaos Theory—and its potential developments, the ultimate traditional boundaries/limits between "complicated systems" and "complex systems" are being/have been breached (and consolidated, even in scientific literature).

The non-linear and unpredictable dynamics that I believe may arise from this will involve a progressive evolution and transformation—achieved by means of *non-linear differentiation* -of all "complicated systems" into "complex systems" and a rediscovery, to some extent paradoxical, of the centrality of the Human/Natural and, consequently, of the re-found/renewed centrality of error, of what is "deviant" or non-conformist, and of unpredictability itself. All this will occur, inevitably, although awareness of the structurally systemic and relational dimension of all social and communicative processes is still wanting. This means that the Human—contrary to what is stated and imagined by much scientific and popular literature on these issues- returns/will return, more and more, to the centre with new powers and new responsibilities, unprecedented in the history of societies and human relationships.

In the future, we shall be obliged, undoubtedly, to deal more and more with "machines" and more or less intelligent (?) "matter", – with ulterior intensification and extension of networks and levels of *connection* between them and us humans—as well as with new automatons/robots which, as mentioned, will be increasingly capable of imitating us and replicating some traits of human behaviour. With the enormously increased and *enhanced* risk of not being able to evaluate fully the consequences of our choices, which are destined to prove more and more powerful and

Meccanica, Bollati Boringhieri, Turin 1994. On these topics see also Searl J.R., *Minds, Brains and Programs*, Harvard University Press, 1984; Searl J.R. (1998), *Mind, Language and Society*, Italian translation, *Mente, Linguaggio e Società*, Raffaello Cortina Ed., Milan 2000.

irreversible. Therefore, the question of epistemological and ethical implications returns once again, more crucial than ever (Figs. 9.2 and 9.3).

The last word, however, will always be that of *Mother Nature* and of that *vitality of Life and spirit* that no technological device, no ecosystem, however hyperconnected and hypercomplex, will ever be able to capture, manage or replicate, much less on the basis of consolidated logics of control and/or confinement.

…

It has been a long journey, a journey which, due to certain problems with my health, has also involved suffering as well as the need to be restarted and repeated several times. Despite all, I have arrived at my destination and am finishing this demanding yet simultaneously exciting effort, marked profoundly by a number of existential events, with a few words by the late Pope Francis, which I consider extremely significant for many reasons, beyond any kind of religious and/or secular identity/affiliation:

> 138. Ecology studies the relationship between living organisms and the environment in which they develop. This necessarily entails reflection and debate about the conditions required for the life and survival of society, and the honesty needed to question certain models of development, production and consumption. It cannot be emphasized enough how everything is interconnected. Time and space are not independent of one another, and even atoms or subatomic particles cannot be considered in isolation. Just as the different aspects of the planet – physical, chemical and biological – are interrelated, so too living species are

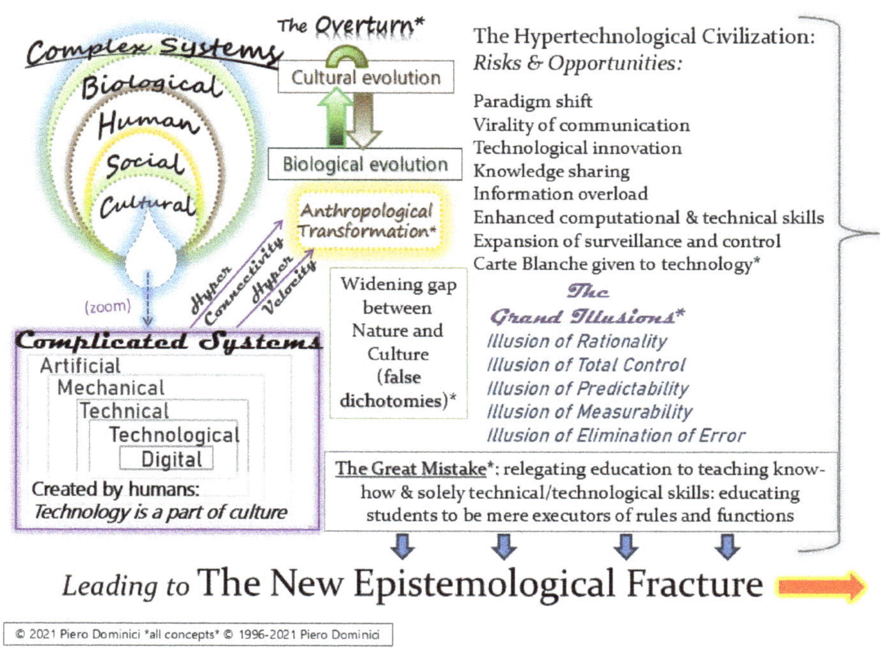

Fig. 9.2 The new epistemological fracture, part 1

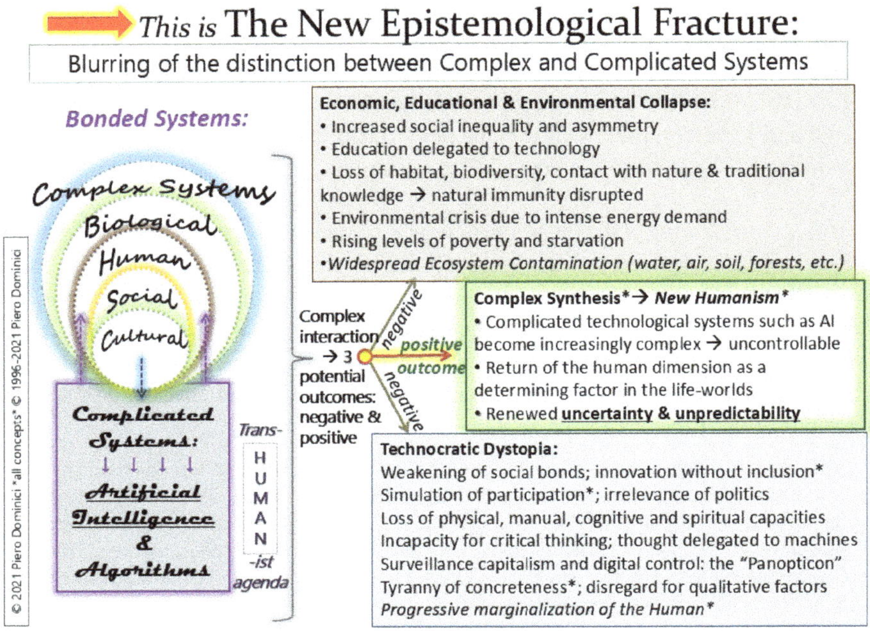

Fig. 9.3 The new epistemological fracture, part 2

> part of a network which we will never fully explore and understand. A good part of our genetic code is shared by many living beings. It follows that *the fragmentation of knowledge and the isolation of bits of information can actually become a form of ignorance, unless they are integrated into a broader vision of reality.*[36]

From this point of view, human beings have not, as yet, fully understood/become aware that the possible achievement of the power to pre-determine their lives does not represent/does not constitute/is not the achievement of a presumed kind of omnipotence, but rather the achievement of a potentially devastating and, in some ways, self-defeating objective: the end of our “being human”, free to make mistakes and be unpredictable, that is, to be ourselves.

For a *non-mechanistic vision of the Social,* the Human and Life. For a “culture of complexity as a culture of responsibility” (1996). Beyond black swans, beyond every black swan, also and above all, to rediscover our fragile, vulnerable, perennially inadequate Humanity, which, at the same time, is also unique, rich, unrepeatable, unpredictable, variable, creative, non-replicable, complex…and hypercomplex!

[36] See Pope Francis, Encyclical Letter *LAUDATO SI’*, Libreria Editrice Vaticana, Vatican City, 2015.

Bibliography

Abramo, M. R. 2002. *Gaston Bachelard e le fisiche del Novecento*. Napoli: Guida.

Addario, N. (a cura di). 1989. *Teorie dei sistemi e teorie dell'azione*. Milano: FrancoAngeli.

Addario, N. 1999. *Autopoiesi della società*. Torino: Rosenberg & Sellier.

Adorno, T. W., and M. Horkheimer. 1947. *Dialektik der Aufklärung. Philosophische Fragmente*; It. trans. *Dialettica dell'Illuminismo*. Torino: Einaudi, 1966.

Agar, N. 2019. *How to Be Human in the Digital Economy*; It. trans. *Non essere una macchina. Come restare umani nell'Era digitale*. Roma: LUISS University Press, 2020.

Ageno, M. 1999. *Le origini dell'irreversibilità*. Torino: Bollati Boringhieri.

Aleo, S., and G. Pica. 2009. *Sistemi giuridici, Complessità, Comunicazione*. Acireale-Roma: Bonanno.

Al-Khalili, J., and J. McFadden. 2014. *Life on the Edge. The Coming of Age of Quantum Biology*; It. trans. *La fisica della vita. La nuova scienza della biologia quantistica*. Torino: Bollati Boringhieri, 2015.

Amagoh, F. 2008. Perspectives on Organizational Change: Systems and Complexity Theories. *The Innovation Journal: The Public Sector Innovation Journal* 13 (3), article 3.

Amagoh, F. 2016. Systems and Complexity Theories of Organizations [w:]. In *Global Encyclopedia of Public Administration, Public Policy, and Governance*, ed. A. Farazmand. Cham: Springer.

Anders, G. 1956. *Die Antiquiertheit des Menschen, I: Über die Seele im Zeitalter der zweiten industriellen Revolution*; It. trans. *L'uomo è antiquato. Vol. I. Considerazioni sull'anima nell'epoca della seconda rivoluzione industriale*. Torino: Bollati Boringhieri, 2003.

Anders, G. 1980. *Die Antiquiertheit des Menschen, II: Über die Zerstörung des Lebens im Zeitalter der dritten industriellen Revolution*; It. trans. *L'uomo è antiquato. Vol. I. Sulla distruzione della vita nell'epoca della terza rivoluzione industriale*. Torino: Bollati Boringhieri, 2003.

Anderson, P. 1972. More Is Different. *Science* 177 (4047): 393–396.

Anderson, P., A. Meyer, K. Eisenhardt, K. Carley, and A. Pettigrew. 1999. Introduction to the Special Issue: Applications of Complexity Theory to Organization Science. *Organization Science* 10 (3): 215–376.

Anolli, L. 2011. *Le sfide della mente interculturale*. Milano: Cortina.

Anselmo, A. 2006. *Edgar Morin dalla sociologia all'epistemologia*. Napoli: Guida.

Apel, K. O. 1973. *Transformation der Philosophie* (2 voll.), vol. 1: *Sprachanalytik, Semiotik, Hermeneutik*, vol. 2: *Das Apriori der Kommunikationsgemeinschaft*; It. trans. *Comunità e comunicazione*, (ed.it. parziale, con intr. di G. Vattimo). Torino: Rosenberg & Sellier, 1977.

Apel, K. O. 1985. *L'etica della responsabilità nell'era della scienza*. Bologna: il Mulino.

Apel, K. O. 1986. *Das Problem der Begründung einer Verantwortungsethik im Zeitalter der Wissenschaft*; It. trans. "Il problema della fondazione di un'etica della responsabilità nell'epoca

P. Dominici, *Beyond Black Swans: Inhabiting Indeterminacy*, Philosophical Studies Series 160, https://doi.org/10.1007/978-3-032-09029-4

della scienza", in E. Berti (a cura di), *Tradizione e attualità della filosofia pratica*, 15–46. Genova: Marietti, 1988.

Apel, K. O. 1992. *Etica della comunicazione*. Milano: Jaca Book.

Apel, K. O. 1997. *Discorso, verità, responsabilità. Le ragioni della fondazione: con Habermas contro Habermas*, (a cura di V. Marzocchi). Milano: Guerini e Associati.

Apel, K. O., and E. Dussel. 1999. *Etica della comunicazione ed etica della liberazione, editoriale Scientifica*, Napoli (a cura di A. Savignano).

Appadurai, A. 1996. *Modernity at Large: Cultural Dimensions of Globalization*; It. trans. *Modernità in polvere*. Roma: Meltemi, 2001.

Ardigò, A. 1988. *Per una sociologia oltre il post-moderno*. Roma-Bari: Laterza.

Ardigò, A. 1989a. La sociologia dell'attore sociale di fronte alla sfida dell'ipercomplessità societaria. *Studi di Sociologia*, Anno 27, Fasc. 3 (luglio-settembre 1989): 267–280.

Ardigò, A. 1989b. *Per una sociologia oltre il post-moderno* (a cura di C. Cipolla and A. Pitasi). Milano: FrancoAngeli, 2020 (nuova edizione).

Ardigò, A. 1990. L'ipercomplessità come problema epistemologico e di organizzazione sociale. L'approccio del sociologo dopo le cibernetiche e la socio-sistemica. In A. Ardigò and G. Mazzoli (a cura di), *L'ipercomplessità tra socio-sistemica e cibernetiche*, 13–40. Milano: FrancoAngeli.

Ardigò, A., and G. Mazzoli, (a cura di). 1986. *Intelligenza artificiale. Conoscenza e società*. Milano: FrancoAngeli.

Ardigò, A., and G. Mazzoli (a cura di) 1990. *L'ipercomplessità tra socio-sistemica e cibernetiche*. Milano: FrancoAngeli.

Arendt, H. 1951. *The Origins of Totalitarianism*; It. trans. *Le origini del totalitarismo*. Milano: Comunità, 1967.

Arendt, H. 1958. *The Human Condition*; It. trans. *Vita activa. La condizione umana*, Milano: Bompiani, 1964.

Arendt, H. 2003. *Responsibility and Judgment*; It. trans. *Responsabilità e giudizio*. Torino: Einaudi, 2004 (a cura di J. Kohn).

Aron, R. 1966. *Trois essais sur l'âge industrielle*; It. trans. *La società industriale*. Milano: Comunità, 1983.

Aron, R. 1969. *Les désillusions du progrès*; It. trans. *Le delusioni del progresso. Saggio sulla dialettica della modernità*. Roma: Armando, 1991

Ashby, W. R. 1956. *An Introduction to Cybernetics*. London: Chapman & Hall.

Askenov, P. 2013. Stanislav Petrov: The Man Who May Have Saved the World. *BBC World News*, September. https://www.bbc.com/news/world-europe-24280831

Axelrod, R. 1984. *The Evolution of Cooperation*; It. trans. *Giochi di reciprocità. L'insorgenza della cooperazione*. Milano: Feltrinelli, 1985.

Axelrod, R. 1997. *The Complexity of Cooperation. Agent-Based Models of Competition and Collaboration*. Princeton: Princeton University Press.

Bachelard, G. 1934. *Le Nouvel Esprit scientifique*; It. trans. *Il nuovo spirito scientifico*. Bari: Laterza, 1978 (nuova ed.).

Bachelard, G. 1995. *La formazione dello spirito scientifico*; It. trans. di E. Castelli. Milano: Cortina.

Bachelard, G. 1998. *La filosofia del non*; trad. di G. Quarta. Roma: Armando.

Bailey, K. D. 1982. *Methods of Social Research*; It. trans. *Metodi della ricerca sociale*. Bologna: il Mulino, 1985.

Baldacci, M. 2012. *Trattato di pedagogia generale*. Roma: Carocci.

Baldacci, M., and E. Colicchi. 2016 (a cura di). *Teoria e prassi in pedagogia*. Roma: Carocci.

Baldini, M. 1986. *Epistemologia e pedagogia dell'errore*. Brescia: La Scuola.

Balibar, É. 2012. *Cittadinanza*. Torino: Bollati Boringhieri.

Ball, P. 2014. *Invisible. The dangerous allure of the Unseen*; It. trans. *L'invisibile. Il fascino pericoloso di quel che non si vede*. Torino: Einaudi, 2016.

Banfield, E. C. 1958. *The Moral Basis of a Backward Society*; It. trans. *Le basi morali di una società arretrata*. Bologna: il Mulino, 1961.

Barabási, A. L. 2002. *Linked. How Everything Is Connected to Everything Else and What It Means for Business, Science, and Everyday Life*; It. trans. *Link. La scienza delle reti*. Torino: Einaudi, 2004.
Barber, B. R. 2007. *Consumed. How Markets Corrupt Children, Infantilize Adults, and Swallow Citizens Whole*; It. trans. *Consumati. Da cittadini a clienti*. Torino: Einaudi, 2010.
Barrow, J. D. 1999. *Impossibilità. I limiti della scienza e la scienza dei limiti*. Milano: Rizzoli.
Barth, R. S. 1973. *Open Education and The American School*. New York: Agathon Press.
Bar-Yam, Y. 1997. *Dynamics of Complex Systems*. Reading: Addison-Wesley.
Bateson, G. 1972. *Steps to an Ecology of Mind*; It. trans. *Verso un'ecologia della mente*. Milano: Adelphi, 1976.
Bateson, G. 1979, *Mind and Nature. A Necessary Unity*; It. trans. *Mente e natura. Un'unità necessaria*. Milano: Adelphi, 1984.
Batty, M., and P. M. Torrens. *Modeling Complexity. The Limits to Prediction*, Working Paper 36. London: Center for Advanced Spatial Analysis, University College London. https://discovery.ucl.ac.uk/id/eprint/158379/1/paper36.pdf.
Bauman, Z. 1998. *Globalization. The Human Consequences*; It. trans. *Dentro la globalizzazione. Le conseguenze sulle persone*. Roma-Bari: Laterza, 1999.
Bauman, Z. 1999. *In Search of Politics*; It. trans. *La solitudine del cittadino globale*. Milano: Feltrinelli, 2000.
Bauman, Z. 2000. *Ponowoczesność. Jako źródlo cierpień*; It. trans. *Il disagio della postmodernità*. Milano: Bruno Mondadori, 2002.
Bauman, Z. 2000a. *Liquid Modernity*; It. trans. *Modernità liquida*. Roma-Bari: Laterza, 2002.
Bauman, Z. 2001. *The Individualized Society*; It. trans. *La società individualizzata. Come cambia la nostra esperienza*. Bologna: il Mulino, 2002
Bauman, Z., and D. Lyon. 2013. *Liquid Surveillance. A Conversation*; It. trans. S*esto potere. La sorveglianza nella modernità liquida*. Bari: Laterza, 2014.
Bayley, K. D. 1994. *Sociology and the New Systems Theory, Toward a Theoretical Synthesis*. Albany: SUNY Press.
Beck, U. 1986. *Risikogesellschaft. Auf dem weg in eine andere Moderne*; It. trans. *La società del rischio. Verso una seconda modernità*. Roma: Carocci, 2000.
Beck, U. 1993. *Die Erfindung des Politischen*; It. trans. *L'era dell'E*. Trieste: Asterios, 2001.
Beck, U. 1994–1997. *I rischi della libertà. L'individuo nell'epoca della globalizzazione*. Bologna: il Mulino, 2000.
Beck, U. 1999. *Schöne neue Arbeitswelt. Vision: Weltbürgergesellschaft*; It. trans. *Il lavoro nell'epoca della fine del lavoro. Tramonto delle sicurezze e nuovo impegno civile*. Torino: Einaudi, 2000.
Beck, U. 2007. *Weltrisikogesellschaft. Auf der Suche nach der verlorenen Sicherheit*; It. trans. *Conditio Humana.Il rischio nell'età globale*. Roma-Bari: Laterza, 2008.
Beck, U., A Giddens., and Lash S. 1994. *Reflexive Modernization*; It. trans. *Modernizzazione riflessiva. Politica, tradizione ed estetica nell'ordine sociale della modernità*. Trieste: Asterios, 1999.
Becker, H. S. 1998. *Tricks of the Trade. How to Think About Your Research While You're Doing It*; It. trans. *I trucchi del mestiere. Come fare ricerca sociale*. Bologna: il Mulino, 2007.
Belardinelli, S. 1996. *Il progetto incompiuto. Agire comunicativo e complessità sociale*. Milano: FrancoAngeli.
Belardinelli, S. 2002. *La normalità e l'eccezione. Il ritorno della natura nella cultura contemporanea*. Soveria Mannelli: Rubbettino.
Bellamy, R. 2008. *Citizenship. A Very Short Introduction*. Oxford: Oxford University Press.
Benasayag, M., and B. Cany. 2021. *Les Nouvelles Figures de l'Agir*; It. trans. *Corpi viventi. Pensare e agire contro la catastrofe*. Milano: Feltrinelli, 2022.
Benes, R., et al. 2017. *Per una pedagogia dell'errore*. Trieste: Asterios.
Beniger, J. R. 1986. *The Control Revolution*; It. trans. *Le origini della società dell'informazione. La rivoluzione del controllo*. Torino; UTET, 1995.

Benjamin, W. 1927–40, 1ª ed. 1982. *Das Passagen-Werk*; It. trans. *Parigi capitale del XIX secolo. I "passages" di Parigi*, in *Opere*, (ed.it. a cura di G. Agamben), vol. XI. Torino: Einaudi, 1986.

Benjamin, W. 1936. Das *Kunstwerk im Zeitalter seiner technischen Reproduzierbarkeit*; It. trans. *L'opera d'arte nell'epoca della sua riproducibilità tecnica*. Torino: Einaudi, 1966.

Benjamin, W. 1950. *Berliner Kindheit um Neunzehnhundert*; It. trans. *Infanzia berlinese*. Torino: Einaudi, 1973.

Benkler, Y. 2006. *The Wealth of Networks. How Social Production Transforms Markets and Freedom*; It. trans. *La ricchezza della Rete. La produzione sociale trasforma il mercato e aumenta le libert*à. Milano: Università Bocconi, 2007.

Bennet, Ch. H. 1985. Dissipation, Information, Computational Complexity and the Definition of Organization. In *Emerging Syntheses*, ed. D. Pines, 215–234. Santa Fe: Santa Fe Institute.

Berger, J. H., and K. Johnson. 2015. *Simple Habits for Complex Times*. Stanford: Stanford University Press.

Berger, P. L., and T. Luckmann. 1966. *The Social Construction of Reality*; It. trans. *La realtà come costruzione sociale*. Bologna: il Mulino, 1969.

Berger, P. L., and T. Luckmann. 1995. *Modernity, Pluralism and the Crisis of Meaning: The Orientation of Modern Men*; It. trans. *Lo smarrimento dell'uomo moderno*. Bologna: il Mulino, 2010.

Bergson H. 1889. *Essai sur les données immédiates de la conscience*; It. trans. *Saggio sui dati immediati della coscienza*, in "Opere 1889–1896", (a cura di P. A. Rovatti). Milano: Mondadori, 1986.

Bergson, H. 1896. *Matière et memoire. Essai sur la relation du corps à l'esprit*; It. trans. *Materia e memoria. Saggio sulla relazione tra il corpo e lo spirito*, in "Opere 1889–1896", (a cura di P. A. Rovatti). Milano: Mondadori, 1986.

Bergson, H. 1907. *L'évolution créatrice*; It. trans. *L'evoluzione creatrice*. Bari: Laterza, 1957.

Berlin, I. 1958. *Two Concepts of Liberty*. Oxford: Clarendon.

Berman, M. 1982. *All That Is Solid Melts into Air. The Experience of Modernity*; It. trans. *L'esperienza della modernità*. Bologna: il Mulino, 1985.

Bernard, C. 1878. *Le science expérimentale*; It. trans. *Un determinismo armoniosamente subordinato. Epistemologia, fisiologia e definizione della vita*. Mimesis: Milano-Udine, 2014.

Berthoz, A. 2009. *La simplexité*; It. trans. *La semplessità*. Torino: Codice, 2011.

Bertolaso, M., and G. Lo Storto. 2020 (a cura di). *Etica digitale. Verità, responsabilità e fiducia nell'era delle macchine intelligenti*. Roma: LUISS.

Bertrando, P., and M. Bianciardi. 2009. *La natura sistemica dell'uomo. Attualità del pensiero di Gregory Bateson*. Milano: Cortina.

Bertuglia, C. S., and F. Vaio. 2003. *Non linearità, caos, complessità*. Torino: Bollati Boringhieri.

Bianchi, P. 2020. *Nello specchio della scuola*. Bologna: il Mulino.

Blastland, M. 2019. *The Hidden Half. How the World Conceals Its Secrets*; It. trans. *La metà nascosta. Le forze invisibili che influenzano ogni cosa*. Torino: Bollati Boringhieri, 2021.

Bo, H.-X., et al. 2021. Posttraumatic Stress Symptoms and Attitude Toward Crisis Mental Health Services Among Clinically Stable Patients with COVID-19 in China. *Psychological Medicine* 51:1052–1053.

Bobbio, N. 1984. *Il futuro della democrazia*. Torino: Einaudi, 1995.

Bobbio, N. 1995. *Eguaglianza e Libertà*. Torino: Einaudi.

Bocchi, G., and M. Ceruti (eds.). 1985. *La sfida della complessità*, (a cura di). Milano: Bruno Mondadori, 2007.

Boccia, Artieri G. 2012. *Stati di connessione. Pubblici, cittadini e consumatori nella (Social) Network Society*. Milano: FrancoAngeli.

Bodei, R. 2016. *Limite*. Bologna: il Mulino.

Boden, M. A. 2018. *Artificial Intelligence. A Very Short Introduction*; It. trans. *L'Intelligenza Artificiale*. Bologna: il Mulino, 2019.

Bohr, N. 1961. *Teoria dell'atomo e conoscenza umana*. Torino: Bollati Boringhieri.

Boltzmann, L. 1905. *Populär öiften*; It. trans. *Modelli matematici, fisica e filosofia*. Torino: Bollati Boringhieri, 1999.
Boniolo, G., ed. 1998. *Filosofia della fisica*. Milano: Bruno Mondadori.
Boorstin, D. J. 1994. *Cleopatra's Nose. Essays on the Unespected*. New York: Vintage Books.
Boorstin, D. J. 1998. *The Seekers. The Story of Man's Continuing Quest to Understand the World*; It. trans. *L'avventura della ricerca*. Milano: Cortina, 2002.
Bostrom, N. 2014. *Superintelligence. Paths, Dangers, Strategies*; It. trans. *Superintelligenza. Tendenze, pericoli, strategie*. Torino: Bollati Boringhieri, 2018.
Boudon, R. 1984. *La place du désordre. Critique des théories du changement sociale*; It. trans. *Il posto del disordine. Critica delle teorie del mutamento sociale*. Bologna: il Mulino, 1985.
Boudon, R. 1999. *Les sens des valeurs*; It. trans. *Il senso dei valori*. Bologna: il Mulino, 2000.
Boudon, R. 2002. *Déclin de la morale? Déclin des valeurs?*; It. trans. *Declino della morale? Declino dei valori?* Bologna: il Mulino, 2003.
Bourdieu, P. 2001. *Science de la science et réflexivité*; It. trans. *Il mestiere di scienziato*. Milano: Feltrinelli, 2003.
Bovone, L. 2000. *Comunicazione. Pratiche, percorsi, soggetti*. Milano: FrancoAngeli.
Bovone, L. 2010. *Tra riflessività e ascolto. L'attualità della sociologia*. Roma: Armando.
Braidotti, R. 2013. *The Posthuman*; It. trans. *Il postumano. La vita oltre l'individuo, oltre la specie, oltre la morte*. Roma: DeriveApprodi, 2014.
Breton, P. 2001. *Il culto di Internet. L'interconnessione globale e la fine del legame sociale*. Torino: Testo&Immagine, 2001.
Brevini, F. 2021. *Abbiamo ancora bisogno degli intellettuali?* Milano: Cortina.
Brooks, S. K., et al. 2020. The Psychological Impact of Quarantine and How to Reduce It: Rapid Review of the Evidence. *Lancet* 395:912–920.
Bruner, J. S. 1961. *The Process of Education*. Cambridge: Harvard University Press.
Bruner, J. S. 1966. *Dopo Dewey. Il processo di apprendimento nelle due culture*. Roma: Armando.
Bruner, J. 1986. *La mente a più dimensioni*. Roma-Bari: Laterza, 2003.
Brynjolfsson, E., and A. McAfee. 2014. *The Second Machine Age*; It. trans. *La nuova rivoluzione delle macchine*. Milano: Feltrinelli, 2015.
Buchanan, M. 2003. *Nexus*. Milano: Mondadori, 2004.
Byung-Chul, H. 2012. *Transparenzgesellschaft*; It. trans. *La società della trasparenza*. Roma: Nottetempo, 2014.
Byung-Chul, H. 2013. *Im Schwarm. Ansichten des Digitalien*; It. trans. *Nello Sciame. Visioni del digitale*. Roma: Nottetempo, 2015.
Calvino, I. 1968. The Spiral. In *Cosmicomics*, English translation by W. Weaver, Harcourt Brace.
Camilli, A. 2013. Gli effetti psicologici della pandemia. *Internazionale* (10 dicembre 2020). Di seguito il link alla versione digitale dell'articolo. https://www.internazionale.it/reportage/annalisa-camilli/2020/12/10/effetti-psicologici-pandemia
Canevaro, A. 2015. *Nascere fragili. Processi educativi e pratiche di cura*. Bologna: EDB.
Canevaro, A. *Scuola inclusiva e mondo più giusto*. Trento: Erickson.
Canevaro, A., and D. Ianes. 2022. *Un'altra didattica è possibile. Esempi e pratiche di ordinaria didattica inclusiva*. Trento: Erickson.
Canfora, L. 2004. *La democrazia. Storia di un'ideologia*. Roma-Bari: Laterza.
Canguilhem, G. 1966. *Le normal et le pathologique*; It. trans. *Il normale e il patologico*. Torino: Einaudi, 1998.
Caniglia, E. 2018. *Neurodiversità*. Roma: Meltemi.
Cao, W., et al. 2020. The Psychological Impact of the COVID-19 Epidemic on College Students in China. *Psychiatry Research* 287:112934.
Capitini, A. 1964. *L'educazione civica nella scuola e nella vita sociale*. Bari: Laterza.
Capitini, A. 1967. *Educazione aperta 1*. Firenze: La Nuova Italia.
Capitini, A. 1968. *Educazione aperta 2*. Firenze: La Nuova Italia.
Capra, F. 1975. *The Tao of Physics*; It. trans. *Il Tao della fisica*. Milano: Adelphi, 1982.

Capra, F. 1996. *The Web of Life*; It. trans. *La rete della vita. Una nuova visione della natura e della scienza*. Milano: Rizzoli, 2001.
Capra, F. 2021. *Patterns of Connection*; It. trans. *Le relazioni nella vita. I percorsi del pensiero sistemico*. Sansepolcro: Aboca, 2022.
Capra, F., and P. L. Luisi. 2014. *The Systems View of Life*. Cambridge: Cambridge University Press.
Carnap, R. 1922–1925. *La costruzione logica del mondo. Pseudoproblemi nella filosofia*. Torino: UTET, 1997 (a cura di Emanuele Severino).
Carnap, R. 1966. *I fondamenti filosofici della fisica. Introduzione alla filosofia della scienza*. Milano: Il Saggiatore.
Cassano, F. 2004. *Homo civicus. La ragionevole follia dei beni comuni*. Bari: Dedalo.
Cassano, F. 2011. *L'umiltà del male*. Roma-Bari: Laterza.
Cassirer, E. 1923–1929. *Philosophie der symbolischen Formen*; It. trans. *Filosofia delle forme simboliche*, III voll. Firenze: La Nuova Italia, 1961–1966.
Cassirer, E. 1937. *Determinismus und Indeterminismus*; It. trans. *Determinismo e indeterminismo nella fisica moderna*. Milano: Mimesis, 2020.
Castel, R. 2003. *L'insécurité sociale. Qu' est-ce qu' être protégé?*; It. trans. *L'insicurezza sociale. Che significa essere protetti?* Torino: Einaudi, 2004.
Castelfranchi, Y., and O. Stock. 2000. *Macchine come noi. La scommessa dell'Intelligenza Artificiale*. Roma-Bari: Laterza.
Castells, M. 1996. *The Rise of the Network Society*; It. trans. *La nascita della società in Rete*, vol. I. Milano: Università Bocconi, 2002.
Castells, M. 1996–1998. *The Information Age, Economy, Society and Culture*, vol. III. Oxford: Blackwell Publishers.
Castells, M. 1997. *The Power of Identity*. Vol. II. Oxford: Blackwell.
Castells, M. 1998. *End of Millenium*. Vol. III. Oxford: Blackwell.
Castells, M. 2001. *The Internet Galaxy. Reflections on the Internet, Business and Society*; It. trans. *Galassia Internet*. Milano: Feltrinelli, 2002.
Castells, M. 2009. *Communication Power*; It. trans. *Comunicazione e potere*. Milano: Università Bocconi, 2009.
Castoriadis, C. 1975. *L'institution imaginaire de la société. II. L'imaginaire social et l'institution*; It. trans. *L'istituzione immaginaria della società*. Torino: Bollati Boringhieri, 1995.
CDC. 2021. *Mental Health and Coping During COVID-19*. https://www.cdc.gov/coronavirus/2019-ncov/daily-life-coping/managing-stress-anxiety.html
Ceccato, S. 1972. *La mente vista da un cibernetico*. Milano-Udine: Mimesis, 2017.
Ceri, P. 2002. *Movimenti globali. La protesta nel XXI secolo*. Roma-Bari: Laterza.
Ceri, P. 2003. *La società vulnerabile. Quale sicurezza, quale libertà*. Roma-Bari: Laterza.
Ceri, P., and P. Borgna. 1998 (a cura di). *La tecnologia per il XXI Secolo. Prospettive di sviluppo e rischi di esclusione*. Torino: Einaudi.
Ceruti, M. 1986. *Il vincolo e la possibilità*. Milano: Feltrinelli.
Ceruti, M. 1995. *Evoluzione senza fondamenti*. Roma-Bari: Laterza.
Ceruti, M. 2018. *Il tempo della complessità*. Milano: Cortina.
Ceruti, M., and F. Bellusci. 2020. *Abitare la complessità*. Milano-Udine: Mimesis.
Cesareo, V., and I. Vaccarini. 2006. *La libertà responsabile. Soggettività e mutamento sociale*. Milano: Vita e Pensiero.
Cesareo, V., and I. Vaccarini. 2012. *L'era del narcisismo*. Milano: FrancoAngeli.
Changeux, J.-P. 1983. *L'Homme Neuronal*; It. trans. *L'uomo neuronale*, Milano: Feltrinelli, 1986.
Changeux, J.-P., and P. Ricoeur. 1998. *Ce qui nous fait penser. La Natura et la Règle*; It. trans. *La natura e la regola. Alle radici del pensiero*. Milano: Cortina, 1999.
Chomsky, N. 1988. *Language and the Problems of Knowledge*; It. trans. *Linguaggio e problemi della conoscenza*. Bologna: il Mulino, 1991.
Cilliers, F. P. 2006. On the Importance of a Certain Slowness. *Emergence: Complexity and Understanding* 8:106–113.
Cingolani, R., and G. Metta. 2015. *Umani e umanoidi*. Bologna: il Mulino.

Cini, M. 1994. *Un paradiso perduto. Dall'universo delle leggi naturali al mondo dei processi evolutivi*. Milano: Feltrinelli.
Cipolla, C. 2014. *Oltre il "peccato originale" della selezione naturale. Alcune proposte teoriche nella società digitale*. Milano: FrancoAngeli.
Cipolla, C. 2015. *Dalla relazione alla connessione nella web society*. Milano: FrancoAngeli.
Cipolla, C. 2019. *Per una scienza sociale eclettica. Riflessioni aforistiche e algoritmiche*. Milano: FrancoAngeli.
Coleman, J. S. 1990. *Foundations of Social Theory*; It. trans. *Fondamenti di teoria sociale*. Bologna: il Mulino, 2005.
Collingridge, D. 1980. *The Social Control of Technology*; It. trans. *Il controllo sociale della tecnologia*. Roma: Editori Riuniti, 1983.
Collins, R. 1988. *Theoretical Sociology*; It. trans. *Teorie sociologiche*. Bologna: il Mulino, 1992.
Cramer, F. 1979. Fundamental Complexity. A Concept in Biological Sciences and Beyond. *Interdisciplinary Sciences Review* 4:132–139.
Cramer, F. 1988. *Chaos und Ordnung. Die Complexe Struktur des Lebendigen*; It. trans. *Caos e Ordine. La complessa struttura del vivente*. Torino: Bollati Boringhieri, 1994.
Crawford, K. 2021. *Atlas of AI. Power, Politics and the Planetary Costs of Artificial Intelligence*; It. trans. *Né intelligente né artificiale. Il lato oscuro dell'IA*. Bologna: il Mulino, 2021.
Crespi, F. (a cura di) 1991. *Etica e scienze sociali*. Torino: Rosenberg & Sellier.
Crespi, F. 1996. *Manuale di Sociologia della cultura*. Roma-Bari: Laterza.
Crouch, C. 2000. *Coping with Post-democracy*; It. trans. *Postdemocrazia*. Roma-Bari: Laterza, 2003.
Crozier, M., and E. Friedberg. 1977. *L'Acteur et le Système. Les contraintes de l'action collective*; It. trans. *Attore sociale e sistema. Sociologia dell'azione organizzata*. Milano: ETAS Libri, 1978.
Cusano, N. 1440. *De Docta Ignorantia*. Milano: Rusconi, 1988.
Dahl, R. A. 1971. *Polyarchy: participation and opposition*; trad. it.: *Poliarchia: partecipazione e opposizione*. Milano: FrancoAngeli, 1980.
Dahl, R. A. 1998. *On Democracy*; It. trans. *Sulla democrazia*. Roma-Bari: Laterza, 2000.
Dahrendorf, R. 1959. *Class and Class Conflict in Industrial Society*; It. trans. *Classi e conflitto di classe nella società industriale*. Bari: Laterza, 1963.
Dahrendorf, R. 1988. *The Modern Social Conflict*; It. trans. Il *conflitto sociale nella modernità: saggio sulla politica della libertà*. Roma: Laterza, 1992.
Dahrendorf, R. 2001. *Dopo la democrazia*. Roma-Bari: Laterza.
Dal Lago, A. 1999. *Non-persone. L'esclusione dei migranti in una società globale*. Milano: Feltrinelli.
Damasio, A. R. 1994. *L'errore di Cartesio. Emozione, ragione e cervello umano*. Milano: Adelphi, 2009.
Damasio, A. R. 2021. *Feeling & Knowing. Making Minds Conscious*; It. trans. *Sentire e conoscere*. Milano: Adelphi, 2022.
Davidson, D. 1980. *Essays on Actions and Events*. Oxford: Oxford University Press.
Davidson, J. D., and W. Rees-Mogg. 1997. *The Sovereign Individual*. New York: Simon & Shuster.
Dawkins, R. 1976. *The Selfish Gene*; It. trans. *Il gene egoista*. Bologna: Zanichelli, 1982.
Dayan, D., and E. Katz. 1992. *Media Events. The Live Broadcasting of History*; It. trans. *Le grandi cerimonie dei media. La Storia in diretta*. Bologna: Baskerville, 1993.
De Biase, L. 2015. *Homo pluralis. Esseri umani nell'era tecnologica*. Torino: Codice.
De Kerckhove, D. 1993. *Brainframes Technology, Mind and Business*; It. trans. *Brainframes. Mente, tecnologia e mercato*. Bologna: Baskerville, 1993.
De Kerckhove, D. 1995. *The Skin of Culture*; It. trans. *La pelle della cultura. Un'indagine sulla nuova realtà elettronica*. Genova: Costa & Nolan, 1996.
de La Boétie, É. 1549–1576. *Discorso della servitù volontaria*. Milano: Feltrinelli, 2014.
de Saint-Exupery, A. 1943. *The Little Prince*. Paris: Reynal and Hitchcock/Gallimard.

de Saussure, F. 1916, postumo; a cura di C. Bally, A. Sechehaye. *Cours de linguistique générale*; It. trans. *Corso di Linguistica generale*. Roma-Bari: Laterza, 1974 (a cura di T. De Mauro).

De Sola Pool, I. 1988. *Technologies of Freedom*; It. trans. *Tecnologie di libertà. Informazione e democrazia nell'era elettronica*. Torino: UTET, 1995.

De Tocqueville, A. 1835–1840. *De la démocratie en Amérique*; It. trans. *Della democrazia in America* in *Scritti politici*, (a cura di N. Matteucci), vol. II. Torino: UTET, 1981.

De Toni, A. F., and L. Comello. 2005. *Prede o ragni. Uomini e organizzazioni nella ragnatela della complessità*. Torino: UTET.

De Toni, A., and G. De Zan. 2015. *Il dilemma della complessità*. Venezia: Marsilio.

Dehaene, S. 2019. *How We Learn*; It. trans. *Imparare. Il talent del cervello, la sfida delle macchine*. Milano: Cortina.

Deleuze, G., and F. Guattari. 1972. *L'Anti-OEdipe*; It. trans. *L'anti-Edipo. Capitalismo e schizofrenia*. Torino: Einaudi, 1975.

Delors, J. 1996. *Learning: The Treasure Within*. Report to UNESCO of the International Commission on Education for the Twenty-first Century, UNESCO, Paris.

Delors, J. 1997. *Nell'educazione un tesoro*. Roma: Armando.

Demetrio, D. 1990. *Educatori di professione*. Firenze: La Nuova Italia.

Demo, H., H. Nes, H. N. Somby, and A. Frizzarin. 2021. In and Out of Classin Inclusive Schools – What's the Meaning? Teachers' Opinion on Push-and Pull-Out in Italy and Norway. *International Journal of Inclusive Education* 27 (14): 1592–1610.

Dennet, D. 2014. *Strumenti per pensare*. Milano: Cortina.

Denney, D. 2005. *Risk and Society*. London: Sage.

Descartes, R. 1637. *Discours de la méthode pour bien conduire sa raison et chercher la vérité dans le sciences*; trad. it., *Discorso sul metodo*. Roma-Bari: Laterza, 2003.

Dewey, J. 1916. *Democracy and Education. An Introduction to the Philosophy of Education*; It. trans. *Democrazia e educazione. Un'introduzione alla filosofia dell'educazione*. Firenze: La Nuova Italia, 1992.

Dewey, J. 1929. *La ricerca della certezza*. Firenze: La Nuova Italia, 1968.

Dewey, J. 1933. *How We Think*; It. trans. *Come pensiamo*. Firenze: La Nuova Italia, 2006.

Di Nicola, P. 1998. *La rete. Metafora dell'appartenenza, analisi strutturale e paradigma di rete*. Milano: FrancoAngeli.

Diamond, J. 1997. *Guns, Germs, and Steel. The Fates of Human Societies*; It. trans. *Armi, acciaio e malattie. Breve storia del mondo negli ultimi tredicimila anni*. Torino: Einaudi, 1998 (cfr. ed. 2006).

Diamond, J. 2005. *Collapse. How Societies Choose to Fail or Succeed*; It. trans. *Collasso. Come le società scelgono di morire o vivere*. Torino: Einaudi, 2005.

Domingos, P. 2015. *The Master Algorithm. How the Quest for the Ultimate Learning Machine Will Remake Our World*; It. trans. *L'Algoritmo definitivo. La macchina che impara da sola e il future del nostro mondo*. Torino: Bollati Boringhieri, 2016.

Dominici, P. 1995–1996. *Per un'etica dei new-media. Elementi per una discussione critica*. Firenze: Firenze Libri, 1998.

Dominici, P. 2003. *La comunicazione nella società ipercomplessa. Istanze per l'agire comunicativo e la condivisione della conoscenza*. Roma: Aracne, 2005.

Dominici, P. 2008. Sfera pubblica e società della conoscenza. In M. Morcellini and B. Mazza (a cura di), *Oltre l'individualismo. Comunicazione, nuovi diritti e capitale sociale*. Milano: Franco Angeli.

Dominici, P. 2010. *La società dell'irresponsabilità*. Milano: FrancoAngeli.

Dominici, P. 2011. *La comunicazione nella società ipercomplessa. Condividere la conoscenza per governare il mutamento*. Milano: FrancoAngeli.

Dominici, P. 2014a. *Dentro la società interconnessa. Prospettive etiche per un nuovo ecosistema della comunicazione*. Milano: FrancoAngeli.

Dominici, P. 2014b. La modernità complessa tra istanze di emancipazione e derive dell'individualismo. *Studi di Sociologia*, n° 3.

Dominici, P. 2015a. Communication and Social Production of Knowledge. A New Contract for the Society of Individuals. *Comunicazioni Sociali*, n° 1.

Dominici, P. 2015b. Il nuovo ecosistema della comunicazione e le sfide della responsabilità/ipercomplessità. *DESK. Rivista Trimestrale di cultura dell'Informazione*, n° 4.

Dominici, P. 2016a. La filosofia come "dispositivo" di risposta alla società asimmetrica e ipercomplessa. In *Il diritto alla filosofia. Quale filosofia nel terzo millennio?*, ed. L. Candiotto and F. Gambetti. Bologna: Diogene Multimedia.

Dominici, P. 2016c. *Il grande equivoco. Ripensare l'educazione (#digitale) per la Società Ipercomplessa* [The Great Mistake. Rethinking Education for the Hypercomplex Society. Milano: Fuori dal Prisma, Il Sole 24 Ore.

Dominici, P. 2016d. L'anello debole e le reti "fuori" dalla Rete: ripensare la cittadinanza nella Società Interconnessa. In *La Rete e il fattore C. Cultura, complessità, collaborazione*, ed. E. Pietrafesa, F. Marzano, T. Medici. Roma: Stati Generali dell'Innovazione.

Dominici, P. 2016e. L'utopia Post-Umanista e la ricerca di un Nuovo Umanesimo per la Società Ipercomplessa. *Comunicazioni Sociali*, n° 3.

Dominici, P. 2017a. The Hypercomplex Society and the Development of a New Global Public Sphere: Elements for a Critical Analysis. *RAZÓN Y PALABRA* 21 (2_97, Abril-junio): 380–405.

Dominici, P. 2017b. L'ipercomplessità, l'educazione e la condizione dei saperi nella Società Interconnessa/iperconnessa. *Il Nodo. Per una pedagogia della Persona*, Anno XXI, 47:81–104.

Dominici, P. 2017c. For an Inclusive Innovation. Healing the Fracture Between the Human and the Technological. *European Journal of Future Research* 6:3, 2018.

Dominici, P. 2017d. Oltre la libertà … di "essere sudditi". In F. Varanini (a cura di), *Corpi, menti, macchine per pensare*, Casa della Cultura, Anno 2, numero 4. Milano.

Dominici, P. 2017e. The Hypertechnological Civilization and the Urgency of a Systemic Approach to Complexity. A New Humanism for the Hypercomplex Society. In *Governing Turbolence. Risk and Opportunities in the Complexity Age*, ed. A. B. F. Neto, E. C. Nieto, G. Chiodi, and S. Petroccia. Cambridge: Cambridge Scholars Publishing.

Dominici, P. 2018a. *La complessità della complessità e l'errore degli errori*. Treccani: Enciclopedia Italiana "Treccani". http://www.treccani.it/magazine/lingua_italiana/speciali/digitale/5_Dominici.html

Dominici, P. 2018b. Hyper-Technological Society? There's No Need for Technicians, but for "Hybrid Figures" (1995). *Morning Future*. https://www.morningfuture.com/en/article/2018/02/16/job-managers-of-complexity-piero-dominici/230/

Dominici, P. 2019a. The Struggle for a Society of Responsibility and Transparency: The Core Question of Education and Culture. In E. Carloni, and D. Paoletti (a cura di), *Preventing Corruption Through Administrative Measures*, European Union Programme Hercule III (2014–2020). Perugia: European Commission, ANAC, Morlacchi, 2019.

Dominici, P. 2019b. *A.A.A. cercansi manager della complessità* [Complexity Manager], Business People.

Dominici, P. 2019c. *Dentro la Società Interconnessa. La cultura della complessità per abitare i confini e le tensioni della civiltà ipertecnologica*. Milano: FrancoAngeli.

Dominici, P. 2019d. Controversies on Hypercomplexity and on Education in the Hypertechnological Era. In *Controversies in the Contemporary World*, ed. A. Fabris and G. Scarafile. Amsterdam/Philadelphia: John Benjamins Publishing Company.

Dominici, P. 2020a. Educating for the Future in the Age of Obsolescence. *CADMUS* 4 (3): 93–109.

Dominici, P. 2020b. Dentro la "civiltà senza errore". Sintesi complesse: l'Umano, il legame sociale e la civiltà ipertecnologica. In *A. De Cesaris (a cura di), Vite digitali. Essere umani nella società del XXI Secolo*. Milano: FrancoAngeli.

Dominici, P. 2020c. Comunicação é cidadania [Communication is Citizenship]. In *Comunicação, Cidadania e Movimentos Sociais: Vivências*, ed. C. Kraus Luvizotto and I. Ferin Cunha, 12–22. Aveiro: Ria Editorial.

Dominici, P. 2020d. Comunicação é complexidade [Communication Is Citizenship]. In *Comunicação, Cidadania e Movimentos Sociais: Perspectivas contemporâneas da participação cidadã*, ed. C. Kraus Luvizotto and I. Ferin Cunha. Aveiro: 1a edição, Ria Editorial.

Dominici, P. 2020e. The Complexity of Communication. The Communication of Complexity. *MATRIZes* 14 (2), May–August, São Paulo, Brasil: 15–19.

Dominici, P. 2021. From Below: Roots and Grassroots of Societal Transformation. *The Social Construction of Change, CADMUS Journal*. http://cadmusjournal.org/article/volume-4/issue-5/essay5-social-construction-change.

Dominici, P. 2022a. The Weak Link of Democracy and the Challenges of Educating Towards Global Citizenship. *Prospects*. Springer Nature, UNESCO.

Dominici, P. 2022b. The Digital Mockingbird: Anthropological Transformation and the "New Nature". *World Futures. The Journal of New Paradigm* 78 (6): 1–29. Routledge, Taylor & Francis, February.

Dominici, P. 2022c. Beyond the Darkness of Our Age. For a Non-Mechanistic View of Complex Organization as Living Organisms. *Rivista Trimestrale di Scienze dell'Amministrazione. Studi di Teoria e ricerca sociale*, n° 2. http://rtsa.eu/RTSA_2_2022_Dominici.pdf?fs=e&s=cl.

Dominici, P. 2022d. Beyond Black Swans. Managing Complexity: A Contradiction in Terms? In *WOSC 2021. Lecture Notes in Networks and Systems*, ed. I. Perko, R. Espejo, V. Lepskiy, and D. A. Novikov, vol. 495. Cham: Springer Nature.

Dominici, P., and J. M. Flores. 2022. La Gran Equivocación: Replantear la educación y la formación virtual para la "sociedad hipercompleja". *Comunicación y Hombre*, Número 18.

Donati, P. 2008. *Oltre il multiculturalismo. La ragione relazionale per un mondo comune*. Roma-Bari: Laterza.

Donati, P. 2011. *Sociologia della riflessività. Come si entra nel dopo-moderno*. Bologna: il Mulino.

Douglas, M. 1985. *Risk Acceptability According to the Social Sciences*. New York: Routledge; It. trans. *Come percepiamo il pericolo. Antropologia del rischio*. Milano: Feltrinelli, 1991.

Douglas, M. 1992. *Risk and Blame: Essays in Cultural Theory*; It. trans. *Rischio e colpa*. Bologna: il Mulino, 1996.

Dreyfus, H. L. 1972. *What Computers Can't Do. The Limits of Artificial Intelligence*; It. trans. *Che cosa non possono fare i computer*. Roma: Armando, 1988.

Dreyfus, H. L. 1992. *What Computers Still Can't Do. A Critique of Artificial Reason*. New York: Harper & Row.

Dunbar, R. 2009. *La scimmia pensante. Storia dell'evoluzione umana*. Bologna: il Mulino.

Dupré, J. 2001. *Human Nature and the Limits of Science*; It. trans. *Natura umana. Perché la scienza non basta*. Roma-Bari: Laterza, 2007.

Durkheim, É. 1893. *De la division du travail social*; trad. it.: *La divisione del lavoro sociale*. Milano: Comunità, 1971.

Dworkin, R. 1978. *Taking Rights Seriously*; It. trans. *I diritti presi sul serio*. Bologna: il Mulino, 1982.

Dyson, F. 1985. *The Origins of Life*. Cambridge: Cambridge University Press.

Easterly, W. 2013. *The Tyranny of Experts. Economists, Dictators, and The Forgetten Rights of the Poor*; It. trans. *La tirannia degli esperti. Economisti, dittatori e diritti negate dei poveri*. Roma-Bari: Laterza, 2015.

Eder, K. 2007. Cognitive Sociology and the Theory of Communicative Action: The Role of Communication and Language in the Making of the Social Bond. *European Journal of Social Theory* 10 (3): 389–408.

Eibl-Eibesfeldt, I. 1988. *Der Mensch – das riskierte Wesen*; It. trans. *L'uomo a rischio*. Torino: Bollati Boringhieri, 1992.

Eisenstadt, S. N. 1999. *Paradoxes of Democracy*; It. trans. *Paradossi della democrazia. Verso democrazie illiberali?* Bologna: il Mulino, 2002.

Elias, N. 1984. *Über die Zeit. Arbeiten zur Wissenssoziologie* II; It. trans. *Saggio sul tempo*. Bologna: il Mulino, 1986.

Elias, N. 1987. *Die Gesellschaft der Individuen*; It. trans. *La società degli individui*. Bologna: il Mulino, 1990.
Ellul, J. 1954. *La technique ou l'enjeu du siècle*; It. trans. *La tecnica rischio del secolo*. Milano: Giuffré, 1969.
Ellul, J. 1984. Tecnologia. In *Enciclopedia del Novecento*. Roma: Istituto dell'Enciclopedia Italiana.
Emery, F. E. (a cura di). 1969. *Systems Thinking*; It. trans. *La teoria dei sistemi. Presupposti, caratteristiche e sviluppi del pensiero sistemico*. Milano: FrancoAngeli, 2001.
Enriques, F. 1906. *Problemi della scienza*. Bologna: Zanichelli, 1985.
Enriques, F. 1932. *Il significato della storia del pensiero scientifico*. Bologna: Zanichelli.
Epstein, J. M., and L. Axtell. 1994. *Growing Artificial Societies. Social Science from the Bottom Up*. Cambridge, MA: The MIT Press.
Eriksen, T. H. 2016. *Overheating. An Anthropology of Accelerated Change*; It. trans. *Fuori controllo. Un'antropologia del cambiamento accelerato*. Torino: Einaudi, 2017.
Esposito, R. 1993. *Nove pensieri sulla politica*. Bologna: il Mulino.
Faggin, F. 2022. *Irriducibile. La coscienza, la vita, i computer e la nostra natura*. Milano: Mondadori.
Featherstone, M. 1995. *Undoing culture: Globalization, postmodernism and identity*; It. trans. *La cultura dislocata. Globalizzazione, postmodernismo, identità*. Roma: SEAM, 1998.
Federici, M. C., and M. Picchio. 2010. *La dimensione incrociata dell'individuo e della società*. Roma: Aracne.
Feng, J., et al. 2021. Psychological Impact During the First Outbreak of COVID-19 on Frontline Health Care Workers in Shanghai. *Frontiers in Public Health* 9:646780.
Ferragina, P., and F. Luccio. 2017. *Il Pensiero computazionale. Dagli algoritmi al coding*. Bologna: il Mulino.
Ferrara, A. 1991. Etica e teoria sociale: tre punti di intersezione. In F. Crespi (a cura di), *Etica e scienze sociali*. Torino: Rosenberg & Sellier.
Ferrara, A. 1998. *Reflective Authenticity, Rethinking the Project of Modernity*; It. trans. *Autenticità riflessiva. Il progetto della modernità dopo la svolta linguistica*. Milano: Feltrinelli, 1999.
Ferrara, A. (a cura di). 2000. *Comunitarismo e liberalismo*. Roma: Editori Riuniti.
Ferrara, A. 2001. *Pensare la società. L'idea di una filosofia sociale*. Roma: Carocci.
Ferraris, M. 2007. *Sans Papier. Ontologia dell'attualità*. Milano: Castelvecchi.
Ferraris, M. 2017. *Postverità e altri enigmi*. Bologna: il Mulino.
Ferraris, M. 2021. *Documanità. Filosofia del mondo nuovo*. Roma-Bari: Laterza.
Ferrarotti, F. 1989. *La sociologia alla riscoperta della qualità*. Bari: Laterza.
Ferrarotti, F. 1990. *Una fede senza dogmi*. Roma-Bari: Laterza.
Ferrarotti, F. 1991. *Trattato di sociologia*. Torino: UTET.
Ferrarotti, F. 1997. *La perfezione del nulla. Promesse e problemi della rivoluzione digitale*. Roma-Bari: Laterza.
Feyerabend, P. K. 1975. *Against Method. Outline of an Anarchistic Theory of Knowledge*; It. trans. *Contro il metodo. Abbozzo di una teoria anarchica della conoscenza*. Milano: Feltrinelli, 1979.
Feynman, R. P. 1963. *Six Easy Pieces*; It. trans. *Sei pezzi facili*. Milano: Adelphi, 2000.
Feynman, R. P. 1985. *QED. The Strange Theory of Light and Matter*; It. trans. *QED. La strana teoria della luce e della materia*. Milano: Adelphi, 1989.
Finn, E. 2017. *What Algorithms Want: Immagination in the Age of Computing*; It. trans. *Che cosa vogliono gli algoritmi. L'immaginazione nell'era dei computer*. Torino: Einaudi, 2018.
Fischer, F. 1990. *Technocracy and the Politics of Expertise*. Newbury Park [etc.]: Sage.
Fischhoff, B. 2002. Risk Perception, Risk Communication, Risk Taking. *Journal of Psychology and Financial Markets* 3:102–111.
Fischhoff, B. 2009. Risk Perception and Communication. In *Oxford Textbook of Public Health*, ed. R. Detels, R. Beaglehole, M. A. Lansang, and M. Gulliford, 5th ed., 940–952.
Fisichella, D. 1997. *L'altro potere: tecnocrazia e gruppi di pressione*. Roma-Bari: Laterza.

Fleck, L.. 2019. *Stili di pensiero. La conoscenza scientifica come creazione sociale*. Italian translation by Francesco Coniglione. Milano: Mimesis.

Flichy, P. 1991. *Une Histoire de la Communication Moderne*; It. trans. *Storia della comunicazione moderna. Sfera pubblica e dimensione privata*. Bologna: Baskerville, 1994.

Focault, M. 1976. *Histoire de la sexualité 1. La Volonté de savoir*; trad. it. *La volontà di sapere*. Milano: Feltrinelli, 1991.

Fodor, J. A. 2000. *The Mind Doesn't Work That Way. The Scope and Limits of Computational Psychology*. The MIT Press; It. trans. *La mente non funziona così*. Roma-Bari: Laterza, 2000.

Fornari, F. 2014. *Il baule di Newton. La sociologia e la sfida della complessità*. Perugia: Morlacchi.

Foucault, M. 1975. *Surveiller et punir. Naissance de la prison*; It. trans. *Sorvegliare e punire. Nascita della prigione*. Torino: Einaudi, 1976.

Foucault, M. 1988. *Technologies of the Self. A Seminar with Michel Foucault*; It. trans. *Tecnologie del Sé. Un seminario con Michel Foucault*. Torino: Bollati Boringhieri, 1992.

Foucault, M. 2004. *Sécurité, territoire, population. Cours au Collège de France 1977–1978*; trad. it. *Sicurezza, territorio, popolazione*, a cura di M. Senellart. Milano: Feltrinelli, 2005.

Foucault, M. 2005. *Nascita della biopolitica. Corso al Collège de France 1978–1979*. Milano: Feltrinelli.

Foucault, M. 2008. *The Birth of Biopolitics: Lectures at the Collége de France, 1978–1979*. Basingstoke: Palgrave Macmillan.

Fourier, J. 1822. *Théorie analytique de la chaleur*, English translation by *Analytical Theory of Heat*. Chicago: Encyclopedia Britannica, 1952.

Francesco, Papa. 2015. *Lettera Enciclica LAUDATO SÍ*. Città del Vaticano: Libreria Editrice Vaticana.

Frauenfelder, E., O. De Sanctis, and E. Corbi. 2011. *Civitas Educationis. Interrogazioni e sfide pedagogiche*. Napoli: Liguori.

Freire, P. 1968. *La pedagogia degli oppressi*. Torino: Gruppo Abele, 2011.

Freire, P. 2004. *Pedagogia dell'autonomia. Saperi necessari per la pratica educativa*. Torino: EGA.

Friedman, J. 2019. *Power Without Knowledge: A Critique of Technocracy*. Oxford: Oxford University Press.

Frisby, D. 1985. *Fragments of Modernity. Theories of Modernity in the Work of Simmel, Kracauer and Benjamin*; It. trans. *Frammenti di modernità. Simmel, Kracauer, Benjamin*. Bologna: il Mulino, 1992.

Fry, H. 2018. *Hello World. How to Be Human in the Age of the Machine*; It. trans. *Hello World. Essere umani nell'era delle macchine*. Torino: Bollati Boringhieri, 2019.

Furedi, F. 2004. *Therapeutic Culture. Cultivating Vulnerability in an Uncertain Age*; It. trans. *Il nuovo conformismo. Troppa psicologia nella vita quotidiana*. Milano: Feltrinelli, 2005.

Gadamer, H. G. 1976. *Vernunft im Zeitalter Wissenschaft*; It. trans. *La ragione nell'età della scienza*. Genova: Il Melangolo, 1999.

Galimberti, U. 1999. *Psiche e techne*. Milano: Feltrinelli.

Galli, C. 2011. *Il disagio della democrazia*. Torino: Einaudi.

Gallino, L. 1978. *Dizionario di Sociologia*. Milano: Utet.

Gallino, L. 1992. *L'incerta alleanza. Modelli di relazioni tra scienze umane e scienze naturali*. Torino: Einaudi.

Gallino, L. 2000. *Globalizzazione e disuguaglianze*. Roma-Bari: Laterza.

Gallino, L. 2001. *Il costo umano della flessibilità*. Roma-Bari: Laterza.

Gammaitoni, L., and A. Vulpiani. 2019. *Perché è difficile prevedere il futuro. Il sogno più sfuggente dell'uomo sotto la lente della fisica*. Bari: Edizioni Dedalo.

Gandolfi, A. 2008. *Formicai, imperi, cervelli. Introduzione alla scienza della complessità*. Torino: Bollati Boringhieri.

Ganesan, B., et al. 2021. Impact of Coronavirus Disease 2019 (COVID-19) Outbreak Quarantine, Isolation, and Lockdown Policies on Mental Health and Suicide. *Frontiers in Psychiatry* 12:565190.

Gardner, H. 1991. *The Unschooled Mind. How Children Think and How Schools Should Teach*; It. trans. *Educare al comprendere. Stereotipi infantile e apprendimento scolastico*. Milano: Feltrinelli, 1993.
Gazzaniga, M. S. 2008. *Human*; It. trans. *Human. Quel che ci rende unici*. Milano: Cortina, 2009.
Gell-Mann, M. 1994. *The Quark and the Jaguar. Adventures in the Simple and the Complex*; It. trans. *Il quark e il giaguaro. Avventura nel semplice e nel complesso*. Torino: Bollati Boringhieri, 1996–2017.
Gell-Mann, M. 1995. *Complexity*. New York: Wiley.
Gembillo, G., and A. Anselmo. 2013a. *Filosofia della Complessità*. Firenze: Le Lettere.
Gembillo, G., and A. Anselmo. 2013b. *Filosofia complessa e sociologia qualitativa*. Roma: Aracne.
Gentili, P. L. 2018. *Untangling Complex Systems: A Grand Challenge for Science*. Boca Raton: CRC Press, Taylor & Francis Group.
Ghirardi, G. C. 2018. *Simmetrie nell'arte e nella scienza*. Roma: Carocci.
Giddens, A. 1971. *Capitalism and Modern Social Theory*; It. trans. *Capitalismo e teoria sociale. Marx, Durkheim, Weber*. Milano: Il Saggiatore, 1975.
Giddens, A. 1990. *The Consequences of Modernity*; It. trans. *Le conseguenze della modernità*. Bologna: il Mulino, 1994.
Giddens, A. 1991. *Modernity and Self-Identity: Self and Society in the Late Modern Age*. London: Polity Press.
Giddens, A. 1994. Vivere in una società post-tradizionale. In U. Beck, A. Giddens, and S. Lash, 101–159.
Giddens, A. 1999. *Runaway World. How Globalization Is Reshaping Our Lives*. London: Profile Books; It. trans. *Il mondo che cambia. Come la globalizzazione ridisegna la nostra vita*. Bologna: il Mulino, 2000.
Giddens, A. 2000. *The Third Way and Its Critics*; It. trans. *Cogliere l'occasione. Le sfide di un mondo che cambia*. Roma: Carocci, 2000.
Giddens, A. 2007. *Europe in the Global Age*; It. trans. *L'Europa nell'età globale*. Roma-Bari: Laterza, 2007.
Gigerenzer, G. 2014. *Risk Savvy. How to Make Good Decisions*; It. trans. *Imparare a rischiare. Come prendere decisioni giuste*. Milano: Cortina, 2015.
Gili, G. 1990. *La teoria della società di massa*. Roma: ESI.
Gili, G. 2001. *Il problema della manipolazione. Peccato originale dei media?* Milano: FrancoAngeli.
Gili, G. 2005. *La credibilità. Quando e perché la comunicazione ha successo*. Soveria Mannelli: Rubettino.
Gleick, J. 1987. *Chaos*; It. trans. *Caos*. Milano: Rizzoli, 1989.
Goffman, E. 1959. *The Presentation of Self in Everyday Life*; It. trans. *La vita quotidiana come rappresentazione*. Bologna: il Mulino, 1969
Goodwin, B. 1994. *How the Leopard Changed Its Spots. The Evolution of Complexity*. Princeton: Princeton Science Library.
Graeber, D. *The Utopia of Rules. On Technology, Stupidity and the Secret Joys of Bureaucracy*. New York: Melville House, 2015.
Graeber, D., and D. Wengrow. 2021. *The Dawn of Everything*; It. trans. *L'alba di tutto. Una nuova storia dell'Umanità*. Milano: Rizzoli, 2022.
Gramsci, A. 1917. *Odio gli indifferenti*. Milano: Chiarelettere, 2011.
Gramsci, A. 1948–1951. *Quaderni del carcere*, 1ª ed. Torino: Einaudi, 1975, 4 voll.
Grandi, R. 1992. *I mass media fra testo e contesto*. Milano: Lupetti, 1994.
Granovetter, M. 1973. The Strength of Weak Ties. *American Journal of Sociology* 78:1360–1380.
Grassi, W. 2018. *Farfalle e uragani. Complessità: la teoria che governa il mondo*. Milano: Hoepli.
Griswold, W. 1994. *Culture and Societies in a Charging World*; It. trans. *Sociologia della cultura*. Bologna: il Mulino, 1997.
Grossman, L. K. 1995. *The Electronic Republic. Reshaping Democracy in the Information Age*; It. trans. *La repubblica elettronica*. Roma: Editori Riuniti, 1997.

Habermas, J. 1962. *Strukturwandel der Oeffentlichkeit*; It. trans. *Storia e critica dell'opinione pubblica*. Bari: Laterza, 1971.

Habermas, J. 1970. The Scientization of Politics and Public Opinion. In *Toward a Rational Society: Student Protest, Science and Politics*. Boston: Beacon.

Habermas, J. 1981. *Theorie des kommunikativen Handelns*, Bd. I *Handlungsrationalität und gesellschaftliche Rationalisierung*, Bd. II *Zur Kritik der funktionalistischen Vernunft*; It. trans. *Teoria dell'agire comunicativo*, Vol. I *Razionalità nell'azione e razionalizzazione sociale*, Vol. II *Critica della ragione funzionalistica*. Bologna: il Mulino, 1986.

Habermas, J. 1985a. *Der philosophische Diskurs der Moderne. Zwölf Vorlesungen*. Frankfurt am Main: SuhrKamp Verlag; It. trans. *Il discorso filosofico della modernità*. Roma-Bari: Laterza, 1987.

Habermas, J. 1985b. *Die Neue Unübersichtlichkeit. Die Krise des Wohlfahrtsstaates und die Erschöpfung utopischer Energien*; It. trans. *La nuova oscurità. Crisi dello stato sociale ed esaurimento delle utopie*. Roma: Lavoro, 1998.

Habermas, J. 1991. *Vergangenheit als Zukunft*; It. trans. *Dopo l'utopia. Il pensiero critico e il mondo d'oggi*. Venezia: Marsilio, 1992.

Habermas, J. 1992. *Faktizität und Geltung. Beiträge zur Diskurstheorie des Rechts und des demokratischen Rechtsstaats*; It. trans. *Fatti e norme. Contributi a una teoria discorsiva del diritto e della democrazia*. Milano: Guerini, 1996.

Habermas, J. 1996. *Die Einbeziehung des Anderen. Studien zur politischen Theorie*; It. trans. *L'inclusione dell'altro. Studi di teoria politica*. Milano: Feltrinelli, 1998.

Habermas, J. 1998. *Die postnationale Konstellation. Politische Essays*; It. trans. *La costellazione postnazionale*. Milano: Feltrinelli, 1999.

Habermas, J. 2001. *Die Zukunft der menschlichen Natur. Auf dem Weg zu einer liberalen Eugenik?* It. trans. *Il future della natura umana. I rischi di una genetica liberale*. Torino: Einaudi, 2002.

Habermas, J. 2004. *Des gespaltene Westen*; It. trans. *L'Occidente diviso*. Roma-Bari: Laterza, 2007.

Habermas, J. 2013. *Im Sog der Technokratie. Kleine Politische Schriften XII*; It. trans. *Nella spirale tecnocratica. Un'arringa per la solidarietà europea*. Roma-Bari: Laterza, 2014.

Habermas, J., and N. Luhmann. 1971. *Theorie der Gesellschaft oder Sozialtechnologie. Was leistet die Systemforschung*; It. trans. *Teoria della società o tecnologia sociale*. Milano: Etas Kompass, 1973.

Haida Nation, Canada. https://www.dfo-mpo.gc.ca/oceans/publications/sk-b-managementplan-plangestion/page04-eng.html.

Haken H. 1977. *Synergetics: An Introduction. Nonequilibrium Phase-Transitions and Self-Organization in Physics, Chemistry, and Biology*, new ed. 1983. Heidelberg: Springer.

Hall, A. D., and R. E. Fagen. 1956. Definition of System. *General Systems* 1:18–28.

Hammersley, M. 2013. *The Myth of Research-Based Policy and Practice*; It. trans. *Il mito dell'evidence-based. Per un uso critico della ricerca sociale applicate*. Milano: Cortina, 2016.

Hannerz, H. 1992. *Cultural Complexity. Studies in the Social Organization of Meaning*; It. trans. *La complessità culturale. L'organizzazione sociale del significato*. Bologna: il Mulino, 1998.

Haraway, D. J. 1991. *Simians, Cyborg and Women: The Reinvention of Nature*. New York: Routledge, (trad. it. di L. Borghi); It. trans. *Manifesto cyborg. Donne, tecnologie e biopolitiche del corpo*. Milano: Feltrinelli, 1995.

Hardt, M., and A. Negri. 2000. *Empire*; It. trans. *Impero. Il nuovo ordine della globalizzazione*. Milano: BUR, 2002.

Hardt, M., and A. Negri. 2004. *Multitude. War and Democracy in the Age of Empire*; It. trans. *Moltitudine. Guerra e democrazia nel nuovo ordine imperiale*. Milano: Rizzoli, 2004.

Hardt, M., and A. Negri. 2009. *Commonwealth*; It. trans. *Comune. Oltre il privato e il pubblico*. Milano: Rizzoli, 2010.

Hartman, S, and P. Degeorges. 2019. "DON'T PANIC": Fear and Acceptance in the Anthropocene. *ISLE – Interdisciplinary Studies in Literature and Environment* 26 (2, Spring): 456–472.

Harvey, D. 1990. *The Condition of Postmodernity*; It. trans. *La crisi della modernità*. Milano: Il Saggiatore, 1993.

Hawkins, M. 1997. *Social Darwinism in European and American Thought, 1860–1945. Nature as Model and Nature as Threat*. Cambridge: Cambridge University Press.
Hedström, P. 2005. *Dissecting the Social. On the Principles of Analytical Sociology*; It. trans. *Anatomia del sociale. Sui principi della sociologia analitica*. Milano: Bruno Mondadori, 2006.
Hegel, G. W. F. 1807. *Phänomenologie des Geistes*. Bamberg-Würzburg; It. trans. *Fenomenologia dello spirito*. Firenze: La Nuova Italia, 1960, 2 voll.
Hegel, G. W. F. 1817. *Enzyklopädie der philosophischen Wissenschaften im Grundrisse*; It. trans. *Enciclopedia delle scienze filosofiche in compendio*. Roma-Bari: Laterza, 1987
Heidegger, M. 1962. *Die Technik und die Kehre*; It. trans. *La svolta*. Genova: Il Melangolo, 1990.
Heidegger, M. 1976. *Brief über den "Humanismus"*; It. trans. *Lettera sull' "Umanismo"*. Milano: Adelphi, 1995.
Heisenberg, W. 1958. *Physics and Philosophy: The Revolution in Modern Science*. Buffalo: Prometheus Books.
Heisenberg, W. 1959, 9* ed. *Wandlungen in den Grundlagen der Naturwissenschaft*; It. trans. *Mutamenti nelle basi della scienza*. Torino: Bollati Boringhieri, 1978.
Heisenberg, W. 1976. *I principi fisici della teoria dei quanti*. Torino: Bollati Boringhieri.
Heisenberg, W. 1999. *Lo sfondo filosofico della fisica moderna*, a cura di G. Gembillo and E. A. Giannetto. Sellerio: Palermo.
Heisenberg, W. 2002. *Indeterminazione e realtà*, a cura di G. Gembillo and G. Gregorio. Napoli: Guida.
Heisenberg, W., et al. 1980. *Discussione sulla fisica moderna*. Torino: Bollati Boringhieri.
Hempel, C. G. 1965. *Aspects of Scientific Explanation*; It. trans. *Aspetti della spiegazione scientifica*. Milano: Il Saggiatore, 1987.
Hess, C., and E. Ostrom. 2007. *Understanding Knowledge as a Commons*; It. trans. *La conoscenza come bene comune. Dalla teoria alla pratica*. Milano: Bruno Mondadori, 2009.
Himanen, P. 2001. *The Hacker Ethic and the Spirit of the Information Age*; It. trans. *L'etica hacker e lo spirito dell'età dell'informazione*. Feltrinelli: Milano, 2001.
Hobsbawn, E. J. 1994. *Age of Extremes. The Short Twentieth Century 1914–1991*; It. trans. *Il Secolo breve. 1914–1991: l'era dei grandi cataclismi*. Milano: Rizzoli, 1995.
Holland, J. H. 1975. *Adaptation in Natural and Artificial Systems*. Ann Arbor: University of Michigan Press.
Honneth, A. 1992. *Kampf um Anerkennung. Grammatik sozialer Konflikte*; It. trans. *Lotta per il riconoscimento. Proposta per un'etica del conflitto*. Milano: Il Saggiatore, 2002.
Hooks, B. 2020. *Teaching to Transgress: Education as a Practice of Freedom*; It. trans. *Insegnare a trasgredire. L'educazione come pratica della libertà*. Milano: Meltemi.
Horkheimer, M. 1947. *Eclipse of Reason*; It. trans. *Eclisse della ragione. Critica della ragione strumentale*. Torino: Einaudi, 1969.
Horkheimer, M., and T. W. Adorno. 1947. *Dialektik der Aufklärung. Philosophische Fragmente*; It. trans. *Dialettica dell'Illuminismo*. Torino: Einaudi, 1966.
Horkheimer, M., and T. W. Adorno (a cura di). 1956. *Soziologische Exkurse*; It. trans. *Lezioni di sociologia*. Torino: Einaudi, 1966.
Hughes, J. A., and W. W. Sharrock. 1997. *The Philosophy of Social Research*; It. trans. *Filosofia della ricerca sociale*. Bologna: il Mulino, 2005.
Huizinga, J. 1939. *Homo ludens*; It. trans. *Homo ludens*. Torino: Einaudi, 1946.
Huntington, S. P. 1993. *The Third Wave. Democratization in the Late Twentieth Century*; It. trans. *La terza ondata. I processi di democratizzazione alla fine del XX secolo*. Bologna: il Mulino, 1995.
Huntington, S. P. 1996. *The Clash of Civilizations and the Remaking of World Order*; It. trans. *Lo scontro delle civiltà e il nuovo ordine mondiale*. Milano: Garzanti, 1997.
Hustvedt, S. 2016. *The Delusions of Certainty*; It. trans. *Le illusioni della certezza*. Torino: Einaudi, 2018.
Huxley, T. H. 1894. *Evolution and Ethics*; It. trans. *Evoluzione ed etica*. Torino: Bollati Boringhieri, 1995.

Illich, I. 1971. *Deschooling Society*; It. trans. *Descolarizzare la società*. Milano: Mimesis, 2019.
Inglehart, R. 1996. *Modernization and Postmodernization. Cultural, Economic and Political Change in 43 Societies*; It. trans. *La società postmoderna. Mutamento, valori e ideologie in 43 paesi*. Roma: Editori Riuniti, 1998.
Ippolita. 2005. *Open non è free. Comunità digitali tra etica hacker e mercato globale*. Milano: Eleuthera.
Israel, G. 2004. *La macchina vivente. Contro le visioni meccanicistiche dell'uomo*. Torino: Bollati Boringhieri.
Israel, G. 2005. The Science of Complexity. Epistemological Problems and Perspectives. *Science in Context* 18, Anno: 1–31.
Izzo, A. 1992. *Storia del pensiero sociologico*. Bologna: il Mulino.
Izzo, A. 1995. *I percorsi della ragione. Il tema della razionalità nella storia del pensiero sociologico*. Roma: Carocci.
Izzo, A. 1999. *L'invincibile perplessità. Fondamenti, storia e problemi della sociologia della conoscenza*. Roma: Armando.
Izzo, A. 2000. *L'anomia. Analisi e storia di un concetto*. Roma-Bari: Laterza.
Jacob, F. 1970. *La logique du vivant*; It. trans. *La logica del vivente*. Torino: Einaudi, 1971.
Jacobs, G., J. Ramanathan, R. Wolff, R. Pricopie, P. Dominici, and A. Zucconi. 2021. A New Paradigm in Global Higher Education for Sustainable Development and Human Security. *CADMUS Journal* IV, November. https://www.cadmusjournal.org/article/volume-4/issue-5/new-paradigm-global-higher-education.
James, W. 1878. *The Brain and the Mind*, Lowell Lectures on 'The Brain and the Mind'. In *Manuscript Lectures. Original Lectures Delivered 1878*, ed. F. Burkhardt, 16–43. Cambridge, MA: Harvard University Press.
James, W. 1887. The Laws of Habit. *The Popular Science Monthly* 30:433–451.
James, W. 1890. *The Principles of Psycology*, 2 voll. New York: Henry Holt & Co.
James, W. 1904. *Essays in Radical Empiricism*; It. trans. *Saggi di empirismo radicale*. Macerata: Quodlibet, 2009.
Jensen, H. J. *Complexity Science. The Study of Emergence*. Cambridge: Cambridge University Press, 2023.
Jonas, H. 1979. *Das Prinzip Verantwortung*; It. trans. *Il principio responsabilità. Un'etica per la civiltà tecnologica*. Torino: Einaudi, 1990.
Josup, M., P. Holme, K. Kanazawa, et al. 2022. Social Physics. *Physics Reports* 948:1–148.
Kahneman, D. 2003. A Perspective on Judgment and Choice. Mapping Bounded Rationality. *American Psychologists* 58:697–720.
Kaku, M. 2018. *The Future of Humanity*; It. trans. *Il futuro dell'umanità*. Milano: Rizzoli, 2018.
Kant, I. 1786. *Was heißt sich im Denken orientieren?*; It. trans. *Che cosa significa orientarsi nel pensiero*. Milano: Adelphi, 1996.
Karlsson, G. 1958. *Social Mechanisms: Studies in Sociological Theory*. Stockholm: Almquist & Wicksell.
Kasperson, J., R. E. Kasperson, N. Pidgeon, and P. Slovic. 2003. The Social Amplification of Risk: Assessing Fifteen Years of Research and Theory. In *The Social Amplification of Risk*, 13–45. Cambridge: Cambridge University Press.
Katz, E. 1957. The Two-Step Flow of Communication: An Up-to-Date Report on an Hypothesis. *Public Opinion Quarterly* 21 (1): 61–78.
Katz, E., and P. Lazarsfeld. 1955. *Personal Influence: The Part Played by People in the Flow of Mass Communications*; It. trans. *L'influenza personale nelle comunicazione di massa*. Torino: Eri, 1968.
Kauffman, S. A. 1971. Gene Regulation Networks. A Theory for Their Global Structure and Behaviours. *Current Topics in Developmental Biology* 6:145–182.
Kauffman, S. A. 1993. *The Origin of Order. Self-organization and Selection in Evolution*. Oxford: Oxford University Press.

Kelly, K. 2010. *What Technology Wants*; It. trans. *Quello che vuole la tecnologia*. Torino: Codice, 2011.
Kiel, L. D. 1994. *Managing Chaos and Complexity in Government*. San Francisco: JoseyBass.
Klein, É. 2018. *Matière à contredire. Essai de philo-physique*; It. trans. *Filosofisica*. Roma: Carocci, 2020.
Klein, S. 2019. *Il tutto e il niente. Sulla bellezza dell'Universo*. Torino: Bollati Boringhieri.
Kohlberg, L. 1981. *Essays on Moral Development, Vol. I: The Philosophy of Moral Development*. San Francisco: Harper & Row, 1981; Vol. II: *The Psychology of Moral Development*. San Francisco: Harper & Row, 1984.
Kohn, A. 1986. *No Contest*; It. trans. *La fine della competizione*. Milano: Baldini & Castoldi, 1999.
Kohn, E. 2013. *How Forests Think: Toward an Anthropology Beyond the Human*; It. trans. *Come pensano le foreste*. Milano: Nottetempo, 2021.
Kojève, A. 1931–1932. *L'idée du determinisme dans la physique classique et dans la physique modern*; It. trans. *L'idea di determinismo nella fisica classica e nella fisica moderna*. Milano: Adelphi, 2018.
Koyré, A. 1944. *Entretiens sur Descartes*; It. trans. *Lezioni su Cartesio*. Milano: Jouvence, 2022.
Koyré, A. 1961. *Les philosophes et la machine. Du monde de l'"à-peu-près" à l'universede la précision*; It. trans. *Dal mondo del pressappoco all'universo della precisione*. Torino: Einaudi, 1967.
Krugman, P. 1996. *The Self-Organizing Economy*. Oxford: Blaclwell; It. trans. *Economia e auto-organizzazione*. Milano: Giuffrè, 2000.
Kuhlmann, M. 2013. What Is Real? *Scientific American* 209 (2): 40.7.
Kuhn, T. 1962. *The Structure of Scientific Revolution*; It. trans. *La struttura delle rivoluzioni scientifiche*. Torino: Einaudi, 1969.
Kumar, K. 1995. *From Post-Industrial to Post-Modern Society. New Theories of the Contemporary World*; It. trans. *Le nuove teorie del mondo contemporaneo. Dalla società post-industriale alla società post-moderna*. Torino: Einaudi, 2000.
Kupiec, J.-J. 2021. *Et si le vivant était anarchique*; It. trans. *La concezione anarchica del vivente*. Milano: Eleuthera, 2021.
Lakatos, I., and A. Musgrave. 1970. *Criticism and the Growth of Knowledge*; It. trans. *Critica e crescita della conoscenza*. Milano: Feltrinelli, 1976.
Landes, D. S. 1969. *The Unbound Prometeus*; It. trans. *Prometeo Liberato. Trasformazioni tecnologiche e sviluppo industriale nell'Europa occidentale dal 1750 ai giorni nostri*. Torino: Einaudi, 1978.
Lanza, R., Pavšič, M., and Berman, B. 2020. *The Grand Biocentric Design*; It. trans. *Il grande disegno biocentrico. Come la vita crea la realtà*. Milano: Il Saggiatore, 2022.
Lash, S. 1990. *The Sociology of Postmodernism*; It. trans. *Modernismo e postmodernismo. I mutamenti culturali delle società complesse*. Roma: Armando, 2000.
Laszlo, E. 1996. *The Systems View of the World: A Holistic Vision for Our Time*. Cresskill: Hampton Press.
Latouche, S. 1989. *L'occidentalisation du monde. Essai sur la signification, la portée et les limites de l'uniformisation planétaire*; It. trans. *L'occidentalizzazione del mondo. Saggio sul significato, la portata e i limiti dell'uniformazione planetaria*. Torino: Bollati Boringhieri, 1992.
Latouche, S. 1995. *La Megamachine. Raison techno-scientifique, raison économique et le mythe du Progrès. Essais à la mémoire de Jacques Ellul*; It. trans. *La Megamacchina. Ragione tecnoscientifica, ragione economica e mito del progresso*. Torino: Bollati Boringhieri, 1995.
Latour, B. 1987. *Science in Action. How to Follow Scientists and Engineers through Society*; It. trans. *La scienza in azione. Introduzione alla sociologia della scienza*. Torino: Comunità, 1998.
Latour, B. 2005. *Reasssembling the Social*; It. trans. *Riassemblare il Sociale*. Milano: Meltemi, 2022.
Lazarsfeld, P., B. Berelson, and H. Gaudet. 1944. *The People Choice. How the Voter Makes Up His Mind in a Presidential Campaign*. New York: Columbia University Press.
Le Moigne, J.-L. 1977. *La théorie du système général. Théorie de la modelisation*. Paris: Presses Universitaires.
Leibniz, G. W. 1967. *Scritti filosofici*, (a cura di D. O. Bianca). Torino: UTET.

Levi, Montalcini R. 1990. *Elogio dell'imperfezione*. Milano: Garzanti.

Lévinas, E. 1961. *Totalité et infini. Essay sur l'extériorité*; It. trans. *Totalità e infinito. Saggio sull'esteriorità*. Milano: Jaca Book, 1977.

Lévinas, E. 1972. *Humanisme de l'autre homme*; It. trans. *Umanesimo dell'altro Uomo*. Genova: Il melangolo, 1985.

Lèvy, P. 1992. *Les Technologies de l'intelligence*; It. trans. *Le tecnologie dell'intelligenza*. Bologna: Synergon.

Lévy, P. 1994. *L'Intelligence collective: pour une anthropologie du cyberspace*; It. trans. *L'intelligenza collettiva. Per un'antropologia del cyberspazio*. Milano: Feltrinelli, 1996.

Lévy, P. 1995. *Qu'est-ce que le virtuel*?; It. trans. *Il virtuale*. Milano: Cortina, 1997.

Lévy, P. 1997. *Cyberculture. Rapport au Conseil de l'Europe*; It. trans. *Cybercultura. Gli usi sociali delle nuove tecnologie*. Milano: Feltrinelli, 1999.

Lingua, G. (a cura di). 2003. *Comunicare senza regole? Etica e mass-media nella società globale*. Milano: Medusa.

Lippman, W. 1922. *Public Opinion*; It. trans. *L'opinione pubblica*. Roma: Donzelli, 1995.

Longo, G. 2014. *Il simbionte. Prove di umanità futura*. Milano: Mimesis.

Lorenz, E. N. 1963. *The Essence of Chaos*. Seattle: University of Wash Press.

Lorenz, E. 1969. Three Approaches to Atmospheric Predictability. *Bulletin of the American Meteorological Society* 50:5.

Losee, J. 1972. *A Historical Introduction to the Philosophy of Science*; It. trans. *Filosofia della Scienza. Un'introduzione*. Milano: Il Saggiatore, 2001.

Lovelock, J. 1979. *Gaia. A New Look at Life on Earth*; It. trans. *Gaia. Nuove idee sull'ecologia*. Torino: Bollati Boringhieri, 1981.

Lovink, G. 2016. *Social Media Abyss*; It. trans. *L'abisso dei social media. Nuove reti oltre l'economia dei like*. Milano: Università Bocconi.

Luhmann, N. 1968. *Vertrauen. Ein Mechanismus der Reduktion sozialer Komplexität*; It. trans. *La fiducia*. Bologna: il Mulino, 2002.

Luhmann, N. 1975. *Macht*; It. trans. *Potere e complessità sociale*. Milano: Il Saggiatore, 1979.

Luhmann, N. 1981a. Communication About Law in Interaction Systems. In *Advances in Social Theory and Methodology: Toward an Integration of Micro- and Macro-Sociologies*, ed. Karin Knorr-Cetina and Aaron V. Cicourel, 234–256. Boston: Routledge and Kegan Paul/London and Henley.

Luhmann, N. 1981b. The Improbability of Communication. *International Social Science Journal* 23 (1): 122–132.

Luhmann, N. 1982. The World Society as a Social System. *International Journal of General Systems* 8:131–138; in R. Felix Geyer and Johannes van der Zouwen, eds. *Dependence and Inequality: A Systems Approach to the Problems of Mexico and other Developing Countries*, 295–306. Oxford, 1982.

Luhmann, N. 1984a. *Soziale Systeme*; It. trans. *Sistemi sociali. Fondamenti di una teoria generale*. Bologna: il Mulino, 1990.

Luhmann, N. 1984b. The Self-Description of Society: Crisis Fashion and Sociological Theory. *International Journal of Comparative Sociology* 25:59–72.

Luhmann, N. 1986. *Okologische Kommunikation. Kann die moderne Gesellschaft sich auf ökologische Gefährdungen einstellen?*; It. trans. *Comunicazione ecologica, Può la società moderna adattarsi alle minacce ecologiche?* Milano: FrancoAngeli, 1989.

Luhmann, N. 1987. The Morality of Risk and the Risk of Morality. *International Review of Sociology* 3:87–101.

Luhmann, N. 1988. *Erkenntnis als Konstruktion*; It. trans. *Conoscenza come costruzione*. Roma: Armando, 2007.

Luhmann, N. 1990. The Autopoiesis of Social Systems. In *Essays on Self-Reference*, ed. N. Luhmann. New York: Colombia University Press.

Luhmann, N. 1991. *Soziologie des Risikos*; It. trans. *Sociologia del rischio*. Milano: Bruno Mondadori, 1996.

Luhmann, N. 1992a. *Beobachtungen der Moderne*; It. trans. *Osservazioni sul moderno*. Roma: Armando, 1995.
Luhmann, N. 1992b. What Is Communication? *Communication Theory* 2:251–259.
Luhmann, N. 1996. The Sociology of the Moral and Ethics. *International Sociology* 11 (1): 27–36.
Luhmann, N. 1997. Globalization or World Society: How to Conceive of Modern Society? *International Review of Sociology* 7 (1): 67–79.
Luhmann, N., and R. De Giorgi. 1991. *Teoria della società*. Milano: FrancoAngeli.
Lupton, D. 1999. *Risk*; It. trans. *Il rischio. Percezione, simboli, culture*. Bologna: il Mulino, 2003.
Lupton, D., ed. 1999b. *Risk and Sociocultural Theory: New Directions and Perspectives*. Cambridge: Cambridge University Press.
Lupton, D. 2001. *Surveillance Society. Monitoring Everyday Life*; It. trans. *La società sorvegliata. Tecnologie di controllo della vita quotidiana*. Milano: Feltrinelli, 2002.
Lupton, D., and J. Tulloch. 2003. *Risk and Everyday Life*. London: Sage.
Lynch, M. P. 2004. *True to Life. Why Truth Matters*; It. trans. *La verità e i suoi nemici*. Milano: Cortina, 2007.
Lyon, D. 1994. *The Electronic Eye. The Rise of Surveillance Society*; It. trans. *L'occhio elettronico. Privacy e filosofia della sorveglianza*. Milano: Feltrinelli, 1997.
Lyon, D. 2001. *Surveillance Society. Monitoring Everyday Life*; It. trans. *La società sorvegliata. Tecnologie di controllo della vita quotidiana*. Milano: Feltrinelli, 2002.
Lyotard, J. F. 1979. *La condition postmoderne. Rapport sur le savoir*; It. trans. *La condizione postmoderna*. Milano: Feltrinelli, 1982.
Mach, E. 1977. *La meccanica esposta nel suo sviluppo storico-critico*. Torino: Bollati Boringhieri.
Mach, E. 1982. *Erkenntinis und Irrtum*; It. trans. *Conoscenza ed errore*. Milano: Mimesis, 2017.
Maffei, L. *In Praise of Rebellion*; It. trans. *Elogio della ribellione*. Bologna: il Mulino, 2016.
Maffettone, S. 2001. *Etica pubblica. La moralità delle istituzioni nel terzo millennio*. Milano: Il Saggiatore.
Magatti, M. 2009. *Libertà immaginaria. Le illusioni del capitalismo tecno-nichilista*. Milano: Feltrinelli.
Magatti, M., and C. Giaccardi. 2014. *Generativi di tutto il mondo unitevi. Manifesto per la società dei liberi*. Milano: Feltrinelli.
Magatti, M., and C. Giaccardi. 2022. *Supersocietà. Ha ancora senso scommettere sulla libertà?* Bologna: il Mulino.
Maldonado, T. 1997. *Critica della ragione informatica*. Milano: Feltrinelli.
Maldonado, T. 2005. *Memoria e conoscenza. Sulle sorti del sapere nella prospettiva digitale*. Milano: Feltrinelli.
Mandelbrot, B. B. 1977. *Fractals: Forms, Chance and Dimensions*. San Francisco: WH Freeman.
Manghi, S. 2000. *Il gatto con le ali. Tre saggi per un'ecologia delle pratiche sociali*. Trieste: Asterios.
Manghi, S. 2004. *La conoscenza ecologica. Attualità di Gregory Bateson*. Milano: Cortina.
Mannheim, K. 1929. *Ideologie und Utopie*; It. trans. *Ideologia e utopia*. Bologna: il Mulino, 1957.
Mannheim, K. 1935. *L'uomo e la società in un'età di ricostruzione*. Milano: Comunità, 1959.
Mannheim, K. 1952. *Essays on the Sociology of Knowledge*; It. trans. *Sociologia della conoscenza*. Bari: Dedalo, 1974.
Margulis, L. 1993. *Symbiosis in Cell Evolution in the Archean and Proterozoic Eons*. New York: Freeman.
Maritain, J. 1947. *La persona e il bene comune*. Brescia: Morcelliana, 2009.
Marshall, T. H. 1950. *Citizenship and Social Class and Other Essays*. Cambridge: Cambridge University Press, 2002.
Marx, K. 1867 e 1885–1894. *Das Kapital. Kritik der politischen Ökonomie*, 3 voll.; It. trans. *Il capitale. Critica dell'economia politica*, a cura di D. Cantimori, R. Panzieri, and M. L. Boggeri. Roma: Editori Riuniti, 1951–1956.
Marx, K., and F. Engels. 1845–1846. *Die Deutsche Ideologie. Kritik der neuesten deutschen Philosophie in ihren Repräsentanten Feuerbach, B. Bauer und Stirner, und des deutschen*

Sozialismus in seinen verschiedenen Propheten; It. trans. *L'ideologia tedesca. Critica della più recente filosofia tedesca nei suoi rappresentanti Feuerbach, B. Bauer e Stirner, e del socialismo tedesco nei suoi diversi profeti*. Roma: Editori Riuniti, 1972

Marx, K., and F. Engels. 1848. *Manifest der Kommunistischen Partei*; It. trans. *Manifesto del Partito Comunista*. Torino: Einaudi, 1998.

Mastrocola, P., and L. Ricolfi. 2022. *Manifesto del libero pensiero*. Milano: La Nave di Teseo.

Mathews, K. M., M. C. White, and R. G. Long. 1999. Why Study the Complexity Sciences in the Social Sciences? *Human Relations* 25 (April 1): 439–461.

Mattelart, A. 1996. *La mondialisation de la communication*; It. trans. *La comunicazione globale*. Roma: Editori Riuniti, 1998.

Maturana, H., and F. Varela. 1972, *Autopoiesis and Cognition. The Realization of the Living*; It. trans. *Autopoiesi e cognizione. La realizzazione del vivente*. Venezia: Marsilio, 1985.

Maturana, H. R., and F. J. Varela. 1985. *The Tree of Knowledge*; It. trans. *L'albero della conoscenza*. Milano: Garzanti, 1987.

Mayer, V., and Cukier K. Schönberger. 2013. *Big Data*; It. trans. *Big data. Una rivoluzione che trasformerà il nostro modo di vivere e già minaccia la nostra libertà*. Milano: Garzanti, 2013.

McCall, R., and J. Burge. 2016. Untangling Wicked Problems. *AI EDAM* 30:200–210.

McDowell, J. 1994. *Mind and World*. Harvard University Press; It. trans. *Mente e mondo*. Torino: Einaudi, 1999.

McFadden, J. 2021. *Life is Simple. How Occam's Razor Set Science Free and Unlocked the Universe*; It. trans. *La vita è semplice. Come il Rasoio di Occam ha liberato la Scienza e modellato l'universo*. Torino: Bollati Boringhieri, 2021.

McLuhan, M., and B. R. Powers. 1989. *The Global Village: Transformations in World Life and Media in the 21rst Century*; It. trans. *Il villaggio globale: XXI secolo*. Varese: SugarCo., 1992.

Mead, G. H. 1923. Scientific Method and the Moral Sciences. *International Journal of Ethics* XXXIII:229–247.

Mead, G. H. 1934. *Mind, Self and Society*; It. trans. *Mente, Sè e Società*. Firenze: Barbera, 1966.

Mead, G. H. 1996. *La voce della coscienza*. Milano: Jaca Book.

Mergel, E., and M. Schützwohl. 2021. A Longitudinal Study on the COVID-19 Pandemic and Its Divergent Effects on Social Participation and Mental Health Across Different Study Groups With and Without Mental Disorders. *Social Psychiatry and Psychiatric Epidemiology* 56 (8): 1459–1468.

Merleau-Ponty, M. 1964. *Le visible et l'invisible*; It. trans. *Il visibile e l'invisibile*. Milano: Bompiani, 1969.

Merloni, F. 2005. *Introduzione all'e-government. Pubbliche Amministrazioni e società dell'informazione*. Torino: Giappichelli.

Merton, R. K. 1965. *On the Shoulders of Giants: A Shandean Postscript*; It. trans. *Sulle spalle dei giganti*. Bologna: il Mulino, 1991.

Meyrowitz, J. 1986. *No Sense of Place: The Impact of Electronic Media on Social Behavior*; It. trans. *Oltre il senso del luogo. L'impatto dei media elettronici sul comportamento sociale*. Bologna: Baskerville, 1993.

Milani, L. 2017. *Tutte le opere* (Ed. critica a cura di A. Melloni). Milano: Mondadori.

Miller, J. G. 1971. *Living system*; It. trans. *Teoria generale dei sistemi viventi*. Milano: FrancoAngeli, 1986[3].

Miller, J. H., and S. E. Page. 2007. *Complex Adaptive Systems: An Introduction to Computational Models of Social Life*. Princeton: Princeton University Press.

Mills, C. W. 1959. *The Sociological Imagination*; It. trans. *L'immaginazione sociologica*. Milano: Il Saggiatore, 1962.

Mills, C. W. 1963. *Power, Politics and People*; It. trans. *Saggi di sociologia della conoscenza*. Milano: Bompiani, 1971.

Minogue, K. 2010. *The Servile Mind. How Democracy Erodes the Moral Life*; It. trans. *La mente servile. La vita morale nell'era della democrazia*. Torino: IBL Libri, 2012.

Minsky, M. 1985. *The Society of Mind*; It. trans. *La Società della Mente*. Milano: Adelphi, 1989.

Mitchell, Waldrop M. 1992. *Complexity. The Emerging Science at the Edge of Order and Chaos.* New York: Simon & Schuster Paperbacks.
Monod, J. 1970. *Le hazard et la nécessité*; It. trans. *Il caso e la necessità. Saggio sulla filosofia naturale della biologia contemporanea.* Milano: Arnoldo Mondadori, 1997.
Montessori, M. 1948. *Come educare il potenziale umano.* Milano: Garzanti, 1992.
Montessori, M. 1970. L'errore e il suo controllo. In *La mente del bambino*, ed. M. Montessori. Milano: Garzanti.
Montessori, M. 2008. *Educare alla libertà.* Milano: Mondadori.
Montuori, A. 2014. *Journeys in Complexity: Autobiographical Accounts by Leading Systems and Complexity Thinkers.* Londra: Routledge.
Morin, E. 1973. *Le paradigme perdu: la nature humaine*; It. trans. *Il paradigma perduto. Che cos'è la natura umana?* Milano: Feltrinelli, 1974.
Morin, E. 1977. *La Méthode I. La Nature de la Nature*; trad. it. *Il metodo 1. La natura della natura.* Milano: Cortina, 2001.
Morin, E. 1980. *La Méthode II. La Vie de la Vie*; trad. it., *Il metodo 2. La vita della vita.* Milano: Cortina, 2004.
Morin, E. 1986. *La Méthode III. La Connaissance de la Connaissance*; trad. it., *Il Metodo 3. La conoscenza della conoscenza.* Milano: Cortina, 2007.
Morin, E. 1990. *Introduction à la pensèe complexe*; It. trans. *Introduzione al pensiero complesso.* Milano: Sperling & Kupfer, 1993.
Morin, E. 1991. *La Méthode IV. Les idées. Leur habitat, leur vie, leurs moeurs, leur organisation*; trad. it., *Il Metodo 4. Le idee: habitat, vita, organizzazione, usi e costumi.* Milano: Cortina, 2008.
Morin, E. 1994. *Mes Démons*; It. trans. *I miei demoni.* Roma: Meltemi.
Morin, E. 1999a. *Les sept savoirs nécessaires à l'éducation du futur*; It. trans. *I sette saperi necessari all'educazione del futuro.* Milano: Cortina, 2001.
Morin, E. 1999b. *La tête bien faite*; It. trans. *La testa ben fatta. Riforma dell'insegnamento e riforma del pensiero.* Milano, Cortina, 2000.
Morin, E. 2001. *La Méthode V. L'Humanité de l'Humanité. Tome 1: L'identité humaine*; trad. it., *Il Metodo 5. L'identità umana.* Milano: Cortina, 2002.
Morin, E. 2002. *Educare gli educatori. Una riforma del pensiero per la democrazia cognitiva.* Roma: EDUP.
Morin, E. 2004. *La Méthode VI. Éthique*; It. trans. *Il metodo. Etica.* Milano: Cortina, 2005.
Morin, E. 2008. *On Complexity.* Cresskill: Hampton Press.
Morin, E. 2014. *Enseigner à vivre*; It. trans., *Insegnare a vivere. Manifesto per cambiare l'Educazione.* Milano: Raffaello Cortina Editore, 2015, p. 7.
Morin, E. 2015. *Penser global. L'homme et son univers*; It. trans. *7 lezioni sul Pensiero globale.* Milano: Cortina, 2016.
Morin, E. 2017. *Connaissance Ignorance Mystère*; It. trans. *Conoscenza Ignoranza Mistero.* Milano: Cortina, 2018.
Morin, E., and M. Ceruti. 2013. *La nostra Europa.* Milano: Cortina.
Morin, E., and A. B. Kern. 1993. *Terre-Patrie*; It. trans. *Terra-Patria.* Milano: Cortina, 1994.
Morin, E., and J.-L. Le Moigne. 1999. *L'intelligence de la complexité.* Paris: L'Harmattan.
Morin, E., É.-R. Ciurana, and D. R. Motta. 2003. *Éduquer pour l'ère planétaire. La pensée complexe comme Méthode d'apprentissage dans l'erreur et l'incertitude humaines*; It. trans. *Educare per l'era planetaria. Il pensiero complesso come metodo d'apprendimento.* Roma: Armando, 2004.
Morozov, E. 2011. *The Net Delusion. The Dark Side of Internet Freedom*; It. trans. *L'ingenuità della rete.Il lato oscuro della libertà di internet.* Torino: Codice.
Mounier, E. 1932. Refaire la Renaissance. *Esprit* 1:5–51.
Mounier, E. 1961. *La pensée de Charles Péguy* (1931). In *Oeuvres*, vol. I, 13–125. Paris.
Mumford, L. 1921. *The Story of Utopias*; It. trans. *Storia dell'utopia.* Roma: Donzelli, 1997 (con prefazione di F. Crespi), (1ª ed. italiana 1969).
Mumford, L. 1934. *Technics and Civilization*; It. trans. *Tecnica e cultura.* Milano: Il Saggiatore, 1961.

Mumford, L. 1967. *The Myth of Machine*; It. trans. *Il mito della macchina*. Milano: Il Saggiatore, 1969.

Nagel, T. 1970. *The Possibility of Altruism*; It. trans. *La possibilità dell'altruismo*. Bologna: il Mulino, 1994.

Nagel, T. 1997. *The Last Word*; It. trans. *L'ultima parola. Contro il relativismo*. Milano: Feltrinelli, 1999.

Nancy, J.-L. 2002. Note sur le terme "biopolitique". In *La création du monde ou la mondialisation*, 134–143. Paris: Galilée.

Neumann von, J. 1958. *The Computer and the Brain*. New Haven: Yale University Press.

Neumann von, J. 1966. *The Theory of Self-Reproducing Automata*. Urbana: University of Illinois Press.

Nicolis, G., and C. Nicolis. 2007. *Foundations of Complex Systems*. Singapore: World Scientific.

Nietzsche, F. 1876. *Die Geburt der Tragödie*; It. trans. *La nascita della tragedia*. Milano: Adelphi, 1972.

Nietzsche, F. 1878. *Human, All Too Human: A Book for Free Spirits*. Cambridge: Cambridge University Press, 1996.

Nietzsche, F. 1887. *Genealogia della morale*. Milano: Adelphi, 1984, d. III, cap. XXV, 147–154.

Noë, A. 2010. *Perché non siamo il nostro cervello*. Milano: Cortina.

Norris, P. 2011. *Democratic Deficits: Critical Citizens Revisited*. Cambridge: Cambridge University Press.

Nourbakhsh, I. R. 2013. *Robot Futures*; It. trans. *Robot fra noi*. Torino: Bollati Boringhieri, 2014.

Nussbaum, M. C. 1997. *Cultivating Humanity. A Classical Defense of Reform in Liberal Education*; It. trans. *Coltivare l'umanità. I classici, il multiculturalismo, l'educazione contemporanea*. Roma: Carocci, 1999.

Nussbaum, M. C. 2002. *Giustizia sociale e dignità umana. Da individui a persone*. Bologna: il Mulino. (intr. di C. Saraceno).

Nussbaum, M. C. 2010. *Not for Profit. Why Democracy Needs the Humanities*. Princeton: Princeton University Press; It. trans. *Non per profitto. Perché le democrazie hanno bisogno della cultura umanistica*. Bologna: il Mulino, 2011.

Nussbaum, M. C. 2018. *The Monarchy of Fear. A Philosopher Looks at Our Political Crisis*. New York: Simon &Schuster; It. trans. *La monarchia della paura. Considerazioni sulla crisi politica attuale*. Bologna: il Mulino, 2020.

OECD. 2011. *Divided We Stand, Why Inequality Keeps Rising*. Paris: OECD Publishing.

OECD. 2019. *Education at a Glance 2019*. Paris: OECD Publishing.

OECD. 2021. *Education at a Glance 2021: OECD Indicators*. Paris: OECD Publishing.

Ohmae, K. 1990. *The Borderless World: Power and Strategy in the Interlinked Economy*; It. trans. *Il mondo senza confini*. Milano: Il Sole 24 Ore Libri, 1991.

Ong, W. 1982. *Orality and Literacy*. London/New York: Routledge.

Onnis, E. 2021. *Metafisica dell'emergenza*. Torino: Rosenberg & Sellier.

Oppenheimer, J. R. 1954. *Science and the Common Understanding*; It. trans. *Scienza e Pensiero comune*. Torino: Bollati Boringhieri, 1965.

Ortega y Gasset, J. 1930. *La rebelión de las masas*; It. trans. *La ribellione delle masse*. Torino: Utet, 1979.

Orwell, G. 1949. 1984, Mondadori, Cles (TN) 1994.

Paccagnella, L. 2010. *Open Access. Conoscenza aperta e società dell'informazione*. Bologna: il Mulino.

Packard, V. 1958. *The Hidden Persuaders*; It. trans. *I persuasori occulti*. Torino: Einaudi, 1958.

Packard, N. 1988. *Adaptation Toward the Edge of Chaos*. University of Illinois at Urbana-Champaign, Center for Complex Systems Research.

Pancani, L., M. Marinucci, N. Aureli, and P. Riva. 2021. Forced Social Isolation and Mental Health: A Study on 1,006 Italians Under COVID-19 Lockdown. *Frontiers in Psychology* 12:663799.

Panebianco, A. 2009. *L'automa e lo spirito. Azioni individuali, istituzioni, imprese collettive*. Bologna: il Mulino.

Panikkar, R. 1989. *The Rhythm of Being: The Gifford Lectures*. New York: Orbis Book Published.
Panikkar, R. 2002. *Pace e interculturalità. Una riflessione filosofica*. Milano: Jaca Book.
Parisi, G. 1992. *Field Theory, Disorder and Simulations*. Singapore: World Scientific Pub. Co Inc.
Parisi, G. 2001. La complessità da un punto di vista di un fisico. *Rendiconti Accademia Nazionale delle Scienze – Memorie di Scienze Fisiche e Naturali* 119° XXV:7–14.
Parisi, G. 2021. *In un volo di storni. Le meraviglie dei sistemi complessi*. Milano: Rizzoli.
Parisi, G., P. Urbani, and F. Zamponi. 2020. *Theory of Simple Glasses*. Cambridge: Cambridge University Press.
Parsons, T. 1951. *The Social System*; It. trans. *Il sistema sociale*, (intr. di L. Gallino). Milano: Comunità, 1965.
Pasquale, F. 2020. *New Laws of Robotics: Defendind Human Expertise in the Age of AI*; It. trans. *Le nuove leggi della robotica*. Roma: LUISS University Press, 2021.
Pellerey, M. 2011. L'approccio per competenze è un pericolo per l'educazione scolastica. *Scuola Democratica*, n.s.:37–54.
Penrose, R. 2016. *Fashion, Faith and Fantasy in the New Physics of the Universe*; It. trans. *L'universo è ancora segreto*. Milano: Rizzoli, 2017.
Perkinson, H. J. 1983. *Didattica dell'errore*. Roma: Armando.
Pfeffer, J., and R. I. Sutton. 2006. Evidenced-Based Management. *Harvard Business Review* 84:62–74.
Piaget, J. 1970a. *Psicologia e pedagogia*. Torino: Loescher.
Piaget, J. 1970b. *La situation des sciences de l'homme dans le système des sciences*; It. trans. *Le scienze dell'uomo*. Roma-Bari: Laterza, 1973.
Piazza, M. *Creature dell'abitudine. Abito, costume, seconda natura da Aristotele alle scienze cognitive*. Bologna: il Mulino.
Pizzorno, A. 1973. L'incompletezza dei sistemi. In *Razionalità sociale e tecnologie dell'informazione. Descrizione e critica dell'utopia tecnocratica*, ed. F. Rositi. Milano: Comunità.
Poincaré, H. 1885. L'Équilibre d'une masse fluide animée d'un mouvement de rotation. *Acta Mathematica* 7:259–380.
Poincaré, J. H. 1902, *La science e l'hypothèse*; It. trans. *La scienza e l'ipotesi*, (con prefazione di P. Odifreddi). Bari: Dedalo, 1989.
Poincaré, J. H. 1908. *Science et méthode*; It. trans. *Scienza e metodo*. Torino: Einaudi, 1997 (a cura di C. Bartocci).
Poincaré, H. 1995. *Geometria e caso*; It. trans. di C. Bartocci. Torino: Bollati Boringhieri.
Polanyi, K. 1944. *The Great Transformation*; It. trans. *La grande trasformazione. Le origini economiche e politiche della nostra epoca*. Torino: Einaudi, 1974.
Polanyi, K. 1987. *La libertà in una società complessa*. Torino: Bollati Boringhieri.
Popitz, H. 1995. *Der Aufbruch zur artifiziellen Gesellschaft*; It. trans. *Verso una società artificiale*. Roma: Editori Riuniti, 1996.
Popper, K. R. 1934. *The Logic of Scientific Discovery*; It. trans. *Logica della scoperta scientifica. Il carattere auto correttivo della scienza*. Torino: Einaudi, 1970.
Popper, K. R. 1945a. *The Open Society and Its Enemies. The Spell of Plato*, vol. I; It. trans. *La società aperta e i suoi nemici. Platone totalitario*. Roma: Armando, 1973.
Popper, K. R. 1945b. *The Open Society and Its Enemies: The High Tide of Prophecy: Hegel, Marx, and the Eftermath*, vol. II; It. trans. *La società aperta e i suoi nemici: Hegel e Marx falsi profeti*. Roma: Armando, 1974.
Popper, K. R. 1969. *Scienza e Filosofia*. Torino: Einaudi.
Popper, K. R. 1994. *The Myth of the Framework: In Defence of Science and Rationality*; It. trans. *Il mito della cornice. Difesa della razionalità e della scienza*. Bologna: il Mulino, 1995.
Prigogine, I. 1980. *From Being to Becoming: Time and Complexity in the Physical Sciences*; It. trans. di Bocchi G., Ceruti M., *Dall'essere al divenire*. Torino: Einaudi, 1986.
Prigogine, I. 1992. *Le leggi del caos*. Bari-Roma: Laterza, 1993.

Prigogine, I. 1996. *La fin des certitudes. Temps, chaos et les lois de la nature*; It. trans. *La fine delle certezze. Il tempo, il caos e le leggi della natura*. Torino: Bollati Boringhieri, 1997.

Prigogine, I., and I. Stengers. 1979. *La Nouvelle Alliance. Métamorphose de la science*; It. trans. *La nuova alleanza. Metamorfosi della scienza*. Torino: Einaudi, 1981.

Prigogine, I., and I. Stengers. 1984. *Order Out of Caos*. New York: Bentham Books.

Privitera, W. 2001. *Sfera pubblica e democratizzazione*. Roma-Bari: Laterza.

Pulcini, E. 2001. *L'individuo senza passioni. Individualismo moderno e perdita del legame sociale*. Torino: Bollati Boringhieri.

Purssell, E., D. Gould, and J. Chudleigh. 2020. Impact of Isolation on Hospitalised Patients Who Are Infectious: Systematic Review with Meta-Analysis. *BMJ Open* 10:e030371.

Putnam, R. D. 2000. *Bowling Alone. The Collapse and Revival of American Community*; It. trans. *Capitale sociale e individualismo. Crisi e rinascita della cultura civica in America*. Bologna: il Mulino, 2004.

Quammen, D. 2018. *L'albero intricato*. Milano: Adelphi, 2020.

Rainie, L., and B. Wellman. 2012. *Networked: The New Social Operating System*; It. trans. *Networked. Il nuovo sistema operativo sociale*. Milano: Guerini, 2012.

Rawls, J. 1971. *A Theory of Justice*; It. trans. *Una teoria della giustizia*. Milano: Feltrinelli, 1982.

Reichenbach, H. 1920. *Relativitätstheorie und Erkenntnis apriori*; It. trans. *Relatività e conoscenza a priori*. Roma-Bari: Laterza, 1984.

Rescher, N. 2000. *Nature and Understanding. The Metaphysics and Method of Science*. Oxford: Clarendon Press.

Revelli, M. 2001. *Oltre il Novecento. La politica, le ideologie e le insidie del lavoro*. Torino: Einaudi.

Revelli, M. 2003. *La politica perduta*. Torino: Einaudi.

Revelli, M. 2013. *Finale di partito*. Torino: Einaudi.

Rheingold, H. 1993. *The Virtual Community*; It. trans. *Comunità virtuali*. Milano: Sperling & Kupfer, 1994.

Rheingold, H. 2002. *Smart Mobs*; It. trans. *Smart mobs. Tecnologie senza fili, la rivoluzione sociale prossima ventura*. Milano: Cortina, 2003.

Ribolzi, L. 2011. Tra conoscenza e competenze: tra integrazione e conflitto. *Scuola Democratica*, n.s.:26–36.

Rickles, D., P. Hawe, and A. Shiell. 2007. A Simple Guide to Chaos and Complexity. *Journal of Epidemiology & Community Health, BMJ* 61 (11): 993–937. https://www.ncbi.nlm.nih.gov/pmc/articles/PMC2465602/.

Riesman, D. 1948. *The Lonely Crowd*; It. trans. *La folla solitaria*. Bologna: il Mulino, 1956.

Rifkin, J. 1995. *The End of Work. The Decline of the Global Labor Force and the Dawn of the Post-Market Era*; It. trans. *La fine del lavoro. Il declino della forza lavoro globale e l'avvento dell'era post-mercato*. Milano: Baldini & Castoldi.

Rifkin, J. 2000. *The Age of Access*; It. trans. *L'era dell'accesso. La rivoluzione della new economy*. Milano: Mondadori, 2000.

Riva, F. 2002. *La rinuncia al sé. Intersoggettività ed etica pubblica*. Roma: Edizioni Lavoro.

Robertson, R. 1992. *Globalization. Social Theory and Global Culture*; It. trans. *Globalizzazione. Teoria sociale e cultura globale*. Trieste: Asterios, 1999.

Robertson, R. 1995. Glocalization: Time-Space and Homogeneity-Heterogeneity. In M. Featherstone, S. Lash, and R. Robertson (a cura di) (1995), *Global Modernities*. London: Sage, cit. pp. 23–44.

Robins, K., and F. Webster. *Times of the Technoculture: From the Information Society to the Virtual Life*; It. trans. *Tecnocultura, Dalla società dell'informazione alla vita virtuale*. Milan: Guerini & Associati, 2003.

Robinson, K. 2015. *Scuola creativa. Manifesto per una nuova educazione*. Trento: Erickson, 2016.

Rodari, G. 2013. *Grammatica della fantasia*. Torino: Einaudi.

Rodotà, S. 1997. *Tecnopolitica. La democrazia e le nuove tecnologie della comunicazione*. Bari: Laterza.

Rodotà, S. 2013. *Iperdemocrazia. Come cambia la sovranità democratica con il web*. Roma-Bari: Laterza.
Rogers, K. H., R. Luton, H. Biggs, R. Biggs, S. Blignaut, A. G. Choles, C. G. Palmer, and P. Tangwe. 2013. Fostering Complexity Thinking in Action Research for Change in Social–Ecological Systems. *Ecology and Society* 18 (2): 31.
Rossi, R., et al. 2020. COVID-19 Pandemic and Lockdown Measures Impact on Mental Health Among the General Population in Italy. *Frontiers in Psychiatry* 11:790.
Rota, G.-C., and J. P. S. Kung. 1980. Probabilità. In *Enciclopedia del Novecento Treccani*. Chivasso: Istituto della Enciclopedia Italiana, 2019.
Rovelli, C. 2020. *Helgoland*. Milano: Adelphi.
Rozenblit, L., and F. Keil. 2002. The Misunderstood Limits of Folk Science: An Illusion of Explanatory Depth. *Cognitive Science* 26:521–562.
Rubin, G. J., and S. Wessely. 2020. The Psychological Effects of Quarantining a City. *BMJ* 368:m313.
Ruelle, D. 1991. *Hasard et Chaos*; It. trans. *Caso e Caos. La scoperta della complessità ai confine tra matematica e fisica*. Torino: Bollati Boringhieri, 1992.
Rullani, E. 2004. *Economia della conoscenza. Creatività e valore nel capitalismo delle reti*. Roma: Carocci.
Rusconi, G. E. 1999. *Possiamo fare a meno di una religione civile?* Roma-Bari: Laterza.
Russell, B. 1916. *Why Men Fight: A Method of Abolishing the International Duel*. HardPress Publishing, 2013.
Sadin, É. 2018. *L'Intelligence artificielle on l'enjeu du siècle*; It. trans. *Critica della ragione artificiale*. Roma: LUISS University Press, 2019.
Salmon, C. 2007. *Storytelling. La machine a fabriquer des histoires et à formatter les esprits*; It. trans. *Storytelling. La fabbrica delle storie*. Roma: Fazi, 2008.
Sassen, S. 1998. *Globalization and Its Discontents*; It. trans. *Globalizzati e scontenti. Il destino delle minoranze nel nuovo ordine mondiale*. Milano: Il Saggiatore, 2002.
Schrödinger, E. 1944. *What Is Life? The Physical Aspect of the Living Cell*; It. trans. *Che cos'è la vita?* Milano: Adelphi, 1995.
Schrödinger, E. 1958. *The Physical Basis of Consciousness, Science and Religion, The Mystery of the Sensual Qualities*; It. trans. *L'immagine del mondo*. Torino: Bollati Boringhieri, 1963 e 2017.
Schulz, K. 2010. *Being Wrong. Adventures in the Margin of Error*; It. trans. *L'arte di sbagliare. Avventure nel margine dell'errore*. Milano: Bompiani, 2019.
Searle, J. R. 1984. *Minds, Brains and Programs*. Cambridge: Harvard University Press.
Searle, J. R. 1998. *Mind, Language and Society*; It. trans. *Mente, Linguaggio e Società*. Milano: Cortina, 2000.
Sellow, T. L., M. W. Seeger, and R. R. Ulmer. *Chaos Theory, Informational Needs and National Disasters*. https://www.orau.gov/cdcynergy/erc/content/activeinformation/resources/chaostheory.pdf.
Sen, A. 1992. *Inequality Reexamined*; It. trans. *La diseguaglianza. Un riesame critico*. Bologna: il Mulino, 1994.
Sennett, R. 1998. *The Corrosion of Character: The Personal Consequences of Work in the New Capitalism*; It. trans. *L'uomo flessibile: le conseguenze del nuovo capitalismo sulla vita personale*. Milano: Feltrinelli, 2000.
Shannon, C. E., and W. Weaver. 1949. *The Mathematical Theory of Communication*. University of Illinois Press; It. trans. *La Teoria matematica delle comunicazioni*. Milano: ETAS, 1971.
Sharma, A., et al. 2020. Impact of Isolation Precautions on Quality of Life: A Meta-Analysis. *The Journal of Hospital Infection* 105:35–42.
Sher, L. 2020a. The Impact of the COVID-19 Pandemic on Suicide Rates. *QJM: An International Journal of Medicine* 113:707–712.
Sher, L. 2020b. COVID-19, Anxiety, Sleep Disturbances and Suicide. *Sleep Medicine* 70:124.

Simmel, G. 1903. *Die Großstädte und das Geistesleben*; It. trans. *Le metropoli e la vita dello spirito*. Roma: Armando, 1990.
Simmel, G. 1918. *Der Konflikt der modernen Kultur*; It. trans. *Il conflitto della cultura moderna, e altri saggi*, (a cura di Carlo Mongardini). Roma: Bulzoni, 1976.
Simon, H. A. 1947. *Administrative Behavior*; It. trans. *Il comportamento amministrativo*. Bologna: il Mulino, 1958.
Simon, H. A. 1959. Theories of Decision-Making in Economics and Behavioral Science. *American Economic Review* 49:253–283.
Simon, H. A. 1962. The Architecture of Complexity. *Proceedings of the American Philosophical Society* 106:467–482.
Simon, H. A. 1979. *Models of Thought*. New Haven: Yale University Press.
Simon, H. A. 1997. *Models of Bounded Rationality*, vol. 3, *Empirically Grounded Economic Reason*; It. trans. *Scienza economica e comportamento umano*. Torino: Comunità, 2000.
Simonigh, C. (a cura di). 2012. *Pensare la complessità. Per un Umanesimo Planetario. Con interventi di Edgar Morin, Gianni Vattimo e Gustavo Zagrebelsky*. Mimesis, Milano.
Sini, C. 2000. *Idoli della conoscenza*. Milano: Cortina.
Sloman, S., and P. Fernbach. 2017. *The Knowledge Illusion. Why We Never Think Alone*; It. trans. *L'illusione della conoscenza*. Milano: Cortina, 2018.
Smith, M. 2005. Book Review: Risk Society and Ethical Responsibility. *Sociology* 39 (3): 543–550.
Snow, C. P. 1959. *The Two Cultures*. Cambridge: Cambridge University Press, 2012 (new ed.).
Sofsky, W. 2005. *Das Prinzip Sicherheit*; It. trans. *Rischio e sicurezza*. Torino: Einaudi, 2005.
Sofsky, W. 2007. *Verteidigung des Privaten*; It. trans. *In difesa del privato*. Torino: Einaudi, 2010.
Sombart, W. 1902. *Der moderne Kapitalismus*; It. trans. *Il capitalismo moderno* (ed. it. parziale a cura di A. Cavalli). Torino: UTET, 1967.
Sorel, G. 1921. *Les Illusions du progrès*; It. trans. *Le illusioni del progresso*. Torino: Bollati Boringhieri, 1993.
Sorokin, P. 1937–1941. *Social and Cultural Dynamics*. Cincinnati: American Book Company.
Sparti, D. 2003. *L'importanza di essere umani. Etica del riconoscimento*. Milano: Feltrinelli.
Spengler, O. 1918. *Der Untergang des Abendlandes. Umrisse einer Morphologie der Weltgeschichte*; It. trans. *Il tramonto dell'Occidente. Lineamenti di una morfologia della storia mondiale*. Milano: Longanesi, 1957.
Spengler, O. 1931. *Der Mensch und die Technik*; It. trans. *L'uomo e la tecnica*. Parma: Guanda, 1992.
Sperber, D. 1996. *Explaining Culture*; It. trans. *Il contagio delle idee*. Milano: Feltrinelli, 1999.
Stephenson, N. 1992. *Snow Crash*. London: Penguin Books, 1993.
Stewart, I. 1989. *Does God Play Dice? The Mathematics of Chaos*. Oxford: Blackwell Pub.
Stiglitz, J. E. 2001. *In un mondo imperfetto. Mercato e democrazia nell'era della globalizzazione*. Roma: Donzelli.
Stiglitz, J. E. 2002. *Globalization and Its Discontents*; It. trans. *La globalizzazione e i suoi oppositori*. Torino: Einaudi.
Strogatz, H. S. 2004. *Sync: How Order Emerges from Chaos in the Universe, Nature, and Daily Life*. New York: Hachette Book Group.
Sunstein, C. 2001. *Republic.com*; It. trans. *Republic.com. Cittadini informati o consumatori di informazioni?* Bologna: il Mulino, 2003.
Sunstein, C. 2005. *Laws of Fear. Beyond the Precautionary Principle*; It. trans. *Il diritto della paura. Oltre il principio di precauzione*. Bologna: il Mulino, 2010.
Swartz, R. M. 1976. Mistakes As an Important Part of Learning Process. *High School Journal* LIX (6): 246–257.
Taleb, N. N. 2004. *Fooled by Randomness*; It. trans. *Giocati dal caso*. Milano: Il Saggiatore, 2014.
Taleb, N. N. 2007. *The Black Swan*; It. trans. *Il cigno nero, come l'improbabile governa la nostra vita*. Milano: Il Saggiatore, 2007.
Taleb, N. N. 2012. *Antifragile*; It. trans. *Antifragile. Prosperare nel disordine*. Milano: il Saggiatore, 2013.

Talevi, D., et al. 2020. Mental Health Outcomes of the CoViD-19 Pandemic. *Rivista di Psichiatria* 55:137–144.

Taylor, C. 1992. *The Malaise of Modernity*; It. trans. *Il disagio della modernità*. Roma-Bari: Laterza, 1994.

Taylor, C. 2007. *A Secular Age*; It. trans. *L'età secolare*. Milano: Feltrinelli, 2009.

Tegmark, M. 2014. *Our Mathematical Universe*; It. trans. *L'universo matematico. La ricerca della natura ultima della realtà*. Torino: Bollati Boringhieri, 2014.

Tegmark, M. 2017. *Life 3.0. Being Human in the Age of Artificial Intelligence*; It. trans. *Vita 3.0. Essere umani nell'era dell'intelligenza artificiale*. Milano: Cortina, 2018.

Tentori, T. 1987. *Il rischio della certezza. Pregiudizio/Potere/Cultura*. Roma: Edizioni Studium.

Thompson, D. A. W. 1917, ed. rivista, 1942. *On Growth and Form*; It. trans. *Crescita e forma. La geometria della natura*. Torino: Bollati Boringhieri, 1992.

Thompson, J. B. 1995. *The Media and Modernity. A Social Theory of the Media*; It. trans. *Mezzi di comunicazione e modernità*. Bologna: il Mulino, 1998.

Todorov, T. 1995. *La vie commune. Essai d'anthropologie générale*; It. trans. *La vita comune. L'uomo è un essere sociale*. Milano: Pratiche, 1998.

Toffler, A. 1980. *The Third Wave*; It. trans. *La Terza Ondata*. Milano: Sperling & Kupfer, 1987.

Tomlinson, J. 1999. *Globalization and Culture*; It. trans. *Sentirsi a casa nel mondo. La cultura come bene globale*. Milano: Feltrinelli, 2001.

Tönnies, F. 1887. *Gemeinschaft und Gesellschaft*; It. trans. *Comunità e società*. Milano: Comunità, 1963.

Touraine, A. 1969. *La société post-industrielle*; It. trans. *La società postindustriale*. Bologna: il Mulino, 1970.

Touraine, A. 1992. *Critique de la modernité*; It. trans. *Critica della modernità*. Milano: Il Saggiatore, 1993.

Touraine, A. 1997. *Eguaglianza e diversità. I nuovi compiti della democrazia*. Roma-Bari: Laterza.

Touraine, A. 2004. *Un nouveau paradigme. Pour comprendre le monde aujourd'hui*; It. trans. *La globalizzazione e la fine del sociale. Per comprendere il mondo contemporaneo*. Milano: Il Saggiatore, 2008.

Tramma, S. 2003. *L'educatore imperfetto. Senso e complessità del lavoro educativo*. Roma: Carocci.

Turing, A. M. 1950. Computing Machinery and Intelligence. *Mind: A Quarterly Review of Psycology and Philosophy* 59:433–460.

Turing, A. M. (a cura di G. Lolli). *Collected Works of A.M. Turing. Mechanical Intelligence*; It. trans. *Intelligenza Meccanica*. Torino: Bollati Boringhieri, 1994.

Turner, J. R., and R. M. Baker. 2019. Complexity Theory: An Overview with Potential Applications for the Social Sciences. *Systems* 7 (1), MDPI Journals. https://www.mdpi.com/2079-8954/7/1/4/html.

Tyler, C. M., et al. 2021. A Study of Older Adults' Mental Health Across 33 Countries During the COVID-19 Pandemic. *International Journal of Environmental Research and Public Health* 18:5090.

UIL. 2019. *4th Global Report on Adult Learning and Education: Leave No One Behind: Participation, Equity and Inclusion*. [PDF] Hamburg: UIL. Available at: https://unesdoc.unesco.org/ark:/48223/pf0000372274.

UNESCO. 2016a. *Education 2030. Incheon Declaration and Framework for Action for the Implementation of Sustainable Development Goal 4: Ensure Inclusive and Equitable Quality Education and Promote Lifelong Learning Opportunities for All*. [PDF] Paris: UNESCO. Available at: https://unesdoc.unesco.org/ark:/48223/pf0000245656/PDF/245656eng.pdf.multi.

UNESCO. 2016b. *Recommendation Concerning Technical and Vocational Education and Training (TVET), 2015*. [PDF] Paris: UNESCO. Available at: https://unesdoc.unesco.org/ark:/48223/pf0000245178.

UNESCO. 2017. *Educazione agli Obiettivi di Sviluppo Sostenibile – Obiettivi di apprendimento*. http://unescoblob.blob.core.windows.net/pdf/UploadCKEditor/MANUALE_ITA.pdf.
UNESCO. 2018a. *Global Education Monitoring Report 2019. Migration, Displacement and Education. Building Bridges, Not Walls.* [online] Paris: UNESCO. Available at: https://unesdoc.unesco.org/ark:/48223/pf0000265866
UNESCO. 2018b. *Building Tomorrow's Digital Skills – What Conclusions Can We Draw from International Comparative Indicators?* [online]. Paris: UNESCO. Available at: https://unesdoc.unesco.org/ark:/48223/pf0000261853.
UNESCO. 2020. *Beyond Disruption: Technology Enabled Learning Futures. 2020 Edition of Mobile Learning Week*. 12–14 October 2020. Report. [PDF] Paris: UNESCO. Available at: https://unesdoc.unesco.org/ark:/48223/pf0000377753/PDF /377753eng.pdf.multi.
UNESCO. 2021a. *Draft Text of the Recommendation on the Ethics of Artificial Intelligence. Annex to the Report of the Social and Human Sciences Commission* (SHS). 41 C/73. Adopted by the UNESCO General Conference, 41st session, 2021. [PDF] Paris: UNESCO. Available at: https://unesdoc.unesco.org/ark:/48223/pf0000379920/PDF/379920eng.pdf.multi.page=14.
UNESCO. 2021b. *Reimagining Our Futures Together: A New Social Contract for Education*. http://www.unesco.org/.
UNESCO. 2022a. *Mobile Learning*. Available at: https://en.unesco.org/themes/ict-education/mobile-learning.
UNESCO. 2022b. *Open Educational Resources (OER)*. [online] Paris: UNESCO. Available at: https://en.unesco.org/themes/building-knowledge-societies/oer.
UNESCO. 2022c. *Making Lifelong Learning a Reality: A Handbook*, 3. Institute for Lifelong Learning. https://uil.unesco.org/making-lifelong-learning-reality-handbook.
UNESCO, UNESCO Bangkok. 2017. *Beyond Access: ICT-Enhanced Innovative Pedagogy in TVET in the Asia-Pacific*. Paris: UNE SCO. [PDF].
Urbinati, N. 2011. *Liberi e uguali. Contro l'ideologia individualista*. Roma-Bari: Laterza.
Vattimo, G. 1989. *La società trasparente*. Milano: Garzanti, 2000 (nuova ed.).
Vattimo, G. 1991. *La fine della modernità*. Milano: Garzanti.
Vattimo, G. 2002. *Tecnica ed esistenza*. Milano: Bruno Mondadori.
Veblen, T. 1899. *The Theory of the Leisure Class*; It. trans. *La teoria della classe agiata*. Milano: Il Saggiatore, 1969.
Veblen, T. (a cura di F. L. Viano, and A. Viano). 2012. *Il posto della scienza*. Torino: Bollati Boringhieri
Veca, S. 1990. *Cittadinanza. Riflessioni filosofiche sull'idea di emancipazione*. Milano: Feltrinelli.
Vespignani, A. 2019. *L'algoritmo e l'oracolo. Come la scienza predice il futuro e ci aiuta a cambiarlo*. Milano: Il Saggiatore.
Vico, G. 1967. *La scienza nuova*, a cura di F. Nicolini. Bari: Laterza.
Vigotskij, L. S. 2007. *Pensiero e linguaggio*. Firenze: Giunti.
von Bertalanffy, L. 1968. *General System Theory: Foundations, Development, Applications*; It. trans. *Teoria generale dei sistemi*. Milano: Isedi, 1975.
von Foerster, H. 1981. *Observing Systems*; It. trans. *Sistemi che osservano*. Roma: Astrolabio, 1987.
von Hayek, F. A. 1964. The Theory of Complex Phenomena. In *The Critical Approach to Science and Philosophy. Essay in Honor of K.R. Popper*, ed. M. Bunge. New York: Free Press.
Waldrop, W. M. 1992. *Complexity. The Emerging Science at the Edge of Order and Chaos*; It. trans. *Complessità. Uomini e idee al confine tra ordine e caos*. Torino: Instar Libri, 1995.
Wallerstein, I. 1974. *The Modern World Systems. Capitalist Agriculture and the Origins of the European World-Economy in the Sixteenth Century*; It. trans. *Il sistema mondiale dell'economia moderna. L'agricoltura capitalista e le origini dell'economia-mondo europea nel XVI secolo*, vol. I. Bologna: il Mulino, 1990.
Wallerstein, I. 1980. *The Modern World Systems II. Mercantilism and the Consolidation of the European World-Economy 1600–1750*; It. trans. *Il Sistema mondiale dell'economia moderna. Il mercantilismo e il consolidamento dell'economia-mondo europea 1600–1750*. Bologna: il Mulino,1990b.

Wallerstein, I. 1995a. *After Liberalism*; It. trans. *Dopo il liberalismo*. Milano: Jaca Book, 1998.

Wallerstein, I. 1995b. *Historical Capitalism with Capitalist Civilization*; It. trans. *Capitalismo storico e civiltà capitalistica*. Trieste: Asterios, 2000.

Watzlawick, P., Beavin J. Helmick, and D. D. Jackson 1967. *Pragmatic of Human Communication. A Study of Interactional Patterns, Pathologies, and Paradoxes*; It. trans. *Pragmatica della comunicazione umana. Studio dei modelli interattivi, delle patologie e dei paradossi*. Roma: Astrolabio, 1971.

Weaver, W. 1948. Science and Complexity. *American Scientist* 36:536.

Weber, M. 1922a. *Gesammelte Aufsätze zur Wissenschaftslehre*; It. trans. *Il metodo delle scienze storico-sociali*. Torino: Einaudi, 1958.

Weber, M. 1922b. *Wirthschaft und Gesellschaft. Grundriß der verstehenden Soziologie*; It. trans. *Economia e società*, vol. II. Milano: Comunità, 1961.

Weick, K. E. 1993. The Collapse of Sensemaking in Organizations: The Mann Gulch Disaster. *Administrative Science Quarterly* 38:628–652.

Weinberger, D. 2019. *Everyday Chaos. Technology, Complexity and How We're Thriving in a New World of Possibility*; It. trans. *Caos quotidiano. Un nuovo mondo di possibilità*. Torino: Codice, 2020.

Wellman, B., and S. Berkovitz. 1988. *Social Structure. A Network Approach*, 37. Cambridge: Cambridge University Press.

WHO. 2020. *Mental Health and Psychosocial Considerations During the COVID-19 Outbreak*. https://www.who.int/docs/default-source/coronaviruse/mental-health-considerations.pdf?sfvrsn=6d3578af_8

Wiener, N. 1948. *Cybernetics: or Control and Communication in the Animal and the Machine*; It. trans. *La cibernetica*. Milano: Il Saggiatore, 1968.

Wiener, N. 1950. *The Human Use of Human Beings*; It. trans. *Introduzione alla cibernetica. L'uso umano degli esseri umani*. Torino: Bollati Boringhieri, 1966.

Wiener, N., A. Siegel, B. Rankin, and W. T. Martin. 1976. *Differential Space, Quantum Systems and Prediction*. Cambridge: MIT Press.

Wievorka, M. 2001. *La différence*; It. trans. *La differenza culturale. Una prospettiva sociologica*. Roma-Bari: Laterza, 2002.

Wilson, E. O. 1998. *Consilience. The Unity of Knowledge*. New York: Alfred A. Knopf.

Wittgenstein, L. 1921. *Logisch-philosophische Abhandlung*, (1922), *Tractatus logico-philosophicus*; It. trans. *Tractatus logico-philosophicus e Quaderni 1914–1916*. Torino: Einaudi, 1964.

Wittgenstein, L. 1953. *Philosophical Investigations*; It. trans. *Ricerche filosofiche*. Torino: Einaudi, 1967.

Yuan, Y., et al. 2021. Prevalence of Post-Traumatic Stress Symptoms and Its Associations with Quality of Life, Demographic and Clinical Characteristics in COVID-19 Survivors During the Post-COVID-19 Era. *Frontiers in Psychiatry* 12:665507.

Zhang, Y., and Z. F. Ma. 2020. Impact of the COVID-19 Pandemic on Mental Health and Quality of Life among Local Residents in Liaoning Province, China: A Cross-Sectional Study. *International Journal of Environmental Research and Public Health* 17 (7): 2381.

Zimmer, C. 2021. *Life's Edge: The Search for What It Means to Be Alive*. Londra: Palgrave Macmillan.

Žižek, S. 1999. *Il Grande Altro. Nazionalismo, godimento, cultura di massa*. Milano: Feltrinelli.

Zolo, D. 1987. *Complessità e democrazia. Per una ricostruzione della teoria democratica*. Torino: Giappichelli.

Zuboff, S. 2023. *The Age of Surveillance Capitalism*; It. trans., *Il capitalismo della sorveglianza. Il futuro dell'umanità nell'era dei nuovi poteri*. Roma: LUISS University Press.

Index

P. Dominici, *Beyond Black Swans: Inhabiting Indeterminacy*, Philosophical Studies Series 160, https://doi.org/10.1007/978-3-032-09029-4

N

O

P

Q

R

GPSR Compliance
The European Union's (EU) General Product Safety Regulation (GPSR) is a set of rules that requires consumer products to be safe and our obligations to ensure this.

If you have any concerns about our products, you can contact us on

ProductSafety@springernature.com

In case Publisher is established outside the EU, the EU authorized representative is:

Springer Nature Customer Service Center GmbH
Europaplatz 3
69115 Heidelberg, Germany

www.ingramcontent.com/pod-product-compliance
Ingram Content Group UK Ltd.
Pitfield, Milton Keynes, MK11 3LW, UK
UKHW020911180926
475648UK00002B/11
* 9 7 8 3 0 3 2 0 9 0 2 8 7 *